by Kimberley Willis with Rob Ludlow

***Raising Chickens* For Dummies®, 2nd Edition**

Published by: **John Wiley & Sons, Inc.,** 111 River Street, Hoboken, NJ 07030-5774, www.wiley.com

Published simultaneously in Canada

Library of Congress Control Number is available from the publisher

ISBN 978-1-118-98278-5 (pbk); ISBN 978-1-118-98280-8 (ebk); ISBN 978-1-118-98279-2 (ebk)

Manufactured in the United States of America

10 9 8 7 6 5 4

Contents at a Glance

Table of Contents

Part IV: Breeding: From Chicken to Egg and Back Again 247

Part V: Special Management Considerations 319

Introduction

Across the United States, from California (where Rob lives) to Michigan (where Kim lives) and beyond — and even in other countries — people are discovering the joy of chickens. Some people want to produce their own food, some are nostalgic and longing for a simpler and more pastoral time; and others were sucked in by some cute chicks. Whatever sparked your interest in chickens, we hope this book helps you become a happy, knowledgeable chicken keeper.

Chickens are a special part of both authors' lives. Every day we listen to the questions and concerns people have about chickens. We take great enjoyment in the chickens we own, too. We're thrilled that more cities and townships are allowing people to keep chickens. But that means there's an ever-growing body of folks who need information about chickens. Because Rob and I can't always be there to answer everyone's chicken questions, we decided it was time for a modern, comprehensive chicken book that provides quick answers to all your chicken questions.

In this, the second edition of *Raising Chickens For Dummies*, we've kept all the good parts of the first edition and done some updating to reflect new technology and knowledge about keeping chickens. We've also expanded some chapters to bring you even more information about chicken keeping.

About This Book

This chicken book is different from some of the others out there. It's easy to find the answer you're looking for because of the way the book is organized. Go ahead, flip through the book and see for yourself. Nice bold headings direct your eyes to just the section you need, and you don't have to read the whole book for a quick answer.

This book gives you a broad overlook of all aspects of keeping chickens, from laying hens to meat chickens, but you don't have to read it all at once or in any particular order. You can start anywhere in the book that interests you. Today you may be interested in learning how to care for some cute, fluff-ball chicks you fell in love with at the hardware store, and you'll find that information here. In 5 months or so, when they begin laying eggs, you'll need information on what to do and how to manage hens. That information is here, too.

And if you get tired of those chicks because they all turn out to be big, fat, noisy roosters, well, we give you good butchering instructions to turn them into chicken fricassee. So put this book on your bookshelf in a prominent place. We're sure you'll refer to it again and again.

We're careful to use modern, scientifically correct information on chicken care, and we direct your attention to sources of additional information when necessary. If you don't want to read the sidebars or the technical points, you don't have to; you'll still get the information you need to become a great chicken keeper. To round out the information, we also throw in lots of good, homey, down-to-earth advice that comes from owning and enjoying our own chickens.

When you're reading this book, you may notice that some web addresses break across two lines of text. If you're reading this book in print and you want to visit one of these web pages, simply type the web address into your computer exactly as it's written in the text, as though the line break doesn't exist. If you're reading this as an e-book, you've got it easy — just click the web address to go directly to the web page.

Foolish Assumptions

To get this book flowing, we had to factor in some assumptions about you, the reader. Here's how we've sized you up:

- You want to find out more about keeping chickens or eating the chickens you have.
- You like animals and want to treat them with kindness, and you need some knowledge of their needs.
- Although you've seen and heard chickens, you aren't an expert on them yet and you need some basic information.
- You don't want to raise chickens on some monster scale, like 500 laying hens or 2,500 broilers. (We assume the readers of this book want information on small home flocks.)
- You have some basic carpentry or craft skills. (Although we include some very basic plans for building chicken housing in this book, we don't have enough room to teach you building skills. So if you don't have the skills, we give you permission to call on a friend who does.)

Icons Used in This Book

Icons are special symbols set in the margins near paragraphs of text in the book. They are meant to draw your attention. Some people use them as a way to access certain pieces of important information, such as tips. This book uses the following icons.

Tips are special time- or money-saving pieces of advice. They come from our years of experience with chickens.

This icon urges you to remember this piece of information because it's important. Sometimes a referral to another chapter for more precise information is nearby.

A warning icon means we're mentioning something that may pose a danger to you or your chickens. Pay attention to warning icons — they contain important information.

This icon means we're providing some technical information that may or may not interest you. You can skip this paragraph if you want, without missing any important information.

Beyond the Book

In addition to the material in the print or e-book you're reading right now, this book comes with some bonus information on the web that you can access from anywhere.

If you want some fast answers on some of the most basic parts of chicken keeping, you can go to the *Raising Chickens for Dummies* Cheat Sheet, at `www.dummies.com/cheatsheet/raisingchickens`. In addition to the Cheat Sheet, you'll find links to some bonus articles not found in the book. For example, we've got bonus articles on feeding chickens organically, making your chicken coop a special place, and showing chickens. These links are found on the page preceding each new part of the book. You also can go to `www.dummies.com/extras/raisingchickens` and find all the bonus articles there.

Where to Go from Here

Time to get reading! May we make some suggestions on where to start? Of course, eventually you'll want to read every scrap of this book, but some things you need to know — now!

Here are some ideas of where you may want to begin, depending on your situation:

- If you're one of those rare people who likes to be well prepared before you start a project like raising chickens, you may want to start with Chapter 1.
- If you're sitting here with the book in one hand and a box of chicks at your feet, turn to Chapter 14 to get more info on raising chicks.
- If you have some chickens and they aren't laying the eggs you expected, you need to flip to Chapter 15.
- If you have chickens and they don't appear to be doing so well, check out Chapter 11 to diagnose and treat whatever your chickens are suffering from.
- If someone gave you some chicks for Easter that turned into ten fighting and crowing roosters, try Chapter 16, which discusses how to turn them into something more valuable — meat for the freezer.

Part I

Getting Started with Raising Chickens

getting started with Raising Chickens

Visit `www.dummies.com` for great Dummies content online.

In this part...

- Find out if owning chickens is right for you. Get all the details of chicken ownership to make an informed decision on starting your own chicken journey.
- Whether you're a seasoned chicken owner or you're new to the chicken world, get information on chicken biology, how chickens interact with one another and other animals, and how to identify illnesses that plague chickens.
- Will it be the cute, loveable Silkie or perhaps the brown-egg-laying Isa Brown? Discover the different breeds and what they offer.
- Get some tips on buying chickens. From starting with adults or chicks, to figuring out costs, to finding the right place to buy your birds, we cover all your bases to get your flock started.

Chapter 1

The Joy of Chickens

In This Chapter

- Checking on local legal restrictions for chicken-keeping
- Considering the commitments you need to make
- Counting the costs
- Being mindful of your neighbors

We love chickens, and we hope you're reading this book because you love chickens, too. So we're going to discuss a very basic issue in this chapter: whether you should actually keep chickens. Chickens make colorful, moving lawn ornaments, and they can even furnish your breakfast. But they do take some attention, some expense, and some good information to care for properly.

So consider this chapter as chicken family planning. If you read the information in this chapter and still believe you're ready to start your chicken family, then you have the whole rest of the book to get all the information you need to begin the adventure.

First Things First: Dealing with the Legal Issues

You may be surprised one day to notice chickens in your suburban neighborhood. Many urban and suburban communities are bowing to public pressure and allowing chicken-keeping. But not every community is so enlightened. The person keeping chickens in your neighborhood may be flouting the law. So before you rush out and buy some chickens, too, check whether any laws in your area prevent you from legally keeping chickens.

Almost all property is classified into zoning areas (some very undeveloped areas may have no zoning). Each type of zoning has laws that state what can and cannot be done to property in that zone. This legislation is a way to regulate growth of a community and keep property use in an area similar.

Zoning classification is the job of local governments. Each local governmental unit then assigns laws governing property use within each zone. These laws vary from community to community, but laws and ordinances can regulate what type and how many animals can be kept, what structures and fences can be built, whether a home business can operate, and many other considerations.

The good news, though, is that many cities are giving in to pressure from citizens who want to keep a few chickens for eggs or pets and are allowing poultry-keeping. In most places, a person who wants to use his or her property in a way that's prohibited by the zoning can ask for a zoning variance. Zoning classification can also change if several property owners request the change and it then is approved. The high population of emigrants in some cities who are used to keeping a few chickens in small quarters has also contributed to the relaxation of some rules.

Knowing what info you need

To know whether you can legally keep chickens, first you need to know the zoning of your property. Then you need to know whether any special regulations in that zoning district affect either chicken-keeping or your ability to build chicken housing.

Some common zoning areas are agricultural, residential, and business. You may also find subcategories such as single-family residential or suburban farms. Here's what those categories generally mean for you:

- **If the zoning is listed as agricultural,** you can probably raise chickens without a problem. With this type of zoning, you'll probably also find a notice about the Right to Farm bill on your paperwork. The Right to Farm bill states that any recognized, legal methods of farming can exist or begin at any time in that zone.
- **If the zoning is listed as residential, residential/agriculture, or some other type of zoning, or if you rent or lease your home,** you'll need to determine just what is allowed. Because these zoning areas can have different rules across the United States, you need to have to check with local officials to find out what that zoning allows you to do. And your landlord may have restrictions in the lease against pets or livestock, so read your lease or talk to your landlord.

If you've lived in your home for several years and you've never raised livestock or chickens, you may want to check the zoning with your township because zoning can change over time.

After you've looked into your zoning, you can ask your government officials about any laws regarding keeping animals and erecting sheds or other kinds of animal housing in your zone. You need to be concerned about two types of laws and ordinances before you begin to raise chickens:

- **Laws concerning the ownership of animals at your home location:** Restrictions may cover the number of birds, the sex of the birds, and where on the property chicken coops can be located. In some areas, the amount of property you have and your closeness to neighbors may determine whether you can keep birds and, if so, how many. Your neighbor may own 5 acres and be allowed to keep chickens, but on your 2-acre lot, poultry may be prohibited. Or you may be allowed to keep so many pets per acre, including chickens. Or you may need to get written permission from neighbors. Many other rules can apply.
- **Laws that restrict the types of housing or pens you can construct:** Do you need a permit to build a chicken coop? Does it need to be inspected?

Finding the info

Just because others in your neighborhood have chickens doesn't mean that it's legal for you to have them. Your neighbors may have had them before a zoning change (people who have animals at the time zoning is changed are generally allowed to keep them), they may have a variance, or they may be illegally keeping chickens.

Not only do you need to find out what *you* are allowed to do, chicken-wise, but you also need to make sure that you get that information from the right people. If you recently purchased your home, your deed and your sales agreement likely have your zoning listed on them. If you can't find a record of how your property is zoned, go to your city, village, or township hall and ask whether you can look at a zoning map. Some places have a copy they can give or sell you; in others, you need to look in a book or at a large wall map.

In larger communities, the planning board or office may handle questions about zoning. In smaller towns or villages, the county clerk or an animal control officer may handle questions about keeping animals. In either case, another government unit may handle the issue of building fences and shelters.

Don't take the word of neighbors, your aunt, or other people not connected to local government that it's okay to raise chickens at your home. If you're in the midst of buying a home, don't even take the word of real estate agents about being able to keep chickens or even about the property zoning. You never know whether the information you're getting is legitimate when it comes from a secondary source, so you're better off avoiding consequences by going straight to the primary source of legal info.

If you can, get a copy of the laws or ordinances so you can refer to them later, if the need arises. You may need them so you can show a neighbor who challenges your right to keep chickens or to remind you of how many chickens you can legally own.

Confronting restrictions

If your city, village, or township doesn't allow chicken-keeping, find out the procedure for amending or changing a law or zoning in your location. Sometimes all you need to do is request a zoning variance. A variance allows you, and only you, to keep chickens, based on your particular circumstances.

Finding out what you have to do

In some areas, getting permission to keep chickens is just a formality; in others, it's a major battle. Some places require you to draft a proposed ordinance or zoning variation for consideration. In either case, you'll probably be required to attend a commission meeting and state your case.

Ask your city clerk, township supervisor, or other local government official whether you need to attend a planning commission meeting, another special committee meeting, or the general city commission meeting. Find out the date, time, and location of the meeting. In some areas, you need to make an appointment to speak at a meeting or bring up issues.

Be patient — some of these changes can take months of discussion and mulling over. If you don't succeed the first time, ask what you can do to change the outcome the next time. Then try again.

Presenting a compelling case

Come to any necessary meeting prepared and organized. Try to anticipate any questions or concerns, and have good answers for them. Be prepared to compromise on some points, such as the number of birds allowed. Research bulletins and other information prepared by university poultry specialists that have guidelines and sample ordinances for keeping chickens in urban settings.

Ask other people in your community who seem involved in local government about the process in your community. They may give you valuable tips on how to approach the officials who have the power to change a law or grant a variance.

If you can afford it, you may consider hiring a lawyer to represent you. However, most people want to handle it on their own, if they can. If you have a city commissioner or other official assigned to your neighborhood, you may want to enlist his or her help.

It helps to find other people in your area who also want to keep chickens and who are willing to come to meetings to support you. Local experts such as a 4-H poultry leader, veterinarian, or agriculture teacher who can speak on the behalf of poultry-keeping may help. You can also draft a proposed law or ordinance and get people to sign a petition in support of it.

Assessing Your Capabilities: Basic Chicken Care and Requirements

Chickens can take as much time and money as you care to spend, but you need to recognize the *minimum* time, space, and money commitments required to keep chickens. In the next sections, we give you an idea of what those minimums are.

Time

When we speak about time here, we're referring to the daily caretaking chores. Naturally, setting up housing for your birds takes some time. If you're building a chicken coop, give yourself plenty of time to finish before you acquire the birds. You will have to judge how much time that entails, depending on the scope of the project, your building skills, and how much time each day you can devote to it. See Chapter 6 for more on constructing your own coop.

Count on a minimum of 15 minutes in the morning and the evening to care for chickens in a small flock, if you don't spend a lot of time just observing their antics. Even if you install automatic feeders and waterers (see Chapter 8), a good chicken-keeper should check on the flock twice a day. If you have laying hens, collect the eggs once a day, which shouldn't take long.

Try to attend to your chickens' needs before they go to bed for the night and after they are up in the morning. Ideally, chickens need 14 hours of light and

10 hours of darkness. In the winter, you can adjust artificial lighting so that it accommodates your schedule. Turning on lights to do chores after chickens are sleeping is very stressful for them.

You will need additional time once a week for basic cleaning chores. If you have just a few chickens, this may be less than an hour. The routine will include such chores as removing manure, adding clean litter, scrubbing water containers, and refilling feed bins. Depending on your chicken-keeping methods, you may need additional time every few months for more intensive cleaning chores.

More chickens doesn't necessarily mean more daily time spent on them until you get to very large numbers. A pen full of 25 meat birds may increase your caretaking time only a few minutes versus a pen of 4 laying hens. But how you keep chickens can increase the time needed to care for them. If you keep chickens for showing and you house them in individual cages, feeding and watering them will take at least five to ten minutes per cage.

Space

Each adult full-size chicken needs at least 2 square feet of floor space for shelter and another 3 square feet in outside run space if it isn't going to be running loose much. So a chicken shelter for four hens needs to be about 2 feet by 4 feet, and the outside pen needs to be another 2 feet by 6 feet, to make your total space used 2 feet by 10 feet (these dimensions don't have to be exact). For more chickens, you need more space, and you need a little space to store feed and maybe a place to store the used litter and manure. Of course, more space for the chickens is always better.

As far as height goes, the chicken coop doesn't have to be more than 3 feet high. But you may want your coop to be tall enough that you can walk upright inside it.

Besides the actual size of the space, you need to think about location, location, location. You probably want your space somewhere other than the front yard, and you probably want the chicken coop to be as far from your neighbors as possible, to lessen the chance that they complain.

Money

Unless you plan on purchasing rare breeds that are in high demand, the cost of purchasing chickens won't break most budgets. Adult hens that are good layers cost less than $10. Chicks of most breeds cost a few dollars each. The cost of adult fancy breeds kept for pets ranges from a few dollars to much, much more, depending on the breed. Sometimes you can even get free chickens!

Housing costs are extremely variable, but they are one-time costs. If you have a corner of a barn or an old shed to convert to housing and your chickens will be free-ranging most of the time, then your housing start-up costs will be very low — maybe less than $50. If you want to build a fancy chicken shed with a large outside run, your cost could be hundreds of dollars. If you want to buy a prebuilt structure for a few chickens, count on a couple hundred dollars.

The best way to plan housing costs is to first decide what your budget can afford. Next, look through Chapters 5 and 6 of this book to learn about types of housing. Then comparison-shop to see what building supplies would cost for your chosen housing (or prebuilt structures) and see how it fits your budget. Don't forget to factor in shipping costs for prebuilt units.

You may have a few other one-time costs for coop furnishings, including feeders, waterers, and nest boxes. For four hens, clever shopping should get you these items for less than $50.

Commercial chicken feed is reasonably priced, generally comparable to common brands of dry dog and cat food. How many chickens you have determines how much you use: Count on about a third to a half pound of feed per adult, full-sized bird per day. We estimate the cost of feed for three to four layers to be less than $20 per month.

Focusing Your Intentions: Specific Considerations

You may be nostalgic for the chickens scratching around in Grandma's yard. You may have heard that chickens control flies and ticks and turn the compost pile. You may have children who want to raise chickens for a 4-H project. Maybe you want to produce your own quality eggs or organic meat. Maybe you just want to provoke the neighbors. People raise chickens for dozens of reasons. But if you aren't sure, it helps to decide in advance just why you want to keep chickens.

Egg layers, meat birds, and pet/show chickens take slightly different housing and care requirements. Having a purpose in mind as you select breeds and develop housing will keep you from making expensive mistakes and will make your chicken-keeping experience more enjoyable.

It's okay to keep chickens for several different purposes — some for eggs and others as show birds, for example — but thinking about your intentions in advance makes good sense.

Want eggs (and, therefore, layers)?

While we're at it, let's define *egg* here. The word *egg* can refer to the female reproductive cell, a tiny bit of genetic material barely visible to the naked eye. In this chapter, *egg* refers to the large, stored food supply around a bit of female genetic material. Because eggs are deposited and detached from the mother while an embryo develops, they're not able to obtain food from her body through veins in the uterus. Their food supply must be enclosed with them as they leave the mother's body.

The egg that we enjoy with breakfast was meant to be food for a developing chick. Luckily for us, a hen continues to deposit eggs regardless of whether they have been fertilized to begin an embryo.

If you want layers, you need housing that includes nest boxes for them to lay their eggs in and a way to easily collect those eggs. Layers appreciate some outdoor space; if you have room for them to do a little roaming around the yard, your eggs will have darker yolks and you will need less feed.

Thinking about home-grown meat?

Don't expect to save lots of money raising your own chickens for meat unless you regularly pay a premium price for organic, free-range chickens at the store. Most homeowners raising chickens for home use wind up paying as much per pound as they would buying chicken on sale at the local big-chain store. But that's not why you want to raise them.

You want to raise your own chickens because you can control what they eat and how they are treated. You want to take responsibility for the way some of your food is produced and take pride in knowing how to do it.

It isn't going to be easy, especially at first. But it isn't so hard that you can't master it. For most people, the hardest part is the butchering, but the good news is that, in almost every area of the country, you can find folks who will do that job for you for a fee.

You can raise chickens that taste just like the chickens you buy in the store, but if you intend to raise free-range or pastured meat chickens, expect to get used to a new flavor. These ways of raising chickens produce a meat that has more muscle or dark meat and a different flavor. For most people, it's a *better* flavor, but it may take some getting used to.

Average people who have a little space and enough time can successfully raise all the chicken they want to eat in a year. And with modern meat-type chickens, you can be eating fried homegrown chicken 10 weeks after you get the chicks — or even sooner. So unlike raising a steer or pigs, you can try raising your own meat in less than 3 months to see if you like it.

The major differences between how you're going to raise meat birds and how they are "factory farmed" are in the amount of space the birds have while growing, their access to the outdoors, and what they eat. You can make sure your birds have a diet based on plant protein, if you like, or organic grains or pasture. Most home-raised chickens are also slaughtered under more humane and cleaner conditions than commercial chickens.

Some people also object to the limited genetics that form the basis of commercial chicken production and the way the broiler hybrids grow meat at the expense of their own health. The meat is fatter and softer, and there's more breast meat than on carcasses of other types of chickens.

But many people are getting used to a new taste in chicken. They're concerned about the inhumane conditions commercial meat chickens are often raised in and the way their food is handled before it reaches them. So they're growing their own or buying locally grown, humanely raised chickens.

Chickens that are raised on grass or given time to roam freely have more dark muscle meat, and the meat is a little firmer and a bit stronger in flavor. Your great-grandparents would recognize the taste of these chickens.

Many of you may be thinking that you want to raise some chickens to eat. You want to control the conditions they're raised in, what they're fed, and how they're butchered. Some people want to butcher chickens in ways that conform to kosher or halal (religious) laws. If you want to raise meat birds, here's what you need to think about:

- **Emotional challenges:** If you're the type of person who gets emotionally attached to animals you care for, or if you have children who are very emotional about animals, think carefully before you purchase meat birds. While traditional meat breeds can make okay pets, the broiler strain birds tend not to live too long and are tricky to care for if left beyond the ideal butchering time.

  We like our birds, and we don't like to kill them. But we love eating our own organically and humanely raised meat. How do we get around the emotional thing? We have someone else do the butchering, at another location. In almost every rural community, someone will butcher poultry for a fee. It adds to the cost of the final product, but it isn't much, and to us, it's well worth it.

That being said, we know how to butcher a bird, and we advise everyone who raises meat birds to learn how to do it. A day may come when you need the skill, and knowing about the process makes you aware of all the factors that go into producing meat, including the fact that a life was sacrificed so you can eat meat. You will appreciate even more the final product and all the skills it takes to produce it. In Chapter 16, we discuss butchering. Read the chapter; then see whether you can do what's necessary, if needed.

- **Do you have enough room?** You need enough space to raise at least 10 to 25 birds to make meat production worthwhile. If you live in an urban area that allows only a few chickens, producing meat probably isn't for you. Even in slightly roomier suburban areas, carefully consider your situation before raising meat birds. In these areas, you probably can't let chickens free-range or pasture them, so you will be raising the meat birds in confinement.

If you live in a rural area, feel that you have plenty of room, and think you can do your own butchering, you probably can try raising your own meat chickens. Start with a small batch and see how you do with the process.

Don't think that raising your own meat chickens will save you money at the grocery store. It almost never does. In fact, the fewer birds you raise, the more costly each one becomes. Economy of scale — for example, being able to buy and use 1,000 pounds of feed instead of two 50-pound sacks — helps costs, but most of us can't do that. You raise your own meat for the satisfaction and the flavor.

Enticed by fun and games or 4-H and FFA?

Showing chickens is a rewarding hobby for adults and an easy way for youngsters in 4-H or FFA to begin raising livestock (and possibly earn a reward!). Chickens can also be a good hobby for mentally handicapped adults. The birds are easy to handle, and care is not too complex. A few chickens can provide hours of entertainment, and collecting eggs is a pleasing reward. If you want pet birds, certain chickens tame easily and come in unusual feather styles and colors.

If you're considering raising chickens as show birds or as pets, consider the following requirements:

- **Space:** For showing, you often need to raise several birds to maturity to pick the best specimen to show. You may need extra room. Pet birds in urban areas need to be confined so they don't bother neighbors or run into traffic.
- **Time:** Show birds are often kept in individual cages, which increases the amount of time needed to care for the chickens.
- **Purchasing cost:** Excellent specimens of some show breeds can be quite expensive.

If you live in a rural area, you can indulge your chicken fantasy to the fullest and maybe get one of everything! Just use common sense and don't get more than you can care for or legally own.

Considering Neighbors

Neighbors are any people who are in sight, sound, and smelling distance of your chickens. Even if it's legal in your urban or suburban area to keep chickens, the law may require your neighbors' approval and continued tolerance. And it pays to keep your neighbors happy anyway. If neighbors don't even know the chickens exist, they won't complain. If they know about them but get free eggs, they probably won't complain, either. A constant battle with neighbors who don't like your chickens may lead to the municipality banning your chickens — or even banning everyone's chickens. Regardless of your situation, the following list gives you some ideas to keep you in your neighbors' good graces:

- **Try to hide housing or blend it into the landscape.** If you can disguise the chicken quarters in the garden or hide them behind the garage, so much the better. Don't locate your chickens close to the property line or the neighbor's patio area, if at all possible.
- **Keep your chicken housing neat and clean.** Your chicken shelter should be neat and immaculately clean.
- **Store or dispose of manure and other wastes properly.** Consider where you're going to store or dispose of manure and other waste. You can't use poultry manure in the garden without some time to age because it burns plants. It makes good compost, but a pile of chicken manure composting may offend some neighbors. You may need to bury waste or haul it away.
- **Even if roosters are legal, consider doing without them.** You may love the sound of a rooster greeting the day, but the noise can be annoying to some people. Contrary to popular belief, you can't stop roosters from

crowing by locking them up until well after dawn. Roosters can and do crow at all times of the day — and even at night. Roosters aren't necessary for full egg production anyway; they're needed only for producing fertile eggs for hatching.

- **If you must have a rooster, try getting a bantam one, even if you have full-size hens.** He will crow, but it won't be as loud. Don't keep more than one rooster; they tend to encourage each other to crow more.
- **Keep your chicken population low.** If you have close neighbors try to restrain your impulses to have more chickens than you really need. We suggest two hens for each family member for egg production. The more chickens you keep, the more likely you will have objections to noise or smells.
- **Confine chickens to your property.** Even if you have a 2-acre suburban lot, you may want to keep your chickens confined to lessen neighbor complaints. Foraging chickens can roam a good distance. Chickens can easily destroy a newly planted vegetable garden, uproot young perennials, and pick the blossoms off the annuals. They can make walking barefoot across the lawn or patio a sticky situation. Mean roosters can scare or even harm small children and pets. And if your neighbor comes out one morning and finds your chickens roosting on the top of his new car, he's not going to be happy.

  Cats rarely bother adult chickens, but even small dogs may chase and kill them. In urban and suburban areas, dogs running loose can be a big problem for chicken owners who allow their chickens to roam. Free-ranging chickens can also be the target of malicious mischief by kids. Even raccoons and coyotes are often numerous in cities and suburban areas. And of course, chickens rarely survive an encounter with a car.

  You can fence your property if you want to and if it's legal to do so, but remember that lightweight hens and bantams can easily fly up on and go over a 4-foot fence. Some heavier birds may also learn to hop the fence. Chickens are also great at wriggling through small holes if the grass looks greener on the other side.
- **Be aggressive about controlling pests.** In urban and suburban areas, you must have an aggressive plan to control pest animals such as rats and mice. If your chickens are seen as the source of these pests, neighbors may complain. Read Chapter 9 for tips on controlling pests.
- **Share the chicken benefits.** Bring some eggs to your neighbors or allow their kids to feed the chickens. A gardening neighbor may like to have your manure and soiled bedding for compost. Just do what you can to make chickens seem like a mutually beneficial endeavor.

- **Never butcher a chicken in view of the neighbors.** Neighbors may go along with you having chickens as pets or for eggs, but they may have strong feelings about raising them for meat. Never butcher any chickens where neighbors can see it. You need a private, clean area, with running water, to butcher. If you butcher at home, you also need a way to dispose of blood, feathers, and other waste. This waste smells and attracts flies and other pests. We strongly advise those of you who raise meat birds and have close neighbors to send your birds out to be butchered.

Finally, don't assume that because you and your neighbors are good friends, they won't care or complain about any chickens kept illegally.

Chapter 2

Basic Chicken Biology and Behavior

In This Chapter

- Checking out basic chicken anatomy
- Determining whether a chicken is healthy
- Getting a glimpse of how chickens behave
- Seeing how chickens interact with other animals

Most people can identify a chicken even if they've never owned one. A few chicken breeds may fool some folks, but by and large, people know what a chicken looks like. However, if you're going to keep chickens, you need to know more than that. To select and raise healthy birds, which involves understanding breed variations, identifying and treating illness, and talking about your chickens (particularly when seeking healthcare advice or discussing breeding), you need to be able to identify a chicken's various parts. In the first part of this chapter, we cover basic chicken biology.

When raising chickens, you also need to know how to tell whether a chicken is healthy, so we describe what a healthy bird looks like. Last but not least, to adequately care for and interact with birds, you need to know a bit about their daily routines — how they eat, sleep, and socialize — as well as their molting and reproduction cycles and behavior. This chapter is your guide.

Familiarizing Yourself with a Chicken's Physique

Domestic breeds of chickens are derived from wild chickens that still crow in the jungles of Southeast Asia. The Red Jungle fowl is thought to be the primary ancestor of domestic breeds, but the Gray Jungle fowl has also contributed some genes. Wild chickens are still numerous in many parts of southern Asia, and chickens have escaped captivity and gone feral or "wild" in many subtropical regions in other parts of the world. So we have a pretty good idea of the original appearance of chickens and their habits.

Our many dog breeds are an example of what man can do by selectively breeding for certain traits. Dog breeds from Chihuahuas to Saint Bernards derived from the wolf. During domestication, not only did the size change, but the color, hair type, and body shape were altered in numerous ways as well. Chickens may not have as many body variations as dogs, but they do have a few, and man has managed to change the color and "hairstyles" of some chickens as well.

Wild hens weigh about 3 pounds, and wild roosters weigh 4 to 4½ pounds. More than 200 breeds of chickens exist today, ranging in size from 1-pound bantams to 15-pound giants. Wild chickens are slender birds with an upright carriage. Some of that slender body shape remains, but modern chicken breeds have many body variations. When you bite into a juicy, plump chicken breast from one of the modern meat breeds, you're experiencing one of those body variations firsthand.

Domestic chickens come in a wide range of colors and patterns of colors. In the next chapter, we discuss some of the chicken breeds that have been developed. Modern chickens generally keep the distinct color differences between male and female, with the males remaining flashy and the females more soberly garbed. However, in some breeds, such as the White Leghorns and many other white and solid-color breeds, both sexes may be the same color.

Even when the chickens are a solid color, differences in the comb and shape of feathers help distinguish roosters from hens. However, in the Sebright and Campine breeds of chickens, the hens look like the roosters, with only a slight difference in the shape of the tail. To see the differences between roosters and hens, refer to Figure 2-1 later in this chapter.

It's really quite amazing how man has been able to manipulate the genetics of wild breeds of chickens and come up with the sizes, colors, and shapes of chickens that exist today.

Figure 2-2: Some different types of chicken combs.

Illustration by Barbara Frake

The comb acts like the radiator of a car, helping to cool the chicken. Blood circulates through the comb's large surface area to release heat. The comb also has some sex appeal for chickens.

The eyes and ears

Moving on down the head, you come to the chicken's eyes. Chickens have small eyes — yellow with black, gray, or reddish-brown pupils — set on either side of the head. Chickens, like many birds, can see colors. A chicken has eyelids and sleeps with its eyes closed.

Chicken ears are small openings on the side of the head. A tuft of feathers may cover the opening. The ears are surrounded by a bare patch of skin that's usually red or white. A fleshy red lobe hangs down at the bottom of the patch. In some breeds, the skin patch and lobe may be blue or black. The size and shape of the lobes vary by breed and sex.

If a chicken has red ear skin, it generally lays brown eggs. If the skin patch around the ear is white, it usually lays white eggs. A chicken may occasionally have blue or black skin elsewhere, but the skin around the ear will still be red or white. This coloring can help you decide whether a mixed-breed hen will lay white or brown eggs, if that's important to you.

Three breeds lay blue or greenish eggs: the Araucana, the Ameraucana, and the Easter Eggers. Those breeds have red ear-skin patches.

The beak and nostrils

Chickens have beaks for mouths. Most breeds have yellow beaks, but a few have dark blue or gray beaks. The lower half of a chicken's beak fits inside the upper half of the beak. When the bird is breathing normally, you should not see a gap where daylight shows between the beak halves. Also, neither beak half should be twisted to one side.

A bird's beak is made of thin, hornlike material and functions to pick up food. Beaks are present on baby chicks, and a thickened area on the end of the beak, called the *egg tooth,* helps them chip their way out of the eggshell. Chickens also use their beaks to groom themselves, running their feathers through their beaks to smooth them.

Chickens don't have teeth, but inside the beak is a triangular-shaped tongue. The tongue has tiny barbs on it that catch and move food to the back of the mouth. Chickens have few taste buds, and their sense of taste is limited.

At the top of the beak are the chicken's two *nostrils,* or nose openings. The nostrils are surrounded by a raised tan patch called the *cere.* In some birds, the nostrils may be partially hidden by the bottom of the comb. Birds with topknots have much larger nostril caverns. The nostrils should be clean and open. A chicken's sense of smell is probably as good as a human's, according to the latest research.

The wattles and the neck

Under the beak are two more fleshy lobes of skin, one on each side. These are called the *wattles.* They're larger in males, and their size and shape differ according to breed. The wattles are usually red, although in some breeds, they can be blue, maroon, black, or other colors.

The neck of the chicken is long and slender. It's made for peeking over tall foliage to look for predators. The neck is covered with small, narrow feathers, called *hackle feathers,* that all point downward. See the section "Finding out about feathers," later in this chapter, for more info.

Checking out the bulk of the body

A chicken's body is rather U-shaped, with the head and tail areas higher than the center. The fleshy area from beneath the neck down to the belly is called the breast. Some breeds — generally those that are raised for meat — have plumper breasts. Birds, of course, do not have mammary glands.

The area of the back between the neck and the tail is called the saddle. Saddle areas can be colorful in male birds. Wings attach to the body on both sides, just below the neck. Chicken wings have a joint in the center, and the bones are shaped a lot like human arm bones.

Finally, you come to the tail. Wild male chickens carry their tails in a slight arch, but the tails tend to be narrow and long, to better slide through the underbrush where the wild chicken lives. Many domesticated male chickens have been bred with exaggerated tail arches; wide, fanlike tails; and extremely long tail feathers, although some breeds have no tail feathers. Male tail feathers are often very colorful. The rooster's tail has no effect on his function, and tail feather differences were selected by man just for show.

Hens, or female chickens, tend to have small tails, either in a fan-type pattern or in a narrow arch pattern. The feathers may be the same color as the body or a contrasting color, but they're not as colorful as the male's.

Looking at the legs and feet

Most chickens have four toes, each of which has a nail. Three point toward the front, and one points to the back. Some breeds have a fifth toe. This toe is in the back, just above the backward-pointing toe, and does not touch the ground. It usually curves upward. All chickens also have a spur, although it's very small in most hens. A spur is like a toe that doesn't touch the ground; instead, it sticks straight out of the back of the chicken's leg. It's hard and bony. Roosters may have large, sharp spurs. Hens rarely have large spurs, and if an old hen gets a bigger spur, they are still smaller than a roosters. Roosters use their spurs for defense and fighting.

Checking out chicken skin

The skin of most chickens is yellow or white, with the exception of the Silkie, which has black skin under all feather colors, and the Egyptian Fayoumis, which has slate-colored skin. The color of a chicken's feathers does not determine the color of the skin. The color of chicken skin can be influenced by what the chicken eats. In chickens that eat a lot of corn and chickens that free-range and eat a lot of greens, yellow skin is a darker, more golden color; white skin turns creamy.

The flesh of chickens is thin, loose, and easy to tear. The feathers grow from follicles all over the chicken's skin. In one breed, the Turken, no feathers grow on the neck. Some think Turkens look like turkeys, but they're not crossed with turkeys.

The lower part of the leg is covered with thick, overlapping plates of skin, called *scales.* As chickens age, the skin on the lower leg looks thicker and rough. Many breeds of chickens have yellow legs, regardless of skin or feather color, but some breeds have white, gray, or black legs.

Finding out about feathers

Feathers cover most of the chicken's body. Most breeds of chickens have bare legs, but some have feathers growing down their legs and even on their toes. Other variations of feathering include *muffs,* puffs of feathers around the ear lobes; *beards,* long, hanging feathers beneath the beak; and *crests* or *topknots,* poofs of feathers on the head that may fall down and cover the eyes.

Some breeds of chickens appear fluffy, and some appear smooth and sleek. Chickens with smooth, sleek feathers are called *hard-feathered,* and birds with loose, fluffy feathers are called *soft-feathered.*

A feather mutation can cause the shaft of the feather to curl or twist, making the feathers on the bird stick out all over in a random fashion. Talk about your bad hair day! These birds are called *Frizzles.* The Frizzle mutation can occur in a number of chicken breeds.

Birds shed their feathers, beginning with the head feathers, once a year, usually in the fall. This shedding period is called the *molt,* and it takes about 7 weeks to complete. The molt period is stressful to chickens. We discuss molt later in this chapter and in more detail in Chapter 10.

Types of feathers

Contour feathers are the outer feathers that form the bird's distinctive shape. They include wing and tail feathers and most of the body feathers.

Down feathers are the layer closest to the body. They provide insulation from cold temperatures. Down feathers lack the barbs and strong central shaft that the outer feathers have, so they remain fluffy. Silkie chickens have body feathers that are as long as the feathers of normal chickens, but their outer feathers also lack barbs, so the Silkie chicken looks furry or fluffy all over.

Feathers also vary according to what part of the chicken they cover. The following list associates these various types of feathers with the chicken's anatomy:

- **On the neck:** The row of narrow feathers around the neck constitutes the *hackles.* Hackle feathers can stand up when the chicken gets angry. These feathers are often a different color than the body feathers, and they may be very colorful in male birds. In most male chickens, the hackle feathers are pointed and iridescent. Female hackle feathers have rounded tips and are duller.
- **On the belly and midsection:** The belly and remaining body areas of the chicken are covered with small, fluffy feathers. In many cases, the underside of the bird is lighter in color.
- **On the wings:** Chickens have three types of feathers on the wings. The top section, closest to the body, consists of small, rounded feathers called *coverts.* The middle feathers are longer and are called *secondaries.* The longest and largest feathers are on the end of the wing and are called *primaries.* Each section overlaps the other just slightly.
- **On the legs:** Chicken thighs are covered with soft, small feathers. In most breeds, the feathers end halfway down the leg, at the hock joint. In some breeds, however, the legs have fluffy feathers right down to and covering the toes.
- **On the tail:** Roosters have long, shiny, attractive tail feathers. In many breeds, the top three or four tail feathers are narrower and may arch above the rest of the tail. These are called *sickle feathers.* Hens have tail feathers, too, but they are short and plainly colored, and they don't arch.

Anatomy of feathers

Feathers are made of keratin, the same stuff that comprises your fingernails and hair. Each feather has a hard, central, stemlike area called a shaft. The bottom of the mature shaft is hollow where it attaches to the skin and is called a quill. Immature feathers have a vein in the shaft, which will bleed profusely if the feather is cut or torn.

Immature feathers are also called pinfeathers because when they start growing, they are tightly rolled and look like pins sticking out of the chicken's skin. They are covered with a thin, white, papery coating that gradually wears off or is groomed off by the chicken running the pinfeather through its beak. When the cover comes off, the feather expands. When the feather expands to its full length, the vein in the shaft dries up.

Chickens can lose a feather at any time and grow a new one, but new feathers are more plentiful during the molting period. The age of a chicken has nothing to do with whether a feather is mature.

On both sides of the shaft are rows of barbs, and on each barb are rows of barbules. The barbules have tiny hooks along the edge that lock them to their neighbors to make a smooth feather. When chickens preen themselves, they are smoothing and locking the feather barbs together.

Feathers grow out of follicles in the chicken's skin. Around each feather follicle in the skin are groups of tiny muscles that allow the feather to be raised and lowered, allowing the bird to fluff itself up.

The color of feathers comes both from pigments in the feather and from the way the keratin that forms the feathers is arranged in layers. Blacks, browns, reds, blues, grays, and yellows generally come from pigments. Iridescent greens and blues usually come from the way light reflects off the layers of keratin. The way the light reflects off the feather is similar to the way light reflects off an opal or pearl. Male chickens generally have more iridescent colors.

A Picture of Health

Although we cover chicken health more carefully in Chapter 10, we briefly discuss what a healthy chicken looks like in this section. Having this information may keep you from mistaking illness or deformity for the normal appearance of a chicken.

Following are some quick pointers for determining whether a chicken is healthy and normal:

- **Eyes:** Chicken eyes should be clear and shiny. When a chicken is alert and active, its eyelids shouldn't be showing. You shouldn't see any discharge or swelling around the eyes.
- **Nose:** Both nostrils should be clear and open, with no discharge from the nostrils.
- **Mouth:** The chicken should breathe with the mouth closed, except in very hot conditions. If cooling the bird doesn't make it breathe with its mouth closed, it is ill.
- **Wings:** The wings of chickens should be carried close to the body in most breeds. A few breeds have wings that point downward. (You need to study breed characteristics to see what is normal for your breed.)

The wings shouldn't droop or look twisted. Sometimes droopy wings are a sign of illness in the bird. A damaged wing that healed wrong won't affect the laying or breeding ability of the bird. However, some birds are hatched with bad wings, which is usually the result of a genetic problem. These birds should not be used for breeding.

- **Feathers:** In general, a chicken shouldn't be missing large patches of feathers. Hens kept with a rooster often have bare patches on the back and the base of the neck near the back. These patches are caused by mating and are normal. You should never see open sores or swelling where the skin is bare. Sometimes feathers are pulled out, particularly tail feathers, when capturing a bird. If the bird appears healthy otherwise and the skin appears smooth and intact, it's probably fine.

  A healthy bird has its feathers smoothed down when it is active. Some breed differences are noteworthy — for example, obviously, a Frizzle with its twisted feathers will never look smooth. A bird with its feathers fluffed out that isn't sleeping or taking a dust bath is probably ill.
- **Feet and toes:** The three front toes of chickens should point straight ahead, and the feet should not turn outward. The hock joints shouldn't touch, and the toes shouldn't point in toward each other. Chicken feet shouldn't be webbed (webbing is skin connecting the toes), although occasionally webbed feet show up as a genetic defect. You shouldn't see any swellings on the legs or toes. Check the bottom of the foot for swelling and raw, open areas.
- **Vent:** The feathers under the tail of the chicken around the *vent,* the common opening for feces, mating, and passing eggs, should not be matted with feces, and you shouldn't see any sores or wounds around it.
- **Mental state:** The chicken should appear alert and avoid strangers if it is in a lighted area. Birds that are inactive and allow easy handling are probably ill. Chickens in the dark, however, are very passive, and this is normal.
- **Activity level:** Here again, differences exist between breeds, but a healthy chicken is rarely still during the daylight hours. Some breeds are more nervous and flighty; others are calm but busy. In very warm weather, all chickens are less active.

On Chicken Behavior

Watching a flock of chickens can be as entertaining as watching teenagers at the mall. Chickens have very complex social interactions and a host of interesting behaviors. And like most domesticated animals, chickens prefer to be kept in groups. A group of chickens is called a *flock.*

Knowing a little about chicken behavior is crucial to keeping chickens. In this section, we briefly discuss some typical chicken behaviors so you can decide whether the chickens you are keeping are normal or just plain crazy.

Hopefully, knowing a little bit about chicken behavior may sway you if you're sitting on the fence about whether to raise chickens. Raising chickens is a fun hobby, even if you're raising them for serious meat or egg production. When the power goes out, you can go back to the times of your forefathers: Sit out on the porch and watch the chickens instead of TV.

Processing information

Being called a birdbrain is supposed to indicate that you're not very smart. Bird brains may be organized more like reptile brains than mammal brains, but plenty of evidence indicates that birds, including chickens, are pretty smart. Scientists have recently discovered that while the "thinking" area of a bird's brain may look different from the same part of a mammal's brain, birds are capable of thought processes that even some species of mammals can't achieve. And chickens' brains are able to repair a considerable amount of damage, something mammal brains can't do.

Birds, including chickens, understand the concept of counting and can be trained to count items to achieve an award. Most mammals who are said to count are actually responding to signals from the trainer. Birds also trick or deceive other birds, and even other animals, which means they must be able to understand the outcome of a future or planned action.

Some birds also mimic the sounds of other birds and animals; few other animals mimic sounds. Chickens, however, cannot be taught to talk as some bird species can, and they don't mimic other animals. Chickens probably fall about midrange on the intelligence scale of birds.

Chicken brains have a large optic area because vision is very important to their survival. A chicken can spot a hawk or hawklike object from a good distance away, and the brain immediately tells the chicken to run for cover or freeze, whichever will be most effective. They also learn to spot and avoid other predators quite quickly. We have known chickens that actually differentiate between different dogs and know whether they're friends or foes.

Chicken eyes are also adept at spotting the tiniest seed or the slightest movement of a bug. We have seen them pick up ant eggs and pick the seed off a bit of dandelion fluff. While human eyes can miss an expertly camouflaged tomato hornworm, even a big fat one, beady chicken eyes quickly spot it.

Communication

Chickens are very vocal creatures, and they communicate with each other frequently. Chickens are rarely quiet for long unless they are sleeping. The range of sounds that chickens make is wide and somewhat open to human interpretation, but we've attempted to define some of the sounds as follows:

- **Crowing:** The loud "cock-a-doodle-do" a rooster makes is the chicken noise most people know best. Roosters crow when they become sexually mature, and they don't do it just in the morning. The crow announces the rooster's presence to the world as ruler of his kingdom: It's a territorial signal. Different roosters have different crows — some are loud, some softer, some hoarse sounding, some shrill, and so on. Roosters crow all day long.
- **Cackling:** Hens make a loud calling noise after they lay an egg. Many times other hens join in. It can go on for a few minutes. Some people call it a signal of pride; others say it's a yell of relief!
- **Chucking or clucking:** Both roosters and hens make a "chuck-chuck" or "cluck-cluck" sound as a conversational noise. It occurs at anytime and can be likened to people talking among themselves in a group. Who knows what they discuss?
- **Perp-perping:** Roosters make a soft "perp-perp" noise to call hens over to a good supply of food. Hens make a similar noise to alert their chicks to a food source.
- **Rebel yelling:** Hey, it's hard to describe these noises, but chickens give out a loud holler of alarm when they spot a hawk or other predator. All the other chickens scatter for cover.
- **Growling:** All chickens can make a growling noise. Hens commonly make this noise when they're sitting on eggs and someone disturbs them. It's a warning sound and may be followed by an attack or a peck.
- **Squawking:** Grab or scare a chicken of either sex, and you'll probably hear this loud sound. Sometimes other chickens run when they hear the noise, and other times they're attracted, depending on the circumstances.
- **Other noises:** The preceding sounds are only some of the more common chicken noises. Baby chicks peep, hens make a sort of crooning sound when they're nesting, and some hens seem to be humming when they're happy and contented. Roosters make aggressive fighting noises. Sit around a chicken coop long enough, and you'll hear the whole range of sounds.

Table manners

Chickens are notorious for eating almost anything. Their taste buds are not well developed, and tastes that we consider bad don't faze them. This can be their downfall if they eat something like Styrofoam, paint chips, fertilizer, or other things that look like food to them. Good chicken-keepers need to protect their charges from eating things like pesticide-coated vegetation, plastic, Styrofoam beads, and other harmful items.

Chickens eat bugs and worms, seeds and vegetation, and meat. They can't break bones into pieces, but they will pick the meat off them. We've seen them eat snakes and small mice. They will pick through the feces of other animals for edible bits, and they'll scratch up the compost pile looking for choice nuggets.

It takes only about 2½ hours for food to pass completely through a chicken's digestive system. The food a chicken picks up in its beak is first sent to the *crop,* a pouchlike area in the neck for storage. The crop is stretchy and allows the chicken to quickly grab sudden food finds and store them for a slower ride through the rest of the digestive system. From the crop, food passes to the stomach, where digestive enzymes are added.

People often worry when they see a huge swelling on their chicken's throat. That swelling is generally a full crop, meaning that the chicken has just been a little greedy, much like a chipmunk when it fills its cheeks. Over a few hours, the "swelling" subsides as food in the crop is passed along the digestive system.

You may also notice chickens picking up small rocks or pieces of gravel, sometimes called *grit.* These go into the gizzard, just beyond the stomach, and help the chicken break down food like human teeth do. When chickens roam freely, they get plenty of grit for digestion. If confined, you may need to provide it — we discuss that in Chapter 8.

Both male and female chickens actively hunt for food a good part of the day. Hens sitting on eggs are an exception: They leave the nest for only brief periods of time to feed. Chickens that are confined still go through the motions of hunting for food, scratching and picking through their bedding, and chasing the occasional fly. If food is plentiful, chickens may rest in the heat of the day or stop to take a dust bath. Chickens don't eat at night or when they're in the dark.

Most breeds of chickens are equally good at finding food given the chance, with a few exceptions. The large, heavy, broiler-type meat birds are like sumo wrestlers — they prefer to park their huge bodies in front of a trough and just sit and eat. They don't do well if they have to rustle their own chow or if their food doesn't consist of high-energy, high-protein items.

Sleeping

When chickens sleep, they really sleep. Total darkness makes chickens go into a kind of stupor. They're an easy mark for predators at this point; they don't defend themselves or try to escape. If you need to catch a chicken, go out with a flashlight a couple hours after darkness has fallen, and you should have no problem, providing you know where they roost. Chickens also sit still through rain or snow if they go to sleep in an unprotected place.

Because they're vulnerable when they sleep, chickens prefer to *roost* (perch) as high off the ground as they can when sleeping. The more "street-savvy" birds also pick a spot with overhead protection from the weather and owls. Chickens like to roost in the same spot every night, so when they're used to roosting in your chicken coop, they'll try to go back home at nightfall even if they've managed to escape that day or are allowed to roam.

Socializing

With chickens, it's all about family. If you don't provide chickens with companions, they will soon make you part of the family. But chickens have very special and firm rules for all family or flock members. Chickens in the wild form small flocks, with 12 to 15 birds being the largest flock. Each wild flock has one rooster.

Ranking begins from the moment chicks hatch or whenever chickens are put together. Hens have their own ranking system, separate from the roosters. Every member of the flock soon knows its place, although some squabbling and downright battles may ensue during the ranking process. Small flocks make chicken life easier. In large flocks of 25 or more chickens and more than one rooster, fighting may periodically resume as both hens and roosters try to maintain the "pecking order."

The dominant hen eats first, gets to pick where she wants to roost or lay eggs, and is allowed to take choice morsels from the lesser-ranked hens. The second-ranked hen bows to none but the first, and so on. In small, well-managed flocks with enough space, the hens are generally calm and orderly as they go about their daily business.

Roosters establish a ranking system, too, if there's more than one in a flock. A group of young roosters without hens will fight, but generally an uneasy truce based on rank will become established. Roosters in the presence of hens fight much more intensely, and the fight may end in death for one of the roosters. If more than one rooster survives in a mixed-sex flock, he becomes a hanger-on — always staying at the edge of the flock and keeping a low profile.

If you have a lot of hens and a lot of space, such as in a free-range situation, each rooster may establish his own separate flock and pretty much ignore the other rooster except for occasional spats. How aggressive a rooster is depends on both the breed and the individuals within a breed. We've had some very aggressive roosters, to the point that they've been a serious nuisance or even a danger to the human caretaker. When a rooster becomes aggressive toward humans, the best option is the soup pot.

A rooster always dominates the hens in his care. Sorry, no women's lib in the chicken world. He gets what he wants when he wants it. And what he doesn't want is a lot of squabbling among his flock. When he's eating, all the hens can eat with him, and no one is allowed to pull rank. If squabbling among hens gets intense at other times, he may step in and resolve the problem.

A rooster can be much smaller and younger than the hens in the flock, but as long as he's mature, he's the ruler of the coop. But it's not all about terrorizing the ladies. The rooster is also their protector and guide, as well as their lover. He stands guard over them as they feed, shows them choice things to eat (usually letting them have the first bites), and even guides them to good nesting spots.

Roosters tend to have a favorite hen — usually, but not always, the dominant hen in the flock — but they treat all their ladies pretty well. They may mate more frequently with the favorite, but all hens get some attention.

Romance

We've established the rooster as the stern but loving protector of his family; now we talk about chicken romance, or the mating process. Roosters have a rather limited courtship ritual, compared to some birds, and the amount of "romancing" varies among individuals, too.

When a rooster wants to mate with a hen, he usually approaches her in a kind of tiptoelike walk and may strut around her a few times. Usually a hen approached this way crouches down and moves her tail to one side as a sign of submission. The rooster jumps on the hen's back, holds on to the back of her neck with his beak, and rapidly thrusts his cloaca against hers a few times. He then dismounts, fluffs his feathers, and walks away. Boastful crowing may also take place soon after mating, although crowing is not reserved just for mating. The hen stands up, fluffs her feathers, and walks away as well. Both may preen their feathers for a few minutes after mating.

A young rooster may mate several hens within a few minutes of each other, but usually mating is spread out throughout the day. A rooster may mate a hen even if he's infertile: Fertility drops as roosters age, and cold weather also causes a drop in fertility.

The celibate hen — living without a rooster

Hens don't need a rooster to complete their lives — or even to lay eggs, for that matter. A hen is hatched with all the eggs she's ever going to have, and she'll lay those eggs for as long as the hen lives (or until she's out of eggs), whether a rooster is around or not. The number of eggs a hen lays over her lifetime varies by breed and the individual. After the third year of life, though, a hen lays very few eggs.

Of course, without a rooster, no babies will be hatched from those eggs, but the eggs that we eat for breakfast don't need to be fertilized to be laid. Hormones control the egg cycle whether a rooster is present or not. And fertilized eggs don't taste differently — nor are they more nutritious — than unfertilized eggs.

Is a hen happier with a rooster around? Our guess is that she probably is because it fits the more natural family lifestyle of chickens. But hens are pretty self-sufficient, and our guess is that if they've never known life with a rooster, they really don't know what they're missing.

New life

At about the time an egg is to be laid, usually early in the morning, hormone levels in the hen rise, and she seeks out a nest and performs some nest-making behaviors: moving nest material around with her beak, turning around in the nest to make a hollow, and sometimes gently crooning a lullaby. Several hens may crowd into a nest box at one time, and they seem to be stimulated by the laying of hens around them.

After the egg is laid, hormone levels generally drop, and the hen hops off the nest and announces her accomplishment with a loud cackling sound. Other hens may join in the celebration. Morning is a noisy time in the henhouse. After laying an egg, the hen generally goes about her daily routine.

A hen that *does* have the instinct to sit on (incubate) eggs and hatch them and is in the mood to do so is called a *broody* hen. Many modern breeds of chickens no longer incubate eggs because the maternal instinct has been bred out of them. Thus, the eggs need to be artificially incubated (put in an incubator) to perpetuate those breeds. The instinct was bred out of them because when a hen is sitting on a clutch of eggs or raising her young, she doesn't lay eggs.

When a hen does go broody, she tries to sneak off and hide her eggs in a secret nest. If she can't, she commandeers one of the nest boxes in the coop. The hen lays about ten eggs before she starts sitting in earnest. Fertilized eggs are fine in their suspended animation stage until she decides to sit. The reason the hen doesn't start sitting on the eggs until about ten have built up is that nature intends for the chicks to hatch all at once so the mom will have an easier time caring for them. It takes 21 days for eggs to hatch.

We discuss the egg formation process in more detail in Chapter 15, and we discuss both natural and artificial incubation processes in Chapter 13.

When the chicks have hatched, their mom leads them out into the world, showing them how to eat and drink and defending them to the best of her ability. At night or when it's cold, the chicks snuggle under her for warmth. A hen can be very aggressive when defending her young, and many tales are told about hens giving their lives to defend their babies. Chicks stay with the mom at least until they're feathered, and often for 4 or 5 months.

Bath time

Another interesting behavior of chickens is their bathing habits. They hate getting wet, but they sure do love a dust bath. Wherever the coop has loose soil — or even loose litter — on the floor, you will find chickens bathing.

Chickens scratch out a body-sized depression in the soil and lie in it, throwing the soil from the hole into their fluffed-out feathers and then shaking to remove it. They seem very happy when doing this, so it must feel good. In nature, this habit helps to control parasites.

In the garden or lawn, these dust-bath bowls can be quite damaging, but you can't do much about it except put up a fence. If your chickens are confined all the time, they'll really appreciate a box of sand to bathe in.

Interacting with Other Poultry and Animals

The previous section discusses how chickens socialize with each other, but many new chicken owners want to know how chickens interact with other poultry or animals. Many people envision a happy barnyard mixture of chickens and other poultry, or perhaps goats and horses. You can have that peaceful environment, in most cases, if you have some knowledge of how chickens interact with other animals.

In this section, we give some general guidelines for interspecies interaction. Individual animals can vary in how tolerant they are. It's up to the chicken owner to carefully supervise animal introductions and intervene when an animal is in danger.

Dogs and cats

Dogs and chickens often don't mix well. In fact dogs can be a chicken-keeper's worst enemies — even your own dog. (It's your dog we talk about here; we discuss the neighbor's dog more in Chapter 9 when we cover fending off predators.)

We think that people can own both dogs and chickens (we own both) if some basic precautions are taken. Be very careful introducing dogs to chickens, even small dogs. Terriers and hunting breeds may be more likely to kill chickens but any breed or mix of breeds is capable of doing harm. Never leave dogs alone with chickens until you have observed the dog interacting with the chickens safely many times under various circumstances.

Some dogs simply ignore chickens, but to many dogs, chickens are just fun. They run and squawk, and it's all very exciting. Even if they don't catch them, dogs should never be allowed to chase chickens. Being chased is very stressful to a chicken: In their minds, a dog is the same as a fox or a wolf. Stressed chickens don't lay well, aren't friendly, and get sick more often. You may think it's cute to see dogs "herding" the chickens, but we assure you, the chickens don't.

Training a dog that shows interest in chasing or harming chickens rarely works unless the dog is a pup and you are a good and consistent trainer. If your dog wants to chase or, worse, kill chickens, the chickens and dogs must be confined separately or you must consider which animals you really want to own. Confining both the chickens and the dog gives you double protection.

Dogs will often climb over fences or dig under them if they are excited so plan accordingly. Electric wire or "hot wires" at the top of a good fence and 8 to 12 inches from the ground at the fence bottom will almost always keep dogs inside and are an easy way to secure large areas for dogs. The electric fence setups can be found at any farm store and they won't seriously harm the dog. Invisible fences set ups for dogs won't work to separate dogs from chickens because the chickens won't get a shock when they cross the line and may wander into the dog's reach.

Cats don't usually pose a risk to adult chickens. They will kill and eat chicks, however, so you need to protect the chicks from them. Wait to introduce cats to chickens until the chickens are full grown. Barn cats usually learn at a young age to avoid chickens because adult chickens can be very mean to kittens. Most barn or outside cats will simply ignore chickens. Cats often go hunting for mice and rats in chicken areas, but only rarely will a feral cat kill a chicken. Small bantams are most at risk, so we recommend keeping any small valuable bantams in pens that even your own cats can't get into.

Ducks and geese

In general, ducks and geese get along well with chickens. However, you may need more room than a small backyard to keep them with chickens. Ducks and geese need to be kept where they have a lot of space to roam and a place to bathe. Ducks and geese can also furnish you with edible eggs, but you need to know some info about these combinations.

You can mix ducklings or goslings (baby geese) with chicks in a brooder (a warm, protected spot for baby poultry — see Chapter 14) without them harming each other. However, you'll need a plan for meals. We usually recommend that chicks start out with medicated feed, but ducklings and goslings shouldn't have medicated starter feed because they're sensitive to the antibiotic used. Since you can't keep them from eating each other's feed, all the babies will need unmedicated feed. This compromise may then lead to more disease problems in the chicks.

We recommend using a higher-protein feed, such as broiler feed, for ducklings — the chicks will be okay with that choice, too. As adults, ducks, geese, and chickens can eat the same feed, although special feed mixes for ducks are available.

Keep in mind that ducks are messy, even when they're ducklings. They'll play in the water, and their droppings are more liquid than chicks, so brooders with ducklings need more frequent cleaning to keep them dry. Ducklings don't need to swim while in the brooder (although they will if they can fit inside the water container), and it's not recommended to let them bathe if you're keeping them with chicks. Goslings aren't quite as messy with water.

Be aware that, with the exception of Muscovy ducks, ducks and geese can be noisy, and neighbors may not welcome them.

You'll need to address one other consideration when keeping ducks with chickens. Don't keep male ducks with chickens without female ducks also being present. Ducks are often aggressive sexually. If they're deprived of their own females, they may mate with hens. Unlike roosters, male ducks (called drakes) have a penis and may hurt hens. We've heard of male geese (ganders) also mounting chicken hens, but it's much less frequent. We've seen roosters mount female ducks, too, but it's not as common. These matings don't result in fertile eggs.

Mother hens and ducks sometimes raise each other's babies when allowed to mingle freely. They may lay in each other's nests and sit on each other's eggs. Chicks don't usually follow a stepmama duck into water, but we have heard of it happening. Baby ducklings can confuse a hen when they pop into water to swim, but it rarely causes a problem.

Turkeys

We've kept turkeys and chickens together for many years and don't have problems with them interacting in a free-range situation. However, when breeding turkeys, having broody chickens around can be a problem. The birds will squabble over nests and disrupt egg incubation. Occasionally, individual turkeys and chickens develop a dislike for each other, and this fighting can cause distress to the other birds. If this happens, you need to separate those individuals.

Regarding "Blackhead"

People are often warned about keeping turkeys with chickens because of a disease called blackhead. Blackhead doesn't affect chickens much, but it can kill turkeys. The disease is rare now, and if you buy your chicks and poults (baby turkeys) from hatcheries, you probably won't have a problem. However, if you want to raise rare heritage breeds of turkeys, you may want to keep them away from chickens, just in case.

You can raise baby turkeys (poults) with baby chicks if you remember that the poults need more protein than chicks. You must feed a turkey starter to them (24 to 26 percent protein), or they will develop crooked legs and wings and may die. When the poults are about a month old, we advise separating the turkey poults and regular chicks if the chicks aren't broiler (meat) chicks (you can leave turkeys with broiler chicks until the broilers are butchered). After they're separated, you need to feed the growing chickens chick grower (lower protein). See Chapter 8 for more about feeding chickens. Turkeys need a high-protein feed until they finish growing.

Meat turkeys take from 3 to 5 months to reach butchering size; turkeys wanted for breeding and pets are grown at about 6 months. At that time, you can mingle the turkeys and chickens, if you want, and use the same feed as you do for the chickens.

You can use broody hens to hatch turkey eggs, even though incubation for turkey eggs is 28 days (7 days longer than for chickens). Turkeys also occasionally hatch chicken eggs. Either way, the chicks will probably be successfully raised by their step mommas. Turkey eggs can be eaten just like chicken eggs; they're generally white with brown freckles.

Turkeys and chickens have been known to mate with each other, although it is rare. Sometimes hybrid birds have hatched from these matings, but those birds are sterile.

Guineas

Guinea fowl can be kept successfully with chickens, but they're much noisier than chickens, and we don't recommend them when neighbors are close. They also fly well, wander extensively, and like to roost in trees at night.

Adult guineas can eat the same food as adult chickens. Baby guineas can be raised with chicks, but they should have a higher-protein feed, such as broiler chicken feed, of about 20 to 24 percent protein.

Guineas and chickens have mated with each other in rare circumstances and produced sterile offspring.

Pheasants and quail

In general, we don't recommend mixing pheasants and quail with chickens. As babies, these birds are fragile and require game bird starter as feed. As adults, they're often aggressive, particularly some pheasant breeds, and often kill chickens. We find that pheasants and quail tend to make chickens act wilder when kept together.

Chickens are sometimes used to hatch pheasant or quail eggs and can raise the babies successfully. Pheasants rarely mate with chickens and, when they do, produce sterile offspring.

Livestock

With the exception of pigs, it's mostly up to individual animals how chickens and large livestock interact. Pigs will eat chickens they can catch, and it's highly recommended that you keep chickens away from pig pens. Cattle, sheep, goats, and horses can occasionally be mean to chickens, but usually they ignore them.

Chickens will pick through manure to eat maggots and undigested food. This practice is great fly control, but if it disgusts you, keep the chickens out of pastures and stalls. It doesn't affect the taste or safety of chicken eggs, by the way.

One issue to worry about when keeping chickens with livestock is the health of the livestock. Don't let chickens poop on livestock feed or hay or in their water. Chicken waste can sicken livestock. Also be extremely careful to keep chicken feed where livestock can't get to it. A few handfuls of chicken feed may kill a horse or goat because it causes bloat or colic.

Chapter 3

A Chicken Isn't Just a Chicken: Your Guide to Breeds

In This Chapter

- Understanding breed basics
- Checking out breeds for eggs and meat
- Looking at show, pet, and rare breeds

A chicken breed is a group of chickens that look similar, have similar genetics, and, when bred together, produce more chickens similar to them. Breeds don't just emphasize visual differences, however. Just like most domesticated animals, chickens have been selected over time for their ability to fit certain needs that humans have for them. Some chickens are better as meat birds, some are better at laying eggs, and some are just better looking.

Chickens have been selected and bred for certain traits for many reasons, including desires to produce birds that

- Lay more eggs
- Lay different-colored eggs
- Grow faster
- Provide more breast meat
- Behave more calmly
- Are resistant to disease

In this chapter, we introduce you to breed terminology and discuss how chicken-keepers categorize breeds. Then we explain the most common breeds within each category and give you a mini-biography about each one — we've kept them short and sweet, promise!

Throughout this chapter, we give you an overview of the most common breeds that the casual chicken-keeper needs to know about. If you're interested in breeding purebred chickens or showing chickens, you'll want to dig a little deeper into breed characteristics and color varieties. The American Poultry Association publishes a book called *American Standard of Perfection* every few years that includes color photos and complete breed descriptions. You may be able to find it at a library or at your county Extension office, or you can purchase it from the American Poultry Association, P.O. Box 306, Burgettstown, PA 15021; phone 724-729-3459; `http://www.amerpoultryassn.com`.

What You Need to Know: A Brief Synopsis

In the next few sections, we define some terms you're likely to come across when researching chicken breeds. After that, we talk about some breeds of chickens that we consider good breeds — although what constitutes a good breed for you is subjective, and some of the breeds we don't mention here are still "good" breeds.

The breeds are organized into groups. The heading for each group directs your attention to the common use of the breeds in that group. Thus, if you want to keep chickens for eggs, you can go right to the breeds under "For Egg Lovers: Laying Breeds," and so on.

Common breed terminology

As you explore the world of chicken breeds, you're likely to come across a handful of commonly used terms: *purebreds*, *hybrids*, *strains*, *mutations*, and *mixed breeds*. In the following sections, we tell you what they mean and what you need to know to get up to speed in no time.

Purebreds

Purebred chickens are chickens that have been bred to similar chickens for a number of generations and that share a genetic similarity. If you breed two chickens from the same breed, you should get offspring that look much like the parents. If you own a backyard flock and want to produce new chicks every year from your flock, or if you're interested in showing chickens, you'll want to purchase purebred chickens.

While more than 200 breeds of chickens are known to exist, with many color varieties within those breeds, fewer than half of these breeds are common. In fact, the vast majority of chickens in existence today belong to one of just a few breeds — White (or Pearl) Leghorns, Rhode Island Reds, Cornish, and Plymouth Rocks — and crosses of those breeds. These breeds comprise the commercial chicken industry for eggs and meat. The other breeds can thank small-scale chicken-keepers like you for their continued existence: Those independent chicken-keepers see value in maintaining genetic diversity and keep them from disappearing.

Hybrids

Many common and well-known chicken "breeds" are not breeds at all, but hybrids. *Hybrids* result from crossing two purebreds (see the preceding section). Animal breeders have long known that crossing two purebred animals of different breeds often produces a hybrid animal with good traits from both parents, along with increased health and productivity.

Hybrids are for end use — that is, they're good for only one purpose. Hybrid chickens may produce the most tender meat in the shortest time or lots of big eggs. These birds can't be bred to each other to produce a new flock: When hybrids are bred together, the results are unpredictable. Some babies look like one parent, and some look like neither. Thus, to maintain a supply of hybrid birds, you must raise two separate purebreds as parent stock.

Strains

Both purebreds and hybrids can be further defined as particular strains. A *strain* is usually one breeder's selection and is based on how that breeder feels the stock should look. Purebred strains represent the basic breed characteristics but may be slightly bigger, more colorful, hardier, and so on. In the case of hybrids, the birds that result from mating two particular purebred birds may also be called strains if they're produced by only a single company or breeder.

Strains are often named, especially if they're produced in large numbers for commercial use. The Cornish and White Rock hybrid has several strains. Some have names, like Cornish X or Vantress, whereas others are identified by numbers. Some strains grow faster, some survive heat better, some have white skin, and so forth. The same genetics firm may offer several strains.

Mutations

Occasionally, a *mutation* pops up, causing a chicken to look or act differently than the birds it was bred from. This scenario is nature accidentally rearranging genetic material. Mutations can be good, bad, or unimportant. Sometimes, with careful breeding, good mutations can turn into breeds.

Mixed breeds

Just as the world of dogs has many mutts, the world of chickens has mutt chickens. Mixed-breed chickens are birds whose ancestry isn't known, and they're a combination of many breeds. Mixed-breed chickens can be a great way to start a home flock. In fact, mixed-breed chickens often result from a chicken owner starting with a variety of purebred chickens and letting them breed. If you just want some average layers or chickens for that country feel, go with mixed breeds. The only problem is that if you don't know the parents of a bird that turns out to be beautiful or very productive, you'll have a hard time breeding more birds like it.

Over time, flocks of mixed-breed birds that are allowed to reproduce indiscriminately tend to produce smaller, less productive, and perhaps less healthy birds. The chickens tend to revert back to the size, color, behavior, and laying habits of their wild ancestors.

How breeds are categorized

From the earliest times, humans and animals have been categorized in some way — hunters, gatherers, herders, and so on — and chickens are no different. To make choosing the breeds you want to raise easier, chickens are grouped into the following categories:

- **Dual-purpose breeds:** Dual-purpose breeds are the kind many people keep around the homestead. They lay reasonably well, are calm and friendly, and have enough meat on their bones that excess birds can be used for eating.
- **Egg layers:** The laying breeds won't sit on their own eggs. In the first year of production, they may lay 290 to 300 eggs and only slightly less for the next 2 years. They don't make good meat birds, although they can be eaten.
- **Meat breeds:** Meat birds were developed to have deeper, larger breasts; a larger frame; and fast growth. Most of the chickens we call meat breeds are actually hybrids, although some purebreds, such as the Cornish, are considered great meat birds. Meat-type birds generally don't lay well.
- **Show breeds:** Some chickens are kept for purely ornamental reasons. They generally don't lay a lot of eggs, although their eggs are certainly edible. Most ornamental-type birds don't make good meals, either, although any chicken can be eaten. These birds are kept for their beautiful colors or unusual feathers.
- **Bantam breeds:** Bantams are miniature versions of bigger chicken breeds or small chickens that never had a larger version. If there's no large version, they're called true bantams. These birds range from 2 to about 5 pounds.

Within almost all breeds, you'll find several color variations. Colors are hard to describe in print. If you're unsure about what a color should look like, refer to a good catalog or reference book, such as the *American Standard of Perfection* mentioned at the beginning of the chapter.

For many breeds, some color variance between male and female chickens is normal. Even if the chickens are a solid color, however, males generally have different tail feather structure, larger combs and wattles, and some iridescence to their feathers. Chapter 2 explains sexual differences.

If You Want It All: Dual-Purpose Breeds

Home flock owners often want chickens that will give them a decent amount of eggs and also be meaty enough so that they can use excess birds as meat birds. The eggs taste the same; you just don't get as many as you do from laying breeds. The meat tastes the same if you raise the dual-purpose birds like you raise meat birds, but their breasts are smaller and the birds grow much more slowly. Some of the birds classified as dual-purpose breeds in the following list were once considered to be meat breeds until the Cornish X Rock cross came along (see the section "Best Breeds for the Table"). Figure 3-1 illustrates common dual-purpose breeds, including these breeds:

- **Barnevelders:** The Barnevelder is an old breed that's making a comeback because of its dark brown eggs. Barnevelders are fair layers and are heavy enough to make a good meat bird, although they're slow growing. They come in black, white, and blue-laced, as well as other colors. These calm, docile birds are fluffy looking and soft feathered.
- **Brahmas:** The Brahma is a large, fluffy-looking bird that lays brown eggs. Brahmas are good sitters and mothers and, as such, are often used to hatch other breeds' eggs. They're good meat birds, but they mature slowly. Their feet are feathered, and they come in several colors and color combos. They withstand cold weather well, and they're calm and easy to handle.
- **Orpingtons:** Orpingtons deserve their popularity as a farm breed. They're large, meaty birds, and they're pretty good layers of brown eggs. They'll also sit on eggs and they're good mothers. Buff or golden-colored Orpingtons are the most popular, but they also come in blue, black, and white. These birds are calm and gentle. They can forage pretty well but don't mind confinement.

Figure 3-1: Common dual-purpose breeds — Barred Plymouth Rock (left) and Wyandotte (right).

Illustration by Barbara Frake

- **Plymouth Rocks:** These birds are an excellent old American breed, good for both eggs and meat. White Plymouth Rocks are used for hybrid meat crosses, but several other color varieties of the breed, including buff, blue, and the popular striped black-and-white birds called Barred Rocks, make good dual-purpose birds. They're pretty good layers of medium-size brown eggs, they'll sit on eggs, and they're excellent home meat birds. They're usually calm, gentle birds, good for free-range or pastured production.
- **Wyandotte:** These large, fluffy-looking birds are excellent as farm flocks. They're pretty good layers of brown eggs (some will sit on eggs), and they make excellent meat birds. They come in several colors, with Silver Laced and Columbian being the most popular. Wyandottes are fairly quick to mature, compared to other dual-purpose breeds. Most are docile, friendly birds.
- **Turken/Transylvanian Naked Neck:** With an odd name like this, you may think this breed was crossed with a turkey — but it isn't. This breed comes in many colors, but the distinguishing feature is a lack of feathers on the neck. The genes that make the neck bare also contribute to a good-sized, meaty breast area in the chicken and make it easier to pluck. These birds are fairly good layers of medium to large light brown eggs. While they may look ugly, Turkens are calm and friendly birds.

For Egg Lovers: Laying Breeds

If you want hens that lay a certain color of eggs, you can read breed descriptions or you can look at the color of the skin patch around the ear. Hens that have white skin around the ear generally lay white eggs. Hens that have

Breeding out motherly instincts

Wild chickens lay eggs for only a short time. They normally lay about 10 eggs, and then they stop laying and incubate the eggs. During the time they're sitting on the nest, and for a couple months afterward while they raise the chicks, they won't lay more eggs.

To get domestic chickens to lay more eggs, breeders chose chickens that were less inclined to sit on their eggs or take care of their chicks. After a while, breeds were developed that seldom sat on eggs and laid a lot more eggs than their ancestors. When people found that they could artificially incubate eggs instead of keeping some motherly hens around to raise the offspring of the more productive layers, selection for laying really took off.

red skin around the ears generally lay brown eggs, in any number of shades. (There's no way to tell what shade of brown from looking at the ear patch.) The breeds that lay greenish-blue eggs usually have red ear-skin patches.

It's important to remember that all colors of eggs have exactly the same nutritional qualities and taste.

White-egg layers

The white-egg layers in the following list (see Figure 3-2 for an illustration) are the most productive, although many others also exist. Although there are individual exceptions, white-egg layers tend to be more nervous and harder to tame than brown-egg layers. That may be why many home flocks consist of the latter. But if you want a lot of white eggs, the following birds are the best breeds to choose from:

- **White or Pearl Leghorn:** This bird accounts for at least 90 percent of the world's white-egg production. It's lightweight and has a large, red, single comb. Leghorns also come in other colors that don't lay as many eggs but are fine for home flocks. Leghorns tend to be nervous and don't do as well in free-range or pastured situations as other breeds. California Whites are a hybrid of Leghorns and a Barred breed that are quieter than Pearl Leghorns. Other hybrids are also available.
- **Ancona:** Anconas lay large white eggs. They're black feathered, and some feathers have a white tip that gives the bird a "dotted" appearance. They're similar in shape to the Leghorn. Anconas are flighty and wild acting. Originally from Italy, they're becoming rare and harder to find.

Figure 3-2: Common breeds that lay white eggs — Minorca (far left), white Leghorn (middle), and Hamburg (right).

Illustration by Barbara Frake

- **Andulusian:** The Andulusian chicken was once much rarer, but this breed has enjoyed an upswing in popularity in recent years. Originally developed in Spain, the bird is known for its beautiful blue-gray shade; the edges of the feathers are outlined in a darker gray. These chickens are good layers of medium to large white eggs. They are lightweight birds and like to fly, which can be a problem when confining them. Andulusians are a purebred breed of chicken, but the blue color doesn't breed true: Offspring can be black or white with gray spots, called a splash. These colors are equally fine as layers — they just can't be shown.
- **Hamburg:** One of the oldest egg-laying breeds, Hamburgs are prolific layers of white eggs. They come in spangled and penciled gold or silver, or solid white or black. Hamburgs of all colors have slate-blue leg shanks and rose combs. They're active birds and good foragers, but they're not especially tame.
- **Minorca:** Minorcas are large birds that lay lots of large to extra-large white eggs. They come in black, white, or buff (golden) colors. Minorcas can have single or rose combs. They're active and good foragers, but they're not easy to tame. The Minorca is another bird that's becoming hard to find.

Brown-egg layers

For home flocks, brown-egg layers are popular (see Figure 3-3). The brown eggs these birds lay can vary from light tan to deep chocolate brown, sometimes even within the same breed. As hens get older, their eggs tend to be lighter in color. Some of the best brown-egg layers follow:

Figure 3-3: Common brown-egg layer breeds — Australorp (left) and Rhode Island Red (right).

Illustration by Barbara Frake

- **Isa Brown:** This hybrid makes up the world's largest population of brown-egg layers. Isa Browns are a genetically patented hybrid chicken. Only a few hatcheries can legally produce and sell the chicks, which means chicks hatched at someone's home or at a hatchery other than a licensed one aren't selling real Isa Browns. Isas are a combination of Rhode Island Red and Rhode Island White chickens — which are considered to be separate breeds, not colors — and possibly other breeds. Isa Brown hens are red-brown in color, with some white under-feathering and occasional white tail feathers; the roosters are white. The hens lay large to extra-large brown eggs that range in color from light to chocolate brown. These calm and gentle birds are easy to work with and are also good foragers. They may be production birds, but they have great personalities and are people-friendly. The disadvantages are that they can't be kept for breeding (they don't breed true), and the roosters don't make good meat birds.
- **Amber Link:** This breed is a close relative to the Isa Brown, with a slightly darker brown egg that tends to be medium to large in size. Amber Links are white with some gold-brown feathers in the tail and wing area. They're productive and hardy, and they're are also calm, gentle birds. You'll probably be able to find them only at hatcheries that sell Isa Browns.
- **Red Star, Black Star, Cherry Egger, and Golden Comet:** All these are variations of the same breeding that produced Isa Browns. Some were developed from New Hampshires or other heavy breeds other than Rhode Island Reds. They're prolific layers of brown eggs, but they won't sit on eggs. They also don't make good meat birds because of their light frames. They're usually calm and friendly birds. If you're not going to breed birds and you just want good egg production, any of these chickens will fill the need.

- **Australorp:** A true breed rather than a hybrid, Australorps lay a lot of medium-size, light-brown eggs. Before Isa Browns, they were the brown-egg-laying champions. Both the hens and the roosters are solid black birds with single combs. They're calm, they mature early, and some will sit on eggs. They were developed in Australia from meat birds, and the roosters make moderately good eating.
- **New Hampshire Red:** New Hampshire Reds are similar to Rhode Island Reds, and the two are often confused with one another. True New Hampshire Reds are lighter red, they have black tail feathers and the neck feathers are lightly marked with black. They're more likely to brood eggs than Rhode Island Reds. New Hampshire Reds are usually calm and friendly but active. The breed has two strains: Some are good brown-egg layers, whereas others don't lay as well but are better meat birds.
- **Rhode Island Red:** Rhode Island Reds were developed in the United States from primarily meat birds, with an eye toward making them productive egg layers as well. They lay a lot of large brown eggs. Both sexes are a deep red-brown color and can have a single or rose comb. They're hardy, active birds that generally aren't too wild, but the roosters tend to be aggressive.
- **Rhode Island White:** Rhode Island Whites were developed from a slightly different background than Rhode Island Reds, which is why they're considered separate breeds and can be hybridized. They lay brown eggs. They birds are white with single or rose combs, and they're calm and hardy.
- **Maran:** These birds aren't recognized as a pure breed in all poultry associations. Sometimes referred to as Cuckoo Marans, *cuckoo* refers to a color type (irregular bands of darker color on a lighter background). The breed actually has several color variations, including silver, golden, black, white, wheat, and copper. Marans were once rare, but they're now popular for their very dark brown eggs (remember, though, eggs of different colors don't taste any different!). Not every Maran lays equally dark eggs. The eggs vary in size from medium to large. Most Marans are good layers, but they're not as good as some of the previously noted breeds. The various strains exhibit a lot of variation in the breed in terms of temperament and whether they will brood.
- **Welsummer:** Welsummers are also popular for their very dark brown eggs that are medium to large in size. The hens are partridge colored (dark feathers with a gold edge), whereas the roosters are black with a red neck and red wing feathers. As members of an old, established breed, Welsummers are friendly, calm birds. They're good at foraging, and some will sit on eggs.

Colored-egg layers

Like the brown-egg layers, the colored-egg layers are also popular with home flock owners (Figure 3-4 illustrates two). Colored-egg layers are a novelty. Despite catalogs that show eggs in a rainbow of colors, their eggs are actually shades of blue and blue-green. Sometimes brown-egg layers whose eggs are a creamy light brown are said to lay yellow eggs. Pink, red, and purple eggs are said to exist, but these colors are generally caused by odd pigment mistakes and often aren't repeated. Some colors also are in the eye of the beholder (or the camera exposure), and what you call pink we may call light reddish-brown.

- **Ameraucana:** The origin of the Ameraucana is debatable: Some say it developed from the Araucana (see the next bullet), whereas others say it developed from other South American breeds that lay blue eggs. Ameraucanas come in many colors. They have puffs of feathers or muffs at the side of the head, and some have beards. They have tails, which help distinguish them from Araucanas. Their temperament varies; however, many will sit on eggs. The eggs are blue or blue-green.
- **Araucana:** This breed of chicken is seldom seen in its purebred form. Many chickens sold as Araucana are actually mixes, so buyer beware. Purebred Araucanas have no tail feathers. Either they have tuffs, small puffs of feathers at the ears, instead of large "muff" clumps or they're clean faced. They don't have beards, and they have pea combs. Most have willow (gray-green) legs. The breed has many color variations. Araucanas lay small blue to blue-green eggs and are not terribly prolific layers. They're calm and make good brooders.

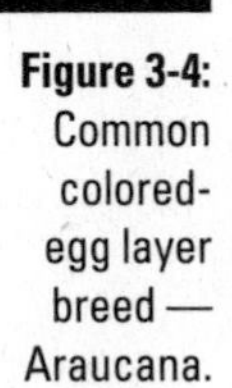

Figure 3-4: Common colored-egg layer breed — Araucana.

Illustration by Barbara Frake

- **Easter Egger:** These birds are mutts in the world of chickens because no one is quite sure of their background. They're usually a combination of the previous breeds and maybe some other South American blue-egg layers or other layers. They can be a bit more prolific in egg laying, but the egg color and temperament of the birds, as well as their adult body color, range greatly. They lay shades of blue, blue-green, and olive eggs.

Best Breeds for the Table

In the not-so-distant past, most meat chickens were young males that were the excess offspring from laying or show birds. They were kept just long enough to make a good meal — and that usually meant about 5 to 6 months of feeding and caring for the young roosters. The excess males in most heavy, generally brown-egg-laying breeds of chickens were used as meat. And in your great-grandparents' day, any hen that quit laying was also used for meat.

About 50 years ago, breeders began to create hybridized strains of chickens specifically for meat (see Figure 3-5). They grew fast and gained weight quickly on less feed than other chickens. They had more breast meat, something consumers seem to want. Over the years, these hybrids — particularly one hybrid, a cross between White Rocks and White Cornish breed chickens — have dominated the commercial meat market. In fact, almost every chicken you can buy in a grocery store is a strain of the previously mentioned cross.

Figure 3-5: Common meat breeds — Jersey Giant (left) and Cornish X Rock (right).

Illustration by Barbara Frake

Most meat birds stem from one hybrid, but home flock owners can consider other good meat breeds as well. Many backyard breeders are trying to develop meat breeds that grow quickly and have good meat yield, but are hardy and active enough to do well in pasture or free-range situations.

The meat of these other meat breeds may taste slightly different if the birds are raised on pasture and take longer to grow. The meat is a little firmer, and there's less breast meat. The taste is described by some as "old-fashioned" chicken flavor, which is hard to describe unless you've eaten both types. If the chickens eat commercial feed and aren't allowed to get too old before butchering, the older type meat birds taste much like hybrid meat birds. The advantage of hybrids is that they grow about twice as quickly and on less feed than conventional meat breeds.

- **Cornish X Rock hybrids:** Almost all chickens sold for meat today are a hybrid of White Cornish and White Rock chickens. They've been tinkered with until we now have an extremely fast-growing bird with huge breasts. They're often ready to butcher at 10 weeks, and they do a remarkable job of converting feed into meat. They have soft, tender-textured, bland-flavored meat and lots of white breast meat. Both sexes make good eating and grow at nearly the same rate. If you have eaten only store-purchased chicken, you certainly know how they taste.

  Different strains exist, even some where feather color has been introduced. Basically, however, commercial meat birds are Cornish-Rock crosses. These birds have special nutritional needs and aren't good for free-range or pastured poultry. They're excellent if you want quick, plump meat birds, but they're not good for anything else.

  These breeds gave up some features to achieve these "meat goals." They're extremely closely related genetically, with three or so large firms controlling the production of parent stock. They have to be managed carefully and fed a high-protein diet to avoid problems with their legs and hearts. Basically, they're basketballs with feathers, they're inactive, and they spend most of their time eating. They can't reproduce normally.
- **Cornish:** The Cornish is an old breed known for its wide stance and big breasts. Several colors other than the white ones used to make meat hybrids exist, including red, buff, and laced patterns. They have a rose comb, and their feathers are tight and sleek. These birds are poor layers of small white eggs; some will brood. They can be aggressive with each other, but they're fairly tame with keepers. Cornish chickens are only fair foragers and are better raised in pens.
- **Jersey Giants:** This American breed comprises the world's largest chickens. They come in black or white. They're meaty birds, but they grow slowly. Jersey Giants are fairly good layers of medium-size brown eggs. They're calm, they're good at foraging, and they make good brooders.

- **Freedom Rangers/Red Rangers:** These birds were developed to be good pasture-fed meat birds, primarily for the organic meat market. They thrive on lower-protein feed and are better foragers than Cornish-Rock crosses. Most are red in color, with a few barred or black spotted feathers; gray and "bronze" colors are also seen. These birds take a few weeks longer than the Cornish-cross broilers to be ready to butcher, but many people think they taste better. They're also less likely to suffer from leg problems or die of heart failure than the Cornish-cross broilers. Rangers can be hard to find, and the breed still has a lot of variability: Some birds mature faster and have more meat than others.

Show Breeds

Some breeds of chickens exist today mainly for pleasure (see Figure 3-6 for a peek at some). They may have been used as layers or meat birds in the past, but better breeds came along and replaced them. The practice of showing chickens keeps many of these breeds alive. Some of the most

Figure 3-6: Common show and pet breeds — Old English Game (top left), Cochin (top right), and Polish (bottom).

Illustration by Barbara Frake

beautiful chickens may not be good layers or meat birds, so we justify keeping them around by raising them to show. They make excellent lawn ornaments and pets, too:

- **Cochin:** Cochins originated in China. They're big, fluffy balls of feathers, with feathers covering the feet, and they come in buff, black, white, and partridge colors. These popular show birds are excellent brooding hens and love to raise families. In fact, they're often used to hatch other breeds' eggs. Their own eggs are small and creamy tan. Cochins are calm and friendly, but they may be picked on in a flock with active breeds.
- **Polish:** Polish chickens are small, silly-looking birds with a floppy crest of feathers that covers their eyes. Their crests may block their vision and make them seem a little shy or stupid, or cause them to be bullied by other chickens. If you're not showing them, trim their crests so they can see better. Some Polish also have beards. These birds come in several colors: One of the most popular is a black-bodied bird with a white crest. Polish lay small, white eggs that they generally won't sit on.
- **Old English Game/Modern Game:** Both of these breeds were once bred for fighting but are now used for show. People either like or hate the look of these birds. They stand very upright, with long necks and legs and tight, sleek feathers. They come in numerous colors. They're active and aggressive birds. Modern Games are larger and heavier. Both types lay small white eggs. Old English Games are good brooders and mothers; Modern Games are less so.
- **Appenzeller Spitzhauben:** This small, sprightly breed has topknot that often looks like a Mohawk and sports a black-and-white polka dot feather pattern. It has become a great show bird for youth, even though the Switzerland-based breed isn't officially recognized in many U.S. poultry shows. The birds are fairly good layers of white eggs.
- **Cubalaya:** This breed came to us from Cuba in the last century and has both full-sized and bantam chickens. The Cubalaya's distinguishing feature is its long, flowing tail that it carries very low. In show chicken terms, this type of tail is called a lobster tail. The birds come in a variety of colors. They also lack a spur, the sharp, hooklike structure on the back of a chicken's leg used for fighting. It's difficult to find a bird with a good tail — and also hard to then keep that tail in good condition. The birds are fair layers of cream-colored eggs and are calm and friendly, although they do like to fly and they range widely.

- **Houdans:** The Houdan is an oddity because it has five toes, compared to most chicken breeds that have four. It's an old breed from France and comes in two colors, white and also black splashed with white, called a mottled Houdan. They also have big, fluffy topknots, similar to the Polish. This type matures quickly and was once used for meat, although it is a smaller bird. It's a pretty good layer of white eggs. Houdans, like other crested, topknot breeds are often picked on by other types of chickens and seem a little stupid. Trimming the feathers so they can see does wonders. These birds are generally calm and friendly.

Perfect for Pets: Bantam Breeds

In chickens, selective breeding using the smallest birds of each generation has produced small breeds called *bantams*. A few breeds have some dwarf genes. People have always enjoyed seeing miniature versions of domestic animals. Bantam chickens are primarily kept for show or as pets, as are most other small versions of domestic animals.

Almost every standard chicken breed has bantam varieties, and some bantam breeds exist only in the small size. When the bantam has a full-size counterpart, the breed description is basically the same, except that the bantams are much smaller. When a chicken has no full-size representative, it is called a *true bantam* (see Figure 3-7).

Bantams are seldom heavy layers, but the small eggs are still good to eat and taste exactly like other chicken eggs. The eggs are the same color as eggs laid by a full-size example of the breed. However, bantams do not make good meat birds. Because most bantam breeds look just like the full-size version of the breed, we describe only true bantams here — breeds that don't have full-size counterparts:

- **Silkie:** The Silkie is probably one of the most popular pet chickens. Silkies may be quite small or a bit larger, but all have distinctive furlike feathers. They're so cute that you may want to cuddle them, and some don't mind being handled. Some have beards, and they come in black, blue, white, buff, and other colors. The skin of Silkies is black, and their combs and earlobes are dark maroon-red, with a turquoise area around the ear. Silkies are the ultimate mothers. They love to sit on eggs and happily raise anyone's babies. They're sometimes picked on by more active breeds. Silkies lay white eggs.

Figure 3-7: True bantams — the Silkie (top) and the Japanese bantam (bottom).

Illustration by Barbara Frake

- **Japanese:** The Japanese bantam has tiny, short legs and a high, arched tail, often in a color that contrasts with the body. It comes in many colors and color combinations and weighs about a pound. Most Japanese bantams are friendly, although the roosters sometimes get mean. Japanese bantams lay white eggs.
- **Serama:** The Serama bantam is the smallest chicken breed in the world (about 1 pound or less) and one of the newest, developed in only the last 50 years. Chickens from this breed can be hard to find and expensive. The tiny birds have an odd V shape to them, with a full breast held very upright and a tail carried straight up. The wings point down and touch the ground. Seramas come in a variety of colors, and there is a Silkie feathered variation, too. The tiny chicks need to be raised with special care, and the birds are not cold hardy; they require heated housing in the winter.
- **Sebright:** The Sebright bantams come in gold and silver as base colors. Each feather is outlined in black, making a pretty pattern. They can be hard to sex because the roosters and hens look so similar. Sebrights have a rose comb (short and crumbled looking) that comes to a spike point in the back. They're active birds and can be aggressive with each other.

- **Antwerp Belgian Bantam:** These birds are very small. They have a rose comb like the Sebright, a muff of feathers around the ear lobe, and a fluffy beard under their beak. They come in several colors and are active and friendly little birds.
- **Mille Fleur:** The color of Mille Fleurs is hard to describe. The base feather color is reddish-gold over the body and wings, with black on the feet and tail. But each feather has a white dot on the end and then a black area, giving the bird a spangled appearance. The feet of Mille Fleurs are covered in feathers, or "booted." The breed has two subvarieties: bearded and nonbearded. Bearded birds have muffs of feathers under the earlobes and a beard that runs under the beak. These beautiful little birds make excellent pets.
- **Porcelain:** These tiny bantams are much like the Mille Fleur, with bearded and nonbearded categories. The color of the Porcelain is pale yellow with a blue area just before the white "spangle" on the feather tip. The tail and feet feathers are blue-gray tipped in white. These birds are quiet and gentle.

Heritage and Rare Breeds

Determining which breeds are rare is difficult because many poultry breeds are getting scarce. The genetics of many breeds need to be preserved because the genetic pool of chickens is becoming limited. Poultry breeds go through fads: For example, the current craze is for chickens that lay very dark brown eggs. Welsummers, Marans, and Penedesencas were once very, very rare in the United States, but they're rapidly gaining in numbers due to the dark brown egg craze.

- **Buckeye:** The Buckeye is an American breed, developed by a woman in Ohio in the late 1800s. Buckeyes are a deep brown color, with some black in the tail. Although they bear some similarities to Rhode Island Reds, Buckeyes make better meat birds because of their deeper, broad-breasted bodies. Good foragers and winter hardy, they're also good layers of brown eggs — perfect homestead chickens that deserve more attention.
- **Chantecler:** The Chantecler was developed in Canada as a homestead chicken for meat and eggs. It comes in white and partridge colors. Chanteclers lay brown eggs, are good foragers, and mature quickly. They will brood eggs. They're not particularly people-friendly birds.

- **Delaware:** Delawares are layers of medium to dark brown, medium-size eggs. They were developed in the United States and are starting to make a comeback after almost fading away as a breed. They're white, with some black ticking in the feathers of the neck and tail. They're calm, they sometimes brood their eggs, and they make decent meat birds, too. They're good layers for cold areas.
- **La Fleche:** The La Fleche is known as the Devil Bird because its comb is shaped like a horn. La Fleches are good layers of large, creamy white eggs. They come in black, blue, white, and cuckoo colors. Fairly good as meat birds, they're quick to mature. They're now quite rare.
- **Faverolle:** Faverolles are pretty and also efficient egg layers. Their eggs tend to be medium in size and creamy rather than pearl white. Faverelles come in two colors: white and salmon. (The salmon color is actually black and white with fawn, or salmon, wing and back areas.) The Faverolle has a "muff" of feathers around the ears, a beard, and lightly feathered feet with a fifth toe. Faverolles are calm and tame, good for home flocks, and will sometimes sit on eggs.
- **Penedesenca:** Still fairly rare in the United States, Penedesencas were developed in Spain and are good layers of small, very dark brown eggs. The birds come in several colors and have an unusual comb shaped like a crown. They're rather wild, they withstand heat well, and they're good free-range birds.
- **Campine:** The Campine is an old breed that originally came from Belgium. It has two color varieties, gold and silver. The gold has golden feathers marked with black bars, and the silver is gray-white with black bars. When the two colors are crossed, they produce chicks that can be sexed by color right after hatching. Female chicks are a red-buff color, and males have a gray cap on their head. Campines were originally developed as a dual-purpose breed and are fairly good layers of white eggs. Both colors are rare now, with silver being the rarest.
- **White-Faced Black Spanish:** This breed has been a little more popular recently and isn't quite as hard to find as it once was. The breed originated in Spain, as the name suggests. These birds are tall and graceful, with glossy black feathers, large bright red combs, and large white earlobes, which contribute to their name. As the birds mature, the rest of the facial area also becomes white. They lay white eggs. These birds are active and can be aggressive with other chickens.
- **Egyptian Fayoumis:** The Egyptian Fayoumis are an old breed originating along the Nile River in the Middle East and brought to the U.S. this century to utilize their traits of quick maturity and resistance to disease. The birds are small, about 2 pounds in weight. Egyptian Fayoumis have slate-blue skin, silver neck feathers, and black-and-white barred body feathers. The birds are pretty good layers of small white eggs. They fly well, tend to be on the wild side, and are quite noisy.

Chapter 4

Buying Chickens

In This Chapter

- Determining the age, sex, and number of birds to buy
- Starting with chicks and adults
- Figuring out where to buy birds and what to look for

Before you become a chicken owner, you need to make some important decisions. A little planning goes a long way toward a good first attempt at raising chickens.

In this chapter, we help you decide whether you want to start with chicks or adult birds and how many and what type of chickens to buy. We talk about how to go about buying chicks and adult birds, including hints on where to buy chickens and how to choose healthy birds.

Planning Your Flock

Chickens are social flock creatures, so you need at least two chickens for them to be totally happy. How many chickens it takes for *you* to be happy is another matter.

Check the rules and regulations of the municipality where you live to see how many chickens you're allowed to keep, if any, before you make any purchase decisions (see Chapter 1).

In the sections that follow, we help you decide how many chickens to buy and whether to start with eggs, chicks, or adults, and we help you decide where to get them.

Deciding what you'll start with: Eggs, chicks, or adults

So you've made the decision to keep chickens. Hopefully you have an idea of why you want them, too. (If not, head back to Chapter 1, because you need to know why you want them before you can start making purchase decisions.) Now you need to decide how to begin your chicken-raising experience, and one way — eggs, chicks, or adults — will probably be best for you. Although we recommend starting with either chicks, started chicks, or fully matured adults, in this section, we cover the pros and cons of all three options.

Starting with fertile eggs

If you're thinking of starting your chicken-raising adventure with fertile eggs, you may want to reconsider. Each fertile egg has an embryo in suspended growth. Fertile eggs are difficult to store correctly and even more difficult to ship — they're likely to break or become chilled or overheated. They must be packed so that they don't experience too much jostling and shaking, because too much shaking kills the embryo. Catalogs and breeders that sell fertile eggs rarely guarantee that any will hatch. Fertile eggs usually cost nearly the same as chicks, too. Finally, incubation with even the best models of incubators can be a tricky process.

The only source for some rare or heritage breeds may be fertile eggs that you have to hatch yourself. If you're faced with this situation, or if you insist on starting with fertile eggs, try to find a local source of eggs: Your chances of hatching the eggs will be better. If you plan to hatch eggs, get your incubator set up and adjusted before the eggs arrive and place them in it promptly upon arrival. Don't expect more than about 50 percent to hatch.

Before buying expensive hatching eggs of rare breeds, you may want to practice hatching less expensive eggs of common breeds to get some experience.

Starting with chicks

Many people start with chicks because they fall in love with some at the farm store or someone gives their kids chicks for Easter. Beginning your chicken-raising experience with baby chicks is probably the most economical and practical way to start, for the following reasons:

- You can choose from a wide assortment of breeds.
- Chicks are less likely to carry disease and parasites than older birds, especially if you buy them from a well-run hatchery.
- You don't have to guess how old the birds are.

- Most hatcheries will sex the chicks for you, so you don't have to guess at sex.
- Chicks usually cost less than older birds.

Many mail-order hatcheries operate across the United States. You can order chicks by catalog or from online sites. Some have dozens of breeds available; others specialize in one or two breeds. Baby chicks survive the trip through the mail pretty well. In the "Mail-order hatcheries" section later in this chapter, we discuss the mail-order process.

When you mail-order baby chicks, you're usually required to order at least a minimum number of chicks. Usually this is 25 chicks, which may be 20 more chicks than you need or want. If you live near a hatchery and can pick up the chicks, you may be able to purchase fewer. You may also have local chicken breeders in your area who will sell you just a few chicks. A local feed store may be able to tell you who sells chicks, too.

Chicks are cute, but for many people, they may not be the best way to begin keeping chickens. Baby chicks do have their drawbacks:

- **They need TLC.** All baby animals take extra time and effort to care for.
- **They're fragile.** Small children and pets can easily damage or kill them.
- **They require special brooding equipment.** They need to be kept warm and protected.
- **Their quality is hard to judge.** Judging the quality of a chick is difficult, so if you want to show chickens, you need to keep a lot of chicks until they're adults so you can then pick the best of the bunch.
- **They need time to mature.** If you want chickens for eggs, know that pullet (female) chicks need at least 5 months to mature and begin to lay eggs. Some breeds need several months longer.
- **Their sex is hard to determine.** While hatcheries can do a pretty good job of sexing chicks, you may not end up with as many hens as you expect, and then you'll have extra roosters to deal with, too. If you buy your chicks from a private breeder, his or her skills at sexing chicks may be poor, and you may not get any hens at all.
- **They need extra protection.** If you just want chickens for living lawn ornaments, it's going to be several months before they can be allowed to roam safely. You may want to consider adult birds if you're the impatient type.

To learn more about what it takes to raise baby chicks, turn to Chapter 14.

Instead of newly hatched chicks, a few hatcheries and some local breeders offer what are called *started birds*, or older chicks. The exact definition of started chicks varies: In some cases, it means chicks that are a few weeks; other times, it means birds that are nearly grown. Started chicks are easier to sex, and you may have a better idea of their show quality. They take less time to reach productive age, a plus if you're buying layers. Meat chickens grow so fast that they're seldom offered as started birds. However, the number of breeds you can purchase as started birds is less than the number of breeds available as baby chicks. The birds cost considerably more to ship because they're heavier and require special handling. If you want started chicks, order them early in the season because the supply is usually limited.

Make sure you know what age the offered birds are, particularly if you're interested in avoiding the use of a brooder.

Starting with adults

You may worry that baby chicks will be too hard to raise. Or maybe you want egg production right away, or you want to be able to assess the show qualities of the birds you buy. In these cases, your best bet is to start with adult birds.

Buying adult birds has both pros and cons. Following are some of the negatives:

- Many people find it impossible to tell how old a chicken is once it's an adult.
- Older hens lay fewer or even no eggs.
- Old roosters may not be fertile.
- Some people have a hard time sexing adult birds, although with most breeds, a little experience soon helps with that.
- Adult birds may have been exposed to many diseases and are more likely to have parasites than chicks. You'll need to examine the birds carefully before you buy.

On the plus side:

- You can quickly assess the quality and color of an adult bird.
- Adult birds require less fussing to get them established in their new home.
- Young adult layers will quickly start providing you with breakfast.

Choosing the Sex

The sex you choose depends on your purpose for raising chickens. Use the following list of reasons to decide whether you need roosters, hens, or both:

- **Show birds or pets:** People rarely keep only roosters unless they just want a pet or they're raising show birds. (In most breeds of chickens, the male is the most colorful and makes the best show bird.) Although roosters can become aggressive as they age, a single pet rooster kept without hens rarely becomes aggressive.
- **Meat birds:** Many people order only *cockerels* (young males) because they grow faster and larger than *pullets* (young females). Cockerels can also be cheaper than pullets in some breeds, but in the broiler strains, they often cost more. You don't have to worry about broiler-type cockerels fighting unless you wait way too long to butcher them.
- **Egg producers:** If you want layers, order sexed pullets or buy adult hens. They cost more, but it's worth it.
- **Breeders:** If you want to breed more than two purebred breeds of chickens or different color varieties, you need more than one rooster. You'll need to plan your housing so that you have two or more separate flocks.

Hens don't need a rooster around to produce eggs or to live a fairly normal life, although they seem to appreciate having a male around. If you live close to neighbors and think they may be bothered by a rooster crowing, you don't have to keep a rooster with your hens. But if you like the sound of a rooster, think roosters are pretty, or feel it's more natural for the chicken family to have one, you don't need more than one rooster with your flock unless you're breeding chickens (see Chapter 12).

Getting the right number of chickens

No matter how many chickens you intend to have eventually, if you're new to chicken-keeping, it pays to start off small. Get some experience caring for the birds and see whether you really want to have more. Even if you have some experience, you may want to go to larger numbers of birds in steps, making sure you have proper housing and enough time to care for the birds at each step.

Because chickens are social and don't do well alone, you need to start with at least two birds: two hens or a rooster and a hen. (Two roosters will fight!) Beyond two birds, the number of birds you choose to raise depends on your needs and situation:

- **Layers:** You can figure that one young hen of an egg-laying strain will lay about six eggs a week, two will lay a dozen eggs, and so on. If the birds aren't from an egg-laying strain but you still want eggs, count on three or four hens for a dozen eggs a week. So figure out how many hens you need based on how many eggs your family uses in a week — just don't forget to figure on more hens if you don't get them from an egg-laying strain.
- **Meat birds:** It really doesn't pay to raise just a few chickens for meat, but if your goal is to produce meat and space is limited, you can raise meat birds in batches of 10 to 25 birds, with each batch of broiler strains taking about 6 to 9 weeks to grow to butchering size. If space and time to care for the birds aren't problems, determine how many chickens your family eats in a week and base your number of meat birds on that.

  If it takes 6 to 9 weeks to raise chickens to butchering age and your family wants two chickens a week, you probably want to buy your meat chickens in batches of 25 and start another group as soon as you butcher the first. Or if you want a rest between batches, raise 50 to 60 meat chicks at a time and start the second batch about three months after the first. Remember that frozen chicken retains good quality for about 6 months.
- **Pet and show birds:** When you're acquiring chickens for pet and show purposes, you're limited only by your housing size and the time and resources you have to care for them. Full-size birds need about 2 square feet of shelter space per bird; bantam breeds need somewhat less. Don't overcrowd your housing.

  If you're going to breed chickens to preserve a breed or produce show stock, plan on at least two hens for each rooster, but not more than ten. In some large breeds with low fertility, you may need a ratio of five or six hens per rooster.

Counting the Costs

Anytime you start a hobby or begin producing food for the household, you run into startup and maintenance costs, and keeping chickens is no exception. However, chickens are more economical to purchase for pets or as food-producing livestock than most other animals. Unless you're looking for expensive rare breeds, most people can start a small flock (4 to 25 chickens) for less than $50. Regardless of whether you're starting with 4 or 25, use the following list of tips to keep costs down when purchasing your flock:

- If you're mail-ordering chicks and need fewer than the minimum number you're required to order, try to find someone to share an order with you. Some feed stores allow people to order chicks in small numbers, and they combine those orders to meet the minimums.
- Some people who want just a few laying hens order a few pullet chicks and then fill the rest of the box with meat-type chicks to obtain the minimum quantity for shipping chicks. Most companies allow this. You raise all the birds together, butchering the meat birds before they take up too much space in your housing. You'll want to buy pullets that are a different color than your broiler birds, so you don't get them confused.
- For meat birds, many people order only cockerels because they grow faster and larger than pullets. Cockerels can also be cheaper than pullets in some breeds, but in the broiler strains, they often cost more. So when ordering Rock-Cornish hybrid chicks, ordering them "as hatched," which means chicks whose sex hasn't been determined, will generally save you money, and in these chicks, both sexes grow equally well.
- Although some people still butcher a few chickens at a time as the need arises, it makes better economic sense to butcher chickens in batches. You use the same amount of electricity for the brooder, and you have to buy bedding and feed and so on, so raising 10 to 25 chicks at a time isn't much more expensive than raising 2 or 3 meat birds.
- When ordering chicks by mail, try to order from a hatchery close to you. The closer the hatchery, the less the shipping costs will be.
- Day-old chicks are the most economical way to buy chickens. Fertile eggs may cost almost the same as chicks, but after the expense of purchasing an incubator, running it, and generally having only half the eggs actually hatch a chick, you reap better cost savings by buying chicks.
- Pay to have chicks vaccinated at the hatchery; it's cheaper for them to do it than for you to buy vaccines or pay a vet. For more about vaccinating chicks, see Chapter 10.
- Buy adult birds in the fall because young birds have just finished growing and people are selling their excess young birds. People are also thinking about winter feed costs, so the birds are less expensive in the fall than in spring when supply is low and demand is high for older birds.
- When purchasing adult hens to lay eggs, do some comparison shopping and be wary of people selling hens at low prices. Old hens that have quit laying eggs are hard to distinguish from young hens. For tips on how to tell whether a hen is laying, see the section "Where to buy adult chickens," later in the chapter. (Old, nonlaying hens eat as much as hens that lay well.)

Starting with Chicks

Chicks come in several colors and even sizes, and when you're looking through a chick catalog or gazing down at a tub full of cute babies at the local store, you may find it hard to pick just a few. But remember, although the chicks are small now, they will quickly need more room. Don't buy more than you can take good care of. Of course, it helps to start out with some good-quality chicks. In the sections that follow, we give you the information you need to find and select healthy chicks from a variety of sources.

Where to get chicks

Baby chicks are available for sale from a number of places, which we discuss in the following sections. You'll have more opportunities in the spring and early summer to find the chickens you want, but some baby chicks are available in all but the coldest months of the year.

Mail-order hatcheries

A baby chick catalog is dangerous, with all its interesting breeds and those cute pictures! You can order chicks by catalog or from online sites from dozens of hatcheries across the United States. Most hatcheries have a list of ship dates. You choose the date for your chicks to be shipped that's closest to the time you want them.

In most hatcheries, eggs are hatched according to demand, so when you order 25 Rhode Island Red chicks, the hatchery adds 25 or so Rhode Island Red eggs to the incubator, which hatch in about 21 days. This description is a simplification of the process, but it serves as a reminder that you need to allow some time between when you order chicks and when you want to receive them.

Don't think that you need to order a lot of extras because many of the chicks will die. Mail-order hatcheries commonly add a free chick or two to shipments to cover shipping losses. If you take care of your chicks correctly, you shouldn't lose many of them.

Baby chicks can be safely shipped by U.S. Mail at most times of the year. They're shipped on the day after they hatch and normally take 24 hours to reach you. Baby chicks can survive well without food or water for 2 or 3 days after hatching. Remnants of the egg yolk are attached to them, and the chicks slowly absorb it all in the first few days after hatching, which makes eating and drinking unnecessary. The chicks are packed closely into boxes so that the combined body heat helps warm them — hatcheries maintain the usual 25-chick minimum order for this reason.

Chicks come in different colors. If you're ordering chicks of multiple breeds but similar colors that may be difficult to tell apart, ask the hatchery if it can separate the breeds with a cardboard divider.

Even in groups of 25, baby chicks can become chilled or overheated during shipping. Order your chicks when the weather is mild in your area. It's also a good idea to order from hatcheries that are within one shipping day of your home. Being in transit for longer than that stresses the chicks and causes more deaths.

Generally, when chicks are shipped to a post office, the office will call and ask you to pick up the chicks, so be prepared to do so. If the carrier knows someone will be home to accept the chicks, he or she may deliver them, too. Believe me, post offices are generally eager to get the chicks to you because the peeping from unhappy chicks, especially a lot of them, can get annoying.

When you pick up your chicks or get them delivered, immediately open the box and inspect them. Let your post office clerk or carrier know if you find a lot of dead chicks. Some hatcheries don't guarantee safe arrival, but many do, and you must fill out a claim at the post office to get a refund or replacement. Count the chicks before you file a claim. Some hatcheries add extra chicks to the order to account for losses.

Some U.S. hatcheries

The following list of hatcheries is not a recommendation of their quality or service and is not inclusive. It's just a starting point to help you find the chicks you're looking for. Hatcheries also sell supplies for chickens, and some sell fertile eggs.

- **Belt Hatchery:** 7272 S. West Ave., Fresno, CA 93706; phone 559-264-2090; `www.belthatchery.com`.
- **Cackle Hatchery:** P.O. Box 529, Lebanon, MO 65536; phone 417-532-4581; `www.cacklehatchery.com`; catalog $3, brochure free.
- **Hoffman Hatchery:** P.O. Box 129, Gratz, PA 17030; phone 717-365-3694; `www.hoffmanhatchery.com`.
- **Ideal Hatchery:** P.O. Box 591, Cameron, TX 76520; phone 254-697-6677; `www.ideal-poultry.com`.
- **Murray McMurray Hatchery:** P.O. Box 458, Webster City, IA 50595; phone 800-456-3280; `www.mcmurrayhatchery.com`; free catalog.
- **Townline Hatchery:** P.O. Box 108, Zeeland, MI 49464; phone 616-772-6514; `www.townlinehatchery.com`.
- **Welp Hatchery:** P.O. Box 77, Bancroft, IA 50517; phone 800-458-4473; `www.welphatchery.com`.

We've ordered chicks through the mail many, many times and have had a bad experience only one time. We had about 50 percent of the chicks arrive dead one time when the box was delayed an extra day because a big snowstorm late in the season delayed mail shipments.

To find reputable hatcheries in your area of the country, ask other chicken owners where they got their birds or visit an online site such as `www.backyardchickens.com` to see what the forum members recommend.

Local breeders

Local breeders have small hatcheries or use hens to hatch eggs, and you visit their establishments to pick out chicks. We're not talking about the guy at the flea market with a box of chicks for sale.

When you visit a local breeder, you can see how the chickens are kept and whether they look healthy and happy. You'll probably be able to purchase just a few chicks and maybe chicks that are already off to a good start. You will have the help and advice of someone experienced with raising chickens.

The disadvantages of purchasing locally are that the breeds you want may not be available and the times of the year when chicks are available may be limited. Some breeders have chicks available most of the time; others want to incubate eggs only when you request chicks, so you may need to plan ahead.

Many breeders aren't able to vaccinate chicks for several diseases the way large hatcheries can, and some may not know how to sex chicks — chick sexing is a specialized skill.

Oh, that guy with the box of chicks at the flea market? You don't want to buy chicks from him because you don't know how they were hatched, you have to take his word for what breed and sex they are, and you're likely to end up with diseased or sickly chicks.

Farm stores

Many farm stores have chicks for sale in the spring. Some take orders for chicks, and others bring in batches of popular breeds to sell. The stores don't make a lot of money off the chicks, but they're hoping you'll purchase starter feed, brooder lamps, water holders, and so on.

Some farm stores take preorders. You look at a list or catalog, decide what chicks you want, and order them at the store. Often you don't have to order the 25-chick minimum, which can be helpful. Several orders are combined to make the shipping minimum. You generally prepay, and a few weeks pass between the day you place your order and the day you pick up the chicks.

The store can care for your babies if you aren't available during post office hours to pick them up, although you shouldn't expect it to care for them longer than a day.

With large, combined orders, customers' individual orders are generally separated by cardboard barriers in the shipping box. Sometimes chicks manage to breach these barriers, and if the different breeds shipped look alike as chicks, some people may find they got the wrong breed or sex as the chicks get older.

Some farm stores also carry chicks for sale to impulse buyers. Check to see whether anyone knows what breed and sex the chicks are. If the store just bought a batch of cute, cheap, mixed chicks, they're probably mostly roosters, and not the type that make good meat birds. You'll want to avoid them.

If the farm stores can tell you that the chicks are broiler chicks or pullets from a laying strain of bird, you'll be able to judge whether they suit your needs. Even if they're a mix of laying breed pullets, for example, they may be a good buy. Look for lively chicks and make sure they're being kept in a warm, clean environment, protected from too much handling. And remember how many you decided you needed — don't overbuy because of the cuteness factor or because they're on sale!

When to buy chicks

If you're mail-ordering chicks, try to order them when weather conditions in your area aren't likely to be too hot or too cold. Bad weather may affect them during shipping, and the cost to run a brooder for chicks is higher when the weather is still very cold.

If you're looking for a rare breed or you must have a certain breed, it pays to shop early in the year. In many cases, if you pay for the chicks, you can request delivery at a later date, but you will have reserved the birds you want. Some rare breeds and some popular breeds sell out early in the spring. Many rare breeds don't lay as many eggs as more common breeds, and the hens available to lay them are fewer also.

Laying birds started as chicks won't begin laying for at least 5 months. You may want to order them as early as you can so that they mature while days are still reasonably long. If they're going to be 5 months old in the middle of winter, they may not start laying eggs until the days start getting longer.

Try to purchase meat birds so that when they come out of the brooder (when they're about 1 month old), it isn't too hot in your area. Heat is stressful for the broiler breeds. It takes less than 3 months to raise broiler-type birds to butchering size.

What to look for

Serious breeders usually have their flocks tested and vaccinated for prominent diseases. If you're buying from a hatchery, make sure the chicks come from certified pullorum-tested flocks. Pullorum is a serious bird disease that will kill all your chicks and endanger anyone else's chickens in your area. Ask what vaccines have been given for other diseases. If the option is offered, have the hatchery vaccinate your chicks for Mareks disease. It costs a bit more, but it's well worth it. Home flock owners have difficulty vaccinating chicks. For more information on vaccines and testing for disease, see Chapter 10.

The following tips mainly apply if you're going to buy your chicks from a breeder or store. Once you have mail-order chicks, you're pretty much stuck with what you have, but the tips we provide may tell you whether something is wrong and you need to call the hatchery.

Chick color

When you're looking in a catalog, you certainly can't judge the health of a chick. You have to depend on the seller to send you good, healthy chicks. But you can check to see whether the catalog description tells you the color of the chick. Not all baby chicks are yellow. Some are brown, brown striped, gray, black, or reddish. Even if you're going to pick chicks from a local breeder, you may want to check out some chick catalogs or websites so you have an idea of what the chicks should look like.

If you don't know what color the chicks are supposed to be, you can generally assume that dark adult birds come from black, brown, or gray chicks. The chicks of chickens with variegated feathers, such as a partridge color, generally have faint dark stripes on a yellow or brown background. White, red, and buff chickens usually have chicks of various shades of yellow. But the color of chicks can vary from breed to breed and even among chicks from the same breed.

When you get your chicks, the color may tell you whether the company sent the right chicks. If the color seems off, check to see whether the hatchery reserved the right to substitute breeds if the breed you wanted wasn't available. Usually you'll be notified if you receive a substitute. If you have questions, call the hatchery. If you're buying locally, you can ask the seller about the chicks' color.

Chick sex

Unless the chicks are from a sex-linked line, it's impossible to tell the sex of newly hatched chicks simply by looking at them — so don't let people tell you that they can. Sex-linked chicks are hybrids of two breeds. The male chicks are one color and the females another, which makes sexing quite easy.

Some people are pretty good at picking out the sex of chicks as they grow by observing their combs and some of the feathers they're growing. If the chicks you're going to buy from a breeder are a few weeks old, the breeder may have a pretty good idea of the sex, but do expect some surprises.

The combs of cockerels generally grow a little faster than the combs of pullets. Cockerel tail feathers may also look different than those of pullets. As they get their adult feathers, sexing young birds becomes much easier.

It's possible to tell the sex of chicks even at a few days of age by looking inside the vent area, but it isn't easy. You have to look inside the cloaca with a strong light, sharp eyes, and some training. Most people, even breeders, have trouble doing it without harming the chick. Hatcheries employ experienced people to sex chicks if they offer sexed chicks other than sex-linked breeds. That's why sexed chicks cost more.

If you're picking the chicks from a group, you pretty much have to take your chances on whether you get males or females.

Type of comb and other breed characteristics

The comb of chicks is small but visible. When you examine the head of a baby chick, you should be able to tell whether the comb will be single, rose, or some other comb formation. You can't tell whether the size and placement are proper until the bird is older.

Chicks of breeds with crests and topknots should have a puffy bump on the head or even a tiny topknot right from hatching. Sometimes this looks like a whorled area on the top of the head. You may also see the beginnings of muffs and beards on chicks from breeds with those traits. Of course, you won't be able to judge their quality until the chicks are grown.

Feather-footed breeds are a bit more difficult to spot as chicks. Some have fluff growing down their legs, but in others, the difference from regular chicks is hard to spot.

If the breed you're looking for is supposed to have five toes, look for the fifth toe on the back of the leg. Don't confuse it with the bump that will become the spur; it should look much like the front toes. The color of a chick's legs should match the breed's characteristic color.

Health

Healthy chicks are active but not too noisy. Of course, they do sleep more than adults, like all baby animals, but if disturbed, they quickly get up and move away.

If a chick is sitting off by itself looking droopy, it may not be healthy. If a chick is touched and it responds very little, it probably isn't healthy. If it's lying on its back with its legs in the air, it's definitely unhealthy!

Chicks that are noisy are unhappy and stressed, from being either cold or hungry and thirsty. When the chicks arrive in a shipping box, the stress is evident from the shrill cheeps. But if you place them in the right temperature with food and water, they should quickly calm down.

If you look at a content group of chicks in a proper brooder, some will be under the heat lamp or near it sleeping peacefully, while others will be eating or drinking or walking around. They will be quiet except for an occasional peep.

Chicks that are panting, with their beaks open, are either too warm or sick. If they appear normal after being cooled down, they should be fine. If the chicks are as far from the heat source as possible, it's probably too hot. If they're piled on each other near the heat source and peeping loudly, it's probably too cold. If chicks are noisy but they aren't obviously hot or cold and food and water are available, something else is wrong. While you can fix the temperature or hunger problem, avoid purchasing chicks if you can't tell what's wrong.

Baby chicks should have two bright, clear eyes, and their rear ends, or vent area, should be clean and not pasted up with feces. Their beaks should be straight, not twisted to one side. Some hatcheries trim the end of the beak to prevent chicks from picking at each other, so don't be alarmed if the beak tip is missing. Their toes should be straight, not bent — or, worse, missing.

Newly hatched chicks have a slight lump on the belly where the egg yolk was, and that's okay. But the belly area should not look sore and red. The chicks shouldn't have any wounds or bloody areas.

Whether you buy chicks, hatch eggs, or adopt adult birds, having a healthy flock begins with choosing healthy birds. Healthy baby chicks will be noisy and active when they arrive in the mail. If many chicks are dead or appear weak and drowsy, contact the shipper right away.

Handling chicks

Many poor chicks have been strangled by the loving grasp of children. Chicken owners need to learn the proper ways of catching and holding chickens of all ages and sizes (Chapter 10 has additional info on the safe handling of chicks).

Children need to be taught how to correctly catch and hold chickens, too, if they're allowed to handle them. Children under age 5 probably shouldn't be allowed to hold chicks without close adult supervision. They should never catch the chicks; instead, have them sit down and hand them the chicks to hold briefly.

Children should never be allowed to kiss chicks or chickens, or rub them on their faces. It makes a cute picture, but it's a dangerous health practice. All chickens — even cute, fluffy ones — can carry salmonella and other nasty bacteria and viruses, even though they appear perfectly healthy. Children shouldn't touch their faces or mouths after handling chickens or eat anything until they've thoroughly washed their hands in hot, soapy water.

Also make sure small children don't rest their faces on brooder or cage edges to get a better look. If they handle feed and water dishes, pick up eggs, or help with other chores, they should promptly wash their hands. You should remember this, too, because children aren't the only ones who can come down with salmonosis or other diseases. Wash your hands before eating, smoking, or putting your hands near your mouth, nose, or eyes.

Starting with Adults

You don't have to start with chicks — you do have options when it comes to chickens. So if you've decided that buying adult chickens is the best route for you, read on to see where to get them and how to choose healthy birds.

Where to buy adult chickens

Sometimes you can purchase adult or nearly adult chickens through the mail, although shipping costs are high. The best place to buy adult birds is from a reputable breeder. If you can attend a poultry show at a state or county fair, you may find good birds for sale. Swap meets and animal auctions are another resource for birds, but use extreme caution in these cases.

Ask at feed stores and check online forums and newspapers for sources of adult birds near you.

What to look for

If you're buying sight unseen — which we don't recommend — make sure you get a guarantee of health and age. If you're buying show birds, get a guarantee that the birds have no show-disqualifying features, unless you and the seller have discussed the bird in question and you're aware of a fault. Because show quality is subjective, you probably won't get a guarantee of the bird's quality.

Sellers should, at the minimum, be able to tell you the breed, sex, age, and correct color of the bird or birds they're selling you.

Checking health

If you're choosing your birds from a seller, look for active, alert birds in clean surroundings. Sick birds may look fluffed up and listless. Nasal discharge and runny eyes are other signs of illness. Unless it's very hot where the chickens are, the birds shouldn't be breathing with open beaks.

The birds shouldn't have wounds, sores, or large bare patches. Hens that have been with a rooster may have a small bare area at the back of the neck and on the back from mating. This area doesn't mean the birds aren't healthy.

Most chickens molt in the fall, so fall isn't a great time to pick chickens because they may look a bit scruffy. When a chicken molts, it loses its feathers and replaces them. This process can take as long as 7 weeks. Although birds that are molting aren't exactly sick, molting puts them under stress, and if that stress is compounded by a change of environment or shipping, the chickens become more susceptible to illness. You can't get a good idea of the birds' feather color and quality, either.

Handle the chicken you intend to buy to see how the flesh feels under the feathers. Birds that are too thin or overly fat can have problems. Check the vent (under the tail) to make sure the feathers aren't caked with diarrhea. Look to see whether all the toes are there, and check the comb for damage. Look through the feathers for parasites.

You may want to check a rooster's temperament if you're buying the bird on his home turf. Enter a coop with him and see whether he acts aggressively toward you. Aggressive roosters are a pain to work with and can even harm small children.

Healthy adult chickens should look like this:

- They have bright, clear eyes.
- They have clean nostrils, with no discharge.
- They breathe with their beaks shut, unless they've just been chased to be caught or it's very hot.
- The comb and wattles are plump and glossy. In roosters, large blackened areas of the comb indicate frostbite, which may cause temporary infertility.

- They don't have any swellings or lumps on the body. Don't mistake a full crop on the neck for a lump.
- Their legs are smooth, with shiny skin. They have four or five toes, depending on the breed. They don't have any swellings or lumps on the bottom of the feet.
- The feathers look smooth, and there are no large patches of bare skin. Look carefully through the feathers for lice.
- They're alert and active.

Determining sex

The feathers and coloring in male and female chickens generally differ greatly. In only a few breeds are the coloration and feathering similar. Breeds that are all white or all black may be slightly harder to sex, but they still have color differences, even if they're subtle.

Roosters generally have longer, arched tail feathers. The hackle feathers on the neck are pointed rather than rounded. Birds that show iridescence in the feathers of the neck and tail are most likely roosters, and you can see this iridescence in solid-color birds if you examine them closely.

A rooster's comb and wattles are larger than those of a hen. Only roosters crow, and you can often get a rooster to crow by crowing at him.

Determining age

Figuring out an adult chicken's age is hard. Hens continue to lay well for about 3 years, and roosters are fertile for about the same amount of time. After that, there's only a slight chance that the birds will be able to reproduce. Some hens lay sporadically for many years. A chicken's average life span is about 8 years, so you may be buying unproductive birds if you aren't careful. You may be okay with that if you're looking for only yard birds, but the birds should be priced lower than younger chickens.

Aged chickens have thick, scaly skin on the legs. The spur is long and wicked-looking on roosters. When a hen quits laying, she may develop a big spur, too.

Hens that are laying have deep-red, glossy, moist-looking combs and wattles. They have widely spaced pubic bones and a moist, large *cloaca* (the area where waste and eggs are passed). When hens are old or not laying, their combs and wattles look dull and dry. Their pubic bones seem close together, and the cloaca looks small and shriveled. Roosters also tend to have duller combs and wattles as they age.

Purchasing chickens for show

If you want good show birds, you need to get a book that describes the proper qualifications for chicken breeds and study it before purchasing birds. The American Poultry Association publishes a large book every few years that details breed standards. Attending some poultry shows and looking carefully at the winning birds also helps. Clubs operate for almost every breed of chicken, and they publish information on show qualifications for that breed.

No matter what you do, you will make some mistakes choosing show birds at first. Picking winning show chickens takes experience and some luck. Good, honest breeders try to get newcomers the best birds possible, so listen to their advice when picking birds.

Transporting your birds safely

Some adult chickens are still shipped by air, but airlines are getting fussier about transporting animals and may not carry them at certain times of the year. You may have to go to the airport to pick up adult birds instead of having them sent through the U.S. Mail to your post office.

If you're going to a breeder to pick up birds, bring a proper carrier. You can buy a carrier specially made for chickens, but any pet carrier works well. You can find pet carriers cheaply at garage sales and flea markets. Check them out to make sure the doors still work well and latch securely before buying. You can easily clean pet carriers, and you can stack them so they take up less room without the birds beneath getting soiled.

In the country, you can still see people throwing chickens into feed sacks to carry them home, but this practice is neither humane nor safe. Small wire cages like those for rabbits are another feasible alternative. You can carry baby chicks in cardboard boxes, providing that they have some ventilation holes, but don't try this with older birds — you're likely to end up with chickens running through the neighborhood.

Don't crowd too many birds into one carrier. Make sure there's enough room for the chickens to lie down, stand up, and turn around. Carriers need good ventilation and secure latches. If the trip is an hour or less, the chickens don't need water or food. If the trip is longer, you'll need a water container that clips onto the cage or carrier. Unless the trip takes longer than 12 hours, don't add feed.

Never, ever leave chickens in closed cars in weather warmer than 50 degrees. Even 10 minutes in a closed car in the summer can be too long. Never leave carriers sitting in the sun, either. Chickens can quickly overheat and die.

If you're transporting chickens in the back of a pickup or trailer, cover part of the cage or carrier to shade it and protect the birds from the wind. The back of a pickup or trailer can get hot in the sun, so pay special attention in warm weather.

Part II
Housing Your Flock

Illustration by Barbara Frake

Check out www.dummies.com/extras/raisingchickens for a bonus article with tips on how to really personalize your coop.

In this part...

- Consider which type of housing best suits your space and your flock.
- Design the perfect coop to reflect your style. Options range from constructing the whole project yourself, to ordering from a catalog, to having the best of both worlds by combining homemade structures and premade pieces. Your choices are limitless.
- When you have your coop in place, you need to keep it clean! Find out the good, the bad, and the dirty on furnishing your coop and keeping it safe and tidy.

Chapter 5

Choosing Your Housing Type

In This Chapter

- Understanding the essentials of chicken housing
- Exploring various types of housing
- Choosing a type of housing to suit your needs

Unless you live in the middle of nowhere and turn your chickens loose to fend for themselves, you need some kind of housing for your chickens. Many years ago, every farm had a chicken house. It was generally a long, narrow building with windows on the east or south side, a door for the caretaker, and a smaller door allowing the chickens to come and go at will. Sometimes the small door opened into a fenced area; other times, the chickens roamed the barnyard.

Perhaps your property has an old chicken house or a shed or barn that you can convert to chicken housing. If you're lucky enough to have one of these structures, you may not need to do much to provide housing. But if you're like many people who've developed the urge to raise chickens, you need to build or buy chicken housing before you can begin chicken-keeping. First, you need to decide what kind of housing suits your needs, and your options are many.

One idea you probably don't want to pursue is to keep chickens in your house. Chickens simply don't make good house pets. They can't be housebroken, and their droppings are wet, messy, and very smelly because they're high in ammonia. Even cute, tiny balls of chick fluff smell bad enough that you soon want to move them. Chicken feathers give off a dust that isn't good for humans to be constantly inhaling. And pet chickens can be destructive to houseplants, furnishings, and clothing, just like a puppy or kitten. On the more serious side, chicks can carry salmonella and other bacteria that can cause serious human illness.

Chickens do best with their own separate housing, preferably with access to the outdoors. In this chapter, as the title implies, we provide you with a handy guide to the issues you need to consider when deciding how to house your chickens.

What a Chicken Needs in a Home

Whether you keep chickens for pleasure or to provide the family with eggs and meat, as the keeper of a small flock, you need to provide humane, comfortable conditions for your birds. Just as most animals are at their best in calm, comfortable surroundings, your chickens will do better in that situation.

Give your chickens everything they need to be comfortable — good feed; clean water; dry, clean surroundings; nest boxes and roosts; and maybe a sandbox to bathe in. They need enough room to move around comfortably — scratching, pecking, flapping their wings, and conversing with friends — and they must have the ability to avoid their enemies.

The needs we discuss in this section are not only for the comfort and health of your birds, but also for your convenience and comfort, too. After all, if you, as the chicken caretaker, have an easy, efficient way to care for your flock and your birds are healthy and happy, you'll be less inclined to give up the venture.

Shelter from wind and rain

Being exposed to wind or being wet can make conditions uncomfortable even if the temperature is ideal. Your chickens need a house that keeps them dry in wet weather. Wet chickens are unhealthy and unhappy. If chickens have access to shelter, they will go there on their own when it rains.

Chickens also need protection from winter winds. Wind chill factors affect chickens the same way they do humans. A windbreak to shelter housing from prevailing winter winds is ideal. Trees or shrubs can form natural windbreaks, or you can create a temporary windbreak by stretching a tarp between posts.

Protection from predators

Chickens are a tasty meal for many predators, and chasing chickens is a fun activity for other animals, including some of the two-legged kind. If you keep chickens, it's your duty to keep them safe from harm. A mean old rooster may seem to be able to take care of himself, but a car or a big dog can make short work of him.

Predators are a big problem for chicken-keepers in the city, the suburbs, and the country. Predators include neighborhood dogs, which are probably the number-one chicken-killer for all kinds of chicken-keepers. Chickens face many other kinds of predators as well, and urban areas have their fair share of them. We discuss predators in much detail in Chapter 9, but suffice it to say that everyone who keeps chickens will, sooner or later, deal with a predator.

Strong, well-built housing goes a long way toward keeping your chickens safe from predators. At the very least, your chickens should have a predator-proof shelter at night, when they're most vulnerable. How much additional predator protection they need depends on your location and the predators.

Temperature control

Chickens are native to warm, tropical areas, and although breeders have adapted many breeds so that they tolerate cold weather better than their ancestors, chickens do need winter protection in cold areas. Likewise, weather that is too hot can also be deadly for chickens.

Chickens are most comfortable when temperatures range between 40 and 85 degrees Fahrenheit. If temperatures regularly drop below freezing (32 degrees Fahrenheit), egg production either greatly slows or ceases altogether. When temperatures drop into the single figures or below zero, some chickens lose part of their comb or even toes to frostbite.

In cold climates, the poultry house temperature should be kept just above freezing, or about 34 to 40 degrees Fahrenheit, with red heat lamps or dust-resistant electric heaters. That temperature is warm enough to keep eggs and water from freezing, keep hens laying, and prevent frostbite, without breaking your budget.

Chicken bodies give off a lot of heat, and a well-insulated house that stays above freezing may not need heating. Some basic winterizing of the coop, like covering single-pane windows with plastic, helps. In Chapter 6, we discuss winterizing the coop in a little more detail.

Some people may not be able to, or want to, heat the coop in cold weather. Chickens will survive the cold if they're dry and out of the wind, but they may not lay eggs well when the temperatures are below freezing (32 degrees Fahrenheit). Some breeds of chickens handle cold better than others, and you'd be wise to choose those breeds if you won't be heating the coop. These chickens include heavily feathered breeds like Cochins and Brahmas. Breeds with small combs close to the head — such as Wyandottes and Chanteclers — have less chance of frostbitten combs. Refer to Chapter 3 for more breed information.

Don't heat the building too much; just getting the temperature above freezing is enough. Less ventilation is needed in the winter, but it's still important to reduce ammonia and moisture buildup.

Chickens need protection from heat, too. Temperatures in excess of 90 degrees Fahrenheit may harm chickens. Long periods of temperatures above 90 may also decrease egg production. Just as humans suffer more in heat combined with high humidity, the temperature at which chickens become stressed may depend on the humidity.

The smaller the housing and the more chickens in it, the more critical cooling in hot weather is. In hot weather areas, your chicken housing may need to be larger so that chickens can spread out. Higher ceilings and better ventilation are also helpful.

The heavy *broiler* breeds, or meat birds, are especially prone to dying from overheating. Broilers should be raised at cooler times of the year, or some form of cooling may need to be available in the housing. Keeping temperatures below 85 degrees Fahrenheit (or even cooler if the humidity is high) is desirable for broilers. Even broilers that are housed in tractor pens or given free range outside suffer in heat. Other types of chickens have an easier time handling temperatures above 85 degrees, as long as they have shade, good airflow, and access to plenty of water.

In northern areas, some deciduous trees can provide shade both outside and to the housing in the summer while letting the winter sun shine through. In southern areas, year-round shade is preferred.

Enough space to move about normally

Housing should allow chickens to engage in normal chicken activities (even if their life is to be short), including flapping their wings, sitting in the sun, taking a dust bath, and chasing a bug. The more space you can provide, the happier your chickens will be.

Because chickens are flock birds and are unhappy living alone, count on keeping at least two birds. And because keeping chickens is addictive, you may want to consider housing that allows you to expand in the future. If housing is too cramped, birds lose the ability to establish a good social structure, and behavioral problems result. Housing that contains the right elements — roosts, nests, and so forth — promotes good social behavior.

Chickens must have sufficient space to be healthy and happy. Crowded conditions increase fighting and disease, prevent normal social behaviors, and lead to problems with chickens laying eggs on the floor and picking at each other. Overcrowding also increases moisture and ammonia levels.

Each average adult chicken should have 2 to 3 square feet of floor space in the shelter and 3 to 6 square feet of floor space in an outdoor run. If you have no outdoor run, double the indoor space allocation. Bantam breeds can get by on a little less floor space. Give your chickens even more space if you can.

In cold winter areas, an extremely large indoor shelter may be a disadvantage because the birds' body heat can't warm it as well. But being cooped up in the winter inside a small shelter has to be boring. If the outdoor run is protected from strong winds and roofed to keep out deep snow, the birds will spend a little time outside on nice days.

The housing for your birds should allow them to stand normally without touching the ceiling. Chickens prefer to sleep on a roost up off the floor, so taller housing gives them the freedom to do that and feels more natural. But don't count the height of housing as part of the square-feet-per-bird requirement.

If you keep chickens for showing, you need to provide plenty of room in the housing and around roosts so that birds won't rub their feathers on the wire or walls of the housing. Birds with broken or frayed feathers are unsuitable for showing.

If you can give your chickens more space, it's a good idea to do so. It's better to err on the side of too much space than not enough. And with spacious quarters, you can increase the size of your flock in the future.

Sufficient lighting

Like most birds, chickens are active during the day. Light — either natural or artificial — is necessary for them to eat and drink. The amount of light (day length) influences the hormones that control egg laying and fertility. Some natural sunlight (as little as an hour!) is needed for chickens' bodies to make vitamin D, but vitamin D is a common feed supplement, and chickens can survive well with just artificial light. We discuss light's effect on laying in greater detail in Chapter 15.

It's always nice to have some natural light in a chicken shelter. It saves money and probably feels as nice to chickens as it does to us. Windows that face south or east are best. If windows face south or west in warm weather, you need to be able to open them to avoid excessive heat buildup in the shelter. Chickens like their nest-box area to be in a dimly lit location — under windows is usually better than across from them. If you have a tiny shelter for two hens, having inside lighting or a window may not be practical. The chickens will be fine as long as they get light in an outside run.

Even with natural light, sometimes artificial light is helpful:

- Being able to do chores after dark makes your job easier.
- Artificial lighting in the coop on gray winter days in the North makes the hens more active, which keeps them eating and interacting normally. It also helps stimulate egg-laying. Fourteen to 16 hours of light should keep hens laying well all winter. The light needs to be strong enough to read a book in all areas of the indoor space.
- Meat birds, particularly Rock-Cornish broilers, benefit from having their housing lighted 18 to 22 hours a day. These large birds need to eat frequently to keep their metabolism going, and chickens don't eat in the dark. A small period of darkness or dim light is now thought to be beneficial because it gives the bird's metabolism a slight rest.

Except for young chicks and broiler-type birds, chickens appreciate a time of darkness, or at least dim light. For layers, 8 to 10 hours of darkness or very dim light (similar to the level provided by a bedroom night light) is good; for other types of birds, 12 to 14 hours doesn't hurt. Dim light inside the coop at night is helpful for these reasons:

- It allows the birds to defend themselves from some predators.
- If the birds are spooked off their roost at night, they can find their way back on it.

Fresh air

Good ventilation is important in any enclosed chicken housing. The smaller the housing, the more important ventilation becomes. Ventilation replaces stale, ammonia-saturated air with fresh air. Fresh air doesn't mean drafts, though. Drafts are leaks of air that pass across the living space. Ventilation generally refers to clean air that's pulled in near the bottom of a structure, warms and rises, and exits near the top.

The amount of ventilation needed depends on the number of chickens and the type of shelter you have. The more birds per square foot, the more moisture and ammonia there will be. Shelters with high ceilings (6 feet or more) allow warm moisture and ammonia-saturated air to rise away from the breathing space of the chickens. Shelters with low ceilings may have the same square footage of floor space per bird, but the air doesn't have far to rise and needs to exit quickly to keep birds from breathing it.

No magic formula can determine how much ventilation you need — too many variables are at work. Ventilation should be adjustable with windows, vents, or fans that you can open or close for different conditions. If you're comfortable breathing in the shelter and if there's little or no ammonia smell and no moisture buildup on walls or ceilings, your ventilation is adequate.

Lack of adequate ventilation can cause a number of problems:

- **The buildup of ammonia fumes in walk-in housing leads to an increase in respiratory diseases in chickens.** Ammonia comes from chicken droppings and is a lung irritant.
- **Moisture buildup increases humidity,** which makes an area feel warmer when it's hot and colder when it's cold.
- **Excessive moisture is conducive to the growth of mold** and also increases bothersome smells.
- **Disease organisms are more easily spread in moist, warm environments.**

In solid structures, doors, windows, or vents provide ventilation. Cool air can enter naturally at the bottom of a structure through a vent. The air then warms and rises to the top, where it exits through other vents or windows.

In a small shelter with only a few birds, the opening the birds use to go outside can serve as the bottom vent and some openings along the upper part of the shelter can exhaust stale air. Larger shelters need vents near the floor, windows that open, and adjustable vents at the top. If many birds live in the shelter and the weather is warm, an exhaust fan near the ceiling may be needed to speed the exchange of air.

Clean surroundings

In both inside and outside parts of the coop, both you and your birds will appreciate clean surroundings. Some type of litter on the floor inside is usually needed. If this litter gets wet, change it immediately. We discuss types of litter in the next chapter.

Good drainage in the outside areas is also important because it avoids moisture buildup, which increases smells. Chickens don't like wet feet and are uncomfortable in muddy surroundings. Some chicken owners prefer to put down a hard-surfaced floor on the outside areas and sweep it or hose it down. The hose method is best for hot, dry climates where the floor dries quickly. Other owners add sand or gravel to the outside area if it turns into a mud pit at certain times of the year. Although the floor may start out as turf, it will soon deteriorate into a dirt floor. Even the tiny feet of chickens can compact a dirt floor, causing water to stand on top of it instead of draining through it. Sand or gravel keeps the birds' feet clean and the smells down.

Wire floors are sometimes used for meat birds for cleanliness, but they can cause breast sores in broiler-type birds. Use a deep litter to cushion these birds, and keep it clean and dry. Because meat birds are prone to leg problems anyway, don't use slippery flooring like tile, paper, and metal.

Chapter 9 discusses chicken-coop housekeeping further, but if you keep cleanliness in mind when designing or choosing coops, your housekeeping will be much easier.

Surveying Your Housing Options

You have many options when it comes to housing your chickens. Some types of housing overlap a bit, but in this section, we try to provide a general overview of the more common types of chicken housing available. In the section "Choosing a Type of Housing," later in this chapter, we help you compare the options according to the most important factors to consider.

To check out a ton of housing ideas and step-by-step photos and explanations for many options, visit the "Coop Designs" section of the `http://www.backyardchickens.com` website. Thanks to backyardchickens.com members Andrea Wilson, Froggi VanRiper, Renee Fraser, Mike Walsh, Erin Moshier, and Christine and Michael Byrne for letting us show off their great designs.

Before diving into a discussion about the many variations of chicken housing, it helps to be familiar with some basic chicken-keeping vocabulary. Throughout the following sections, and the book as a whole, we use these definitions to identify the various components of a chicken's humble abode:

- **Coop:** *Coop* is a generic name for anything chickens live in, just as *home* is a generalized term for a place humans live. The term *coop* covers both inside (sheltered from the weather) and outside (unsheltered from the weather) enclosed areas for chickens and is interchangeable with *housing* or *pen*.
- **Housing:** Anything you confine your chickens in is considered *housing*. It includes their shelter and any outside space you may have enclosed for them. *Housing* is interchangeable with *coop* and *pen*.
- **Pen:** *Pen* refers to an enclosure for chickens. A pen is similar to a run (see the next bullet), but a run can refer only to an outdoor enclosure. Pens can be indoors, sheltered from the weather, or outdoors, unsheltered from the weather. Pens can have a shelter in them. The term *pen* is also interchangeable with *housing* and *coop*.

- **Shelter:** *Shelter* refers to the place that protects chickens from the weather; it is the indoor part of their confined area. Shelters may be part of housing, coops, or pens.
- **Run:** *Run* refers only to an outside enclosure — the part without protection from the weather — for chickens. A run is usually connected to a shelter (the area that protects chickens from the weather).

Raising chickens in cages

People who want just a few chickens and who have minimal space to keep them may want to use cages. Cages vary from small all-wire cages to larger units that include a shelter and a small outside pen. These types of cages are often sold online or at feed stores. Bantam chickens do better in cages than larger breeds because they require less room to be happy and healthy.

If you use cages, make them as large as possible, and make sure the chickens have room to walk around, flap their wings, and raise their heads normally. Each chicken in a home flock needs at least 2 square feet of floor space. If the chickens are layers, make sure there's enough room for a nest box (see Chapter 7 for more on furnishing the coop).

Even if you don't house your chickens regularly in cages, it's good to have a few cages around that chickens can be kept in comfortably. They come in handy when you have an ill or injured bird that needs to be separated from the flock, a new bird, birds that have returned from shows and need to be quarantined, or a hen sitting on eggs or caring for chicks.

Cages also sometimes temporarily house chickens for breeding or getting the birds used to conditions they will experience when caged at shows. Unless you're separating chickens to keep them nice for showing, or unless they're injured or need quarantining, keep two or more chickens in each cage.

Wire cages

Cages are typically made of wire mesh and can be square, rectangular, or any shape you want. They can be simple and bare inside except for food and water dishes, or you can make them more comfortable by adding roosts and nest boxes (for hens). Cages can be any size, as long as you provide a minimum of 2 square feet of floor space for each chicken and the cages are high enough for the chickens to stand upright in. Two types of flooring are available:

- **Solid flooring:** Made of wood or metal, solid flooring can be removed for cleaning. It can consist of a removable pan or a slide-out wood shelf. This type of flooring is more natural and comfortable for the chickens, but it's a pain for the caretaker because it requires scraping and washing at least once a week to keep it clean. You may have to remove the birds from the cage while you work.
- **Wire flooring:** A pan under the wire floor makes cleanup easy. The mesh openings on floors should be no more than 1-x-½ inch. On the downside, birds sometimes catch their toenails in wire floors and may injure their toes or legs trying to escape. Heavy breeds of roosters are susceptible to bumblefoot, a bacterial infection of the feet caused by small cuts and pressure to the foot pads inflicted by wire flooring.

We recommend using solid floors for meat chickens and heavy breeds (more than 5 pounds), and wire mesh floors for chickens weighing 5 pounds or less and bantams, except for chickens kept for showing. All breeds of chickens kept for showing need solid floors in cages, to minimize the chance of foot injury. Even a broken toe can disqualify a show bird.

You can use wire cages for chickens inside or outside. When inside, place the cages in locations with good ventilation and lighting. A small door above the nest box in the cage wall makes gathering eggs easy.

If wire cages are placed outside, at least part of the cage needs to be covered so that the chickens can get out of the weather. You can cover cages with tarps or pieces of wood. You can hang cages on nails or hooks in the sides of buildings, or suspend them with chains from a frame or even tree limbs.

People who keep chickens in cages may want to have a fenced area outside where the chickens can occasionally enjoy a bit of exercise, maybe while you're cleaning their cages. This outdoor area may simply be your backyard or a special area you've fenced off just for them. While desirable, an outside area isn't absolutely necessary.

All-in-one "fancy" cages

You can find cute, small chicken coops on the market (as well as plans to build them yourself) that feature a small house, similar to a doghouse, with either a small, outside pen attached or a pen that surrounds the shelter. The pen can be on the ground or raised on legs (see Figure 5-1). The caretaker gathers any eggs from outside the shelter through a small door and, in some cases, feeds and waters the chickens there. In other cases, feed and water are placed outside the shelter portion.

Figure 5-1: An all-in-one chicken coop has a combined shelter and run.

Illustration by Barbara Frake

These units are good for two or three chickens, generally layers or pets. They're ideal for city situations. The drawback is lack of space for the chickens. Sometimes chickens kept in these units can be turned loose a few hours a day for some supervised foraging, which makes this type of housing a little more pleasant for the chickens.

These all-in-one units tend to be expensive, especially if they have to be shipped to you, and many of them have features that appeal more to humans than to chickens. Before purchasing one of these units, look it over carefully and think about the provisions all chicken housing needs to include (see the section "What a Chicken Needs in a Home," at the beginning of this chapter).

If you have the time and skills, you may be able to build a small, neat shelter and run that will work better for you and the chickens and that costs less than pre-fab all-in-one housing. Even if you don't have any building skills, you may be able to hire a local person to build a nice coop for less than the cost of pre-fab housing.

Keeping birds cage-free, but indoors only

You can section off a portion of a garage, shed, or barn, or chickens can use all the floor space in an enclosed building. The building needs good ventilation, good lighting, and flooring that's easy to clean. This housing option is a good way to raise a group of broilers that you intend to butcher at 8 to 12 weeks, particularly in cold weather.

If you have no space for outdoor pens, or if other issues make outdoor pens difficult to use, you can also keep laying hens and pet birds indoors. As long as they're not crowded, birds kept loose in an indoor area are able to socially interact and behave naturally, and they're happier than when confined to small cages.

You may see eggs in grocery stores that are labeled "cage-free." This usually means that the hens producing those eggs are kept in large indoor buildings, loose on the floor. (The "free-range" label on a carton of eggs legally means that the hens have access to the outdoors, but they can be confined and don't need to be running around freely.) These eggs generally cost more, to reflect the additional cost of housing birds outside cramped cages. While cage-free conditions may be more humane for the birds, in some operations, the birds are packed as tightly as they are in cages.

Pairing a shelter with a run

One of the oldest and most successful methods of keeping chickens is to provide them with a shelter they can retreat to at night or in bad weather and an outside enclosure that protects them from predators yet allows them access to fresh air and sunshine. Figure 5-2 illustrates this type of housing.

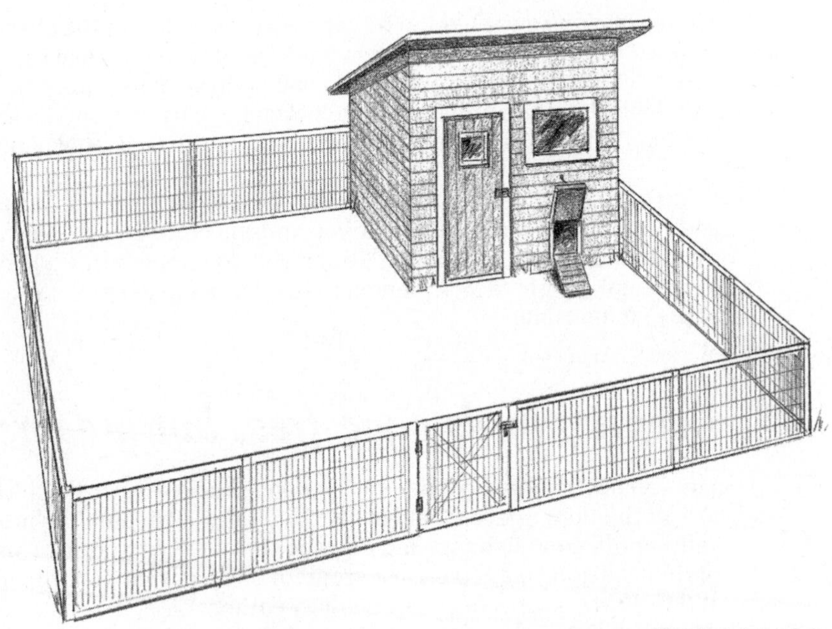

Figure 5-2: A shelter with an outside run.

Illustration by Barbara Frake

Walk-in shelters allow the caretaker to feed and water the birds and collect eggs inside. The shelter area can be a partitioned-off part of a barn or garage or a stand-alone building. A smaller door for chickens usually connects to the outside enclosure. It's a good idea to have human access to the outside enclosure also. The best outdoor enclosures allow plenty of room for exercise. You can let your chickens out for a little free-range roaming, too, but the enclosed outside area and the shelter give you the option of confining them when that isn't safe or desirable.

Walk-in shelters with attached runs can be as small or as large as you need them to be, and they can be built with the intention of expanding in the future as well. Most people who begin with other types of housing end up with this type, unless their space is very limited. It's the most comfortable housing for both humans and chickens.

Offering shelter with free-range access

Some people have a shelter where they can feed and water chickens and from which the chickens can come and go at will. To get an idea of this setup, picture Figure 5-2 minus the fencing. Chickens may have their own shelter or be allowed to roost in any barn or shed on the property. This setup works best when the chickens can roam large spaces without infringing on neighbors and where predators aren't a problem.

Chickens, of course, love this arrangement. But it involves some drawbacks for both them and their caretakers:

- Chickens kept this way are wilder and tend to go off and hide their eggs, particularly when they want to raise a batch of chicks.
- On the range, the chickens are quite vulnerable to predators from the air and land.
- Chickens can wreak havoc on gardens and landscapes, tasting each tomato as it ripens, eating seeds you just planted, scratching out a dust bath in the marigolds, and picking all the flowers off your roses. They can also make a mess on porches and patios.

Free-range birds can cause other headaches, too. A friend of Kim's arrived at her job early one morning, just as it was getting light. A co-worker yelled at her as she was exiting her truck to ask whether the rooster running through the parking lot was hers. (He had seen it jump off her truck's bed.) Sure enough, the woman recognized a rooster from home. He had ridden 30 miles in the bed of the pickup, where he had rested for the night. She spent 3 days on her lunch hour and after work chasing the rooster.

The flavor of free-range meat birds takes some getting used to. Their meat is slightly darker and stronger tasting, and their muscle mass is greater. If they're allowed to get too old before butchering, free-range chickens become tough. Raising a decent-size, free-range meat bird takes several weeks longer than raising a bird that's not allowed to roam. Modern broiler-type Cornish Cross chickens do not do well given free range. They are sedentary birds that like their feed put in front of them, and they require higher levels of protein than they can get from foraging to grow well.

On the other hand, raising free-range chickens has some noteworthy benefits:

- Chickens that roam may get most of their diet free.
- Free-range chickens seem to have less trouble with lice and mites.
- The caretakers of free-range chickens have less cleaning to do.
- Chickens undoubtedly prefer this method over any other.
- Watching the antics of chickens roaming around the yard can be an entertaining and enjoyable experience.

We don't recommend allowing layers free range until they have established a good laying pattern. (We talk about why that's important in Chapter 15.) It's okay, however, to allow them some access to the outdoors in an enclosed area.

Mobile housing methods: Pastured poultry

Technically, the pastured poultry method doesn't allow chickens to roam freely, eating whatever they want. Instead, it's a system of confining chickens on a piece of grass or pasture that needs to be managed or cared for so that it provides the maximum benefits for chickens. Most pastured poultry are also fed some commercial feed or grains.

Chickens that are pastured can be kept in small, moveable housing units with built-in shelters or can be confined in large, fenced areas that are moved, with a free-standing shelter that is also moved with the chickens. Chickens can also be rotated through a series of permanently fenced areas. Each area may have its own shelter, or one shelter may move with the chickens among the areas.

Chickens on pastures are intended to get most of their feed from eating vegetation. When the birds have eaten down one piece of pasture, they're moved to a clean piece. They grow a little more slowly this way, and the meat of pastured poultry tastes differently than the meat of birds that are raised in other housing and fed a completely grain-based diet. Laying hens can be raised this way, as can meat birds. Often replacement pullets are raised on pasture until they're ready to begin laying eggs. The chickens in these units can get a good deal of their food from foraging, as long as the pasture they're put on is good. In mild weather, chickens enjoy this situation.

Pastured poultry is often thought of as easy poultry-keeping, but it requires some intensive management to be successful. While the costs for feed may be greatly reduced and the birds get a natural, healthy diet, work is involved with managing the pasture and moving the enclosures. Poultry manure is hard on grasses, as is the scratching of chickens. Pastured poultry need to be rotated to a clean pen before the vegetation in their enclosure is seriously damaged.

How often the birds have to be rotated depends on how large the enclosure is and how many birds it holds. The type of pasture grasses and the weather are also factors. In mild, wet weather, grass grows vigorously, and the enclosures may need to be moved less often unless mud and smells develop that warrant moving housing more frequently. Dry, hot weather may mean shorter intervals between moves or the need to supplement with other feed. When grass is heavily matted with manure, has turned brown, or has disappeared from large areas, it's time to move the birds. Over time, you gain a feel for what's right for your conditions and birds.

Your county Extension agent can help you determine what forages grow best in your area for a pastured poultry operation and can help you decide whether your current pasture can support poultry. If you're going to use movable housing, avoid putting the chickens in the housing until the frost is out of the ground and you're past the muddy time of year in your area.

Many types of housing can be used to pasture poultry. For small numbers of birds, you can use a chicken tractor-type structure, an A-frame, or hoop housing. But for larger numbers of birds, larger enclosures with free-standing shelters work better. If the birds on pasture are layers, the shelters need nest boxes that can be easily accessed. Hens like to have a shelter high enough to roost off the ground at night. Meat birds are content with a low structure, 3 to 4 feet high.

If you put meat birds in movable pens on pasture, you must take great care not to crush or run over the birds when moving the pens. The Rock-Cornish crosses often move very slowly.

Shelter with an open bottom: The chicken tractor

A *chicken tractor* is a type of housing, usually roofed with an open bottom that is moved from place to place when someone wants the soil in various areas turned over and fertilized (see Figure 5-3). Sometimes these coops are

Figure 5-3: An example of a chicken tractor.

Illustration by Barbara Frake

placed over good pasture just long enough for the chickens to eat it down, and then they're moved to a different spot. The chicken tractor–type coops can be built in a number of shapes and sizes. The smaller tractor coops may be carried or slid along the ground, or small wheels may be attached to the bottom frame to make moving the tractor easy. Larger chicken tractors are heavy, and you may need a real tractor to move them.

Tractor-type coops can be used to raise meat birds, although the hybrid broilers may do better in more conventional housing. Tractor coops can also be used with young pullets being raised to laying age or even with laying hens. In the case of hens, nest boxes must be included in the coop and the hens must have easy access to them. Tractor coops can also be used with pet or show chickens.

Hoop and A-frame housing

Hoop and A-frame houses are variations of tractor housing. Figure 5-4 shows a sample A-frame coop, and Figure 5-5 shows a hoop type of housing.

As the name suggests, the hoop house is like a large tunnel. It can be covered with clear, opaque, or black plastic; shade cloth; metal; canvas; or another material. A-frames are similarly covered, but the frame comes to a point. Both A-frames and hoops may have to be anchored to keep them from blowing over in high wind.

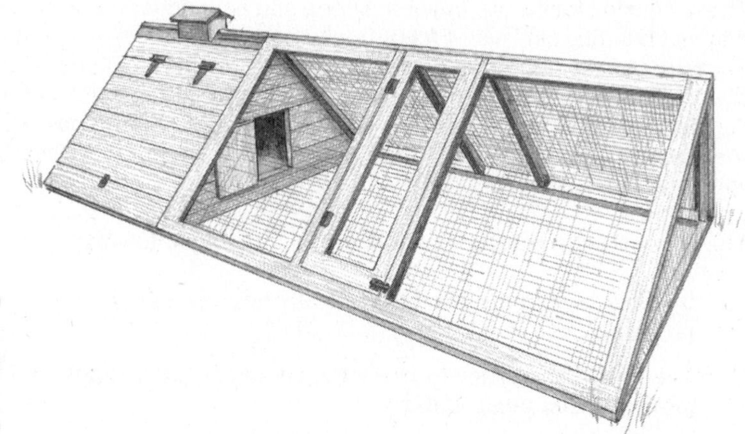

Figure 5-4: An A-frame coop.

Illustration by Barbara Frake

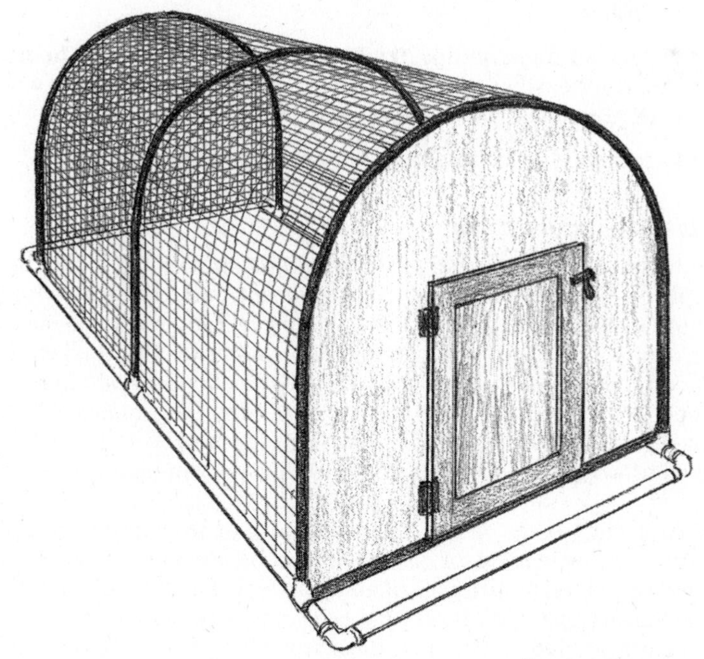

Figure 5-5: A hoop coop.

Illustration by Barbara Frake

These housing forms are quick to set up and economical. They require only minimal building skills, and few supplies are needed because of their shape. They are usually lighter and easier to move than more conventional coops. Many are used to provide temporary housing for larger groups of meat birds. Smaller versions can house a few laying hens, either out on pasture during the summer or, if built so that part of the structure protects them from winter weather, year-round.

The drawbacks to this type of housing include the following:

- **It works best in mild weather.** Really hot, cold, or wet weather causes problems.
- **You must have space to move the coops to a clean spot, and you must move the housing regularly.**
- **The chickens may be vulnerable to predators.** The coops are often placed far away from the human housing, so strong wire must be used on the sides, or predators will break in. Predators may also dig under the frame.
- **In most of these setups, it's hard to catch chickens if the need arises.** You can't easily get inside some of the smaller units, and when you lift a side, you may have chickens everywhere.
- **Feed and water containers may be hard to access.**

Mobile shelter with run

Another method of keeping pastured poultry works much like the tractors, hoops, or A-frames previously noted, except that a separate stand-alone shelter is moved along with some sort of generally unroofed enclosure. (See Figure 5-6.) Each enclosure needs a shelter for the chickens and possibly an additional shaded area, depending on the climate. A shelter may be a completely covered hoop or A-frame structure, or a small shed. It has to be light enough to be moved to a new enclosure; however, it may need to be staked or weighted to keep it from blowing away in high winds.

The large enclosures can be made with regular woven wire, plastic fencing similar to snow fencing, or electrified netting. Posts that can be easily moved should be placed at intervals of about 10 feet. The type of enclosure material you need largely depends on what type of predators you expect to encounter. Use heavy, welded wire if loose dogs are in the area or if larger livestock will be grazing around the chicken enclosures.

The minimum area for pasture enclosures is 5 to 6 square feet per bird, with a shelter size of about 2 square feet per bird. Remember that the spot the shelter sits on can't be used for pasture, and add footage accordingly. Larger amounts of pasture space per bird are better.

Meet your Extension agent

Almost every county in the United States has a county Extension agent. These agents (or educators, as they're sometimes called) are associated with a college in your state called a land grant university. The land grant university's job is to bring research-based knowledge to the general public through Extension offices. You can ask about not only chickens and crops, but also many other issues, including home buying, cooking or nutrition problems, parenting, and so on. Most services are free or cost a nominal amount.

If your Extension educator doesn't have an answer, he or she has resources at the college to turn to for help. You can find your county Extension office by looking in the phone book's Government section, usually under County Government. You can also go to `www.csrees.usda.gov` and click Local Extension Office under the Quick Links heading. When the U.S. map appears, click on your state. A county map will appear; click on your county for contact information.

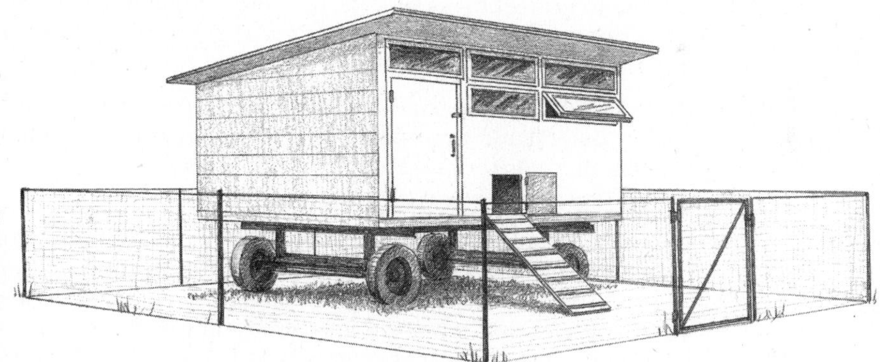

Figure 5-6: Mobile shelter with run.

Illustration by Barbara Frake

Choosing a Type of Housing

Choosing a type of chicken housing to use goes hand in hand with other chicken-keeping considerations. You can keep chickens successfully in many ways, depending on the type of chickens you plan to raise, the proximity of your neighbors, weather conditions and potential predators in your area, and how much space you want for yourself.

- **Housing pets, layers, and meat birds:** If you're keeping a few hens as pets or layers, you're probably best served with a small all-in-one unit, cages, or a small A-frame or hoop house. If you're keeping more layers, a shelter with either a run or free range is ideal.

For meat birds, loose indoor housing or specially designed cages are best. True meat breeds don't exercise much, although you can use a small outside run attached to a shelter to raise them. Because meat birds are here on a temporary basis, the housing can be temporary, too, but it must protect them from predators and weather.

If you have enough land, using the pastured system may be right for you. Chicken tractors, hoops, or A-frames can house meat birds in warm weather and then be stored away over the winter.

- **Housing show birds:** Show birds need larger quarters to prevent feather breakage and fraying. If the breed you're keeping has really long tail feathers, the roosts must be high off the floor. Bedding or run surfaces must be easy to clean.

  Many show specimens are raised on wire floors so that their feet and feathers don't get dirty. Poles are often attached to shelters and runs a few inches from the wall, to prevent the birds from rubbing against the wire or wood walls.

  Be sure to protect dark-colored show birds from direct sunlight, which fades their color or turns it brassy. Avoid grass runs with light-colored birds, to prevent feather stains.

  Many show birds, especially roosters, are best kept in individual cages so they don't fight. Some show breeders use removable dividers in pens to separate birds they want to mate, but roosters may attack each other through these barriers if they're not solid.

- **Considering neighbors and family:** If you're in an urban or suburban area with close neighbors, choose neat, attractive housing that's easy to clean. Instead of building an A-frame out of old pallets, consider partitioning off part of the garage and locating a neat enclosure behind it, or hiding a small, neat shed in the garden, with the chickens confined to the housing most of the time.

  If you have small children, you probably want a coop that they can't get into without help from you. Loose chickens scratching up the garden of a nearby resident or walking down the road may not endear you to your neighbors.

- **Keeping your chickens safe and healthy:** Predators are a big concern in some areas, whereas weather extremes are more important in others. The safest and healthiest housing is generally a well-built shelter with a well-built run. Small, combo shelter-run units may be safe, but they're not a healthy way to house chickens if they're overcrowded or not kept clean. Packing a lot of birds into a small but cute coop makes them stressed and susceptible to disease and behavior issues.

  Chicken tractors, hoops, and A-frames can be safe and healthy in good weather, but they're a nightmare in bad weather. Windstorms, heavy rain, lightning, mud, and cold all take a toll.

It does no good to let your chickens have free range if they can wander into heavy traffic, if dogs or alligators can pick them off, or if kids can treat them meanly. They may be getting all the vitamins and minerals they need from foraging, only to become a healthy meal for a hawk.

Any form of housing, whether inside or out, should be built or purchased with potential predators in mind. If you have dogs, coyotes, or raccoons in your area, use heavy welded wire rather than chicken wire or plastic fencing on your runs or pasture enclosures. Be sure to close doors to shelters at night.

If two-legged thieves or vandals pose a threat, put a padlock on your shelter doors. You may also want to consider night lighting or a security camera.

- **Factoring yourself into the space equation:** Essentially, you need to decide whether you're okay with having a shelter that you must stand outside to service and clean or whether you want a shelter you can walk into. If you want to enter the shelter to collect eggs or clean, you'll probably want ceilings that are taller than you, unless you enjoy stooping. Chasing a bird around in a crouched position isn't fun. Walking on your knees or lying on your belly to clean out a coop will make you wish you didn't keep chickens.

  Small shelters that you can't enter require a removable or drop floor for cleaning. These small coops must eventually be totally cleaned, and that can be a problem if parts of them are hard to reach. If you're keeping only two or three chickens and know you'll be diligent about cleaning their small quarters, then a small, pre-fab housing unit or any housing that you service from outside may work for you. These units fit into small areas where other housing might be a problem — that's the biggest advantage they have to offer.

Whether you will service the housing from inside or outside depends on the style of housing you choose. The preceding points and the following chart can help you see the pros and cons of each type of chicken housing and decide which type appeals to you and fits your needs. You can always alter housing for your convenience. For example, in a walk-in shelter, you can add a small door just above the nest boxes so you can reach in from outside and quickly collect eggs when you don't want to go inside.

Ultimately, you must choose a style to begin chicken-keeping, but don't be afraid to change your housing if you find that it no longer works for you. If you find that trying to clean a small pre-fab unit from the outside is tedious work, graduate to a larger, more practical coop.

Now that you know the basic decision-making factors, consult Table 5-1 to blend your personal considerations and choose a type of housing that's right for you.

Table 5-1 Choosing Suitable Housing for Your Chickens

Type of Shelter	*2–4 Chickens*	*4–10 Chickens*	*10+ Chickens*	*Layers*	*Meat*	*Pet/ Show*	*Cold Weather*	*Other*
Cages, wire or fancy	Yes	Yes	No	Yes	No	Yes	If inside	More work for caretaker; inhibit natural behavior
Indoor, cage-free	Yes, but not practical	Yes	Yes	Yes	Yes	Yes	Yes	Good for meat birds, not so humane for others
Mobile housing	Yes	Yes	Yes, if large enough	Yes	Only fair	Yes	Maybe with the right material and climate	Must be moved often; difficult to access
Shelter with outside run	Possible	Yes	Yes	Yes	Only fair	Yes	Yes	Not best option for meat, but best for all others

Type of Shelter	2–4 Chickens	4–10 Chickens	10+ Chickens	Layers	Meat	Pet/ Show	Cold Weather	Other
Shelter with free range	Yes	Yes	Yes	Yes	Not best method	Yes	Yes, if shelter is large	Chickens vulnerable to predators; may make neighbors unhappy; eggs may be hidden; birds find a lot of their food
Pastured poultry	Not practical	Not practical	Yes	Yes	Must use the right meat strains	No	No	Pasture must be managed; birds need less grain; birds may be vulnerable to predators

Chapter 6

Designing and Constructing a Coop

In This Chapter

- Deciding whether to build, buy, or do some of both
- Finding your building supplies
- Putting up fencing for an outdoor run

Unless you were lucky enough to inherit a nice chicken coop when you bought your property, you need to consider creating some type of housing for your birds. In Chapter 5, we talk about the different types of housing options. This chapter goes a little deeper into the actual construction of a chicken coop.

Before you pick up a hammer or a catalog, you need to complete some specific tasks. The first order of business is to make absolutely certain that keeping chickens at your home is legal (see Chapter 1 for more info on consulting the law). You don't want to put forth the time and effort to build a chicken coop only to be told to remove it.

It's better to spend a lot of time planning than a lot of time fixing mistakes. A good plan makes for a good chance of success in your chicken-raising adventure.

If you're even remotely handy with a hammer and nails, you can build a coop from the ground up. If you're a total klutz and no one wants to see you with a saw in your hand, you can buy ready-made housing. Or you can combine a little building with a little buying to create a one-of-a-kind chicken Taj Mahal.

To Repurpose, to Build, or to Buy? That Is the Question

The challenge is obvious — your chickens need a home. Chicken housing need not be fancy or expensive, but it has to give your birds a place where they

- Stay protected from predators.
- Can stay warm in the winter and cool in the summer.
- Can lay their eggs.
- Can prop up their feet and relax after a hard day of being chickens.

The housing also needs to allow you easy access to your birds. It should be easy to clean and sanitize, and it should also provide good ventilation for the health of your chickens. Chapter 5 discusses different styles of housing. Now you need to decide whether you want to build that housing, buy it, or use some combination of the two.

Checking on housing restrictions

Depending on your area, you may have restrictions on the size or location of your building. It's better to find out your limitations first than deal with unpleasant consequences later. Common building restrictions include the distance a structure has to be from lot lines, from existing structures, and from the road. Restrictions also may govern the size and type of foundation a structure needs.

Make sure the official you talk to knows that the structure will be used to house animals. Different rules may apply for animal housing than for other types of structures. Building a shed on a sturdy skid frame might avoid higher property taxes and building codes, but make sure your community allows moveable structures before building this way.

Start by calling your city or township building department. If you can't find a number, contact the city or township hall and ask who to call. Find out whether you need a permit to build, see what the local building restrictions are, and determine where you need to go to purchase any necessary permits. Get any necessary permits and display them as required.

Many people think they can build over a weekend when city or county officials are out of their offices, and no one will ever notice. But if you get caught, you may have to move all the chickens out of the housing while you fix any problems, and you may have to pay a big fine. It's not worth the hassle.

Making do with what you've got

Without a doubt, the easiest option for housing is combining prebuilt structures with a few of your own personal renovation touches. This method also saves money, in most cases. If you have a shed, garage, or barn that you can convert to an indoor shelter, you can purchase dog kennel panels to create an outside run. You can also use kennel panels on the inside of the structure to divide the chicken space from areas intended for other uses. Conversely, if you already have a large fenced area, you can buy a small shed and convert it to a shelter. Sheds of this type are generally available at most hardware and home improvement stores.

If you're looking for something charming, children's playhouses make good indoor shelters. And if you live in an area with few neighborhood rules and covenants, you can even consider turning an old box trailer or a horse trailer into a chicken shelter with the addition of an outside run.

A greenhouse frame can also work as a chicken shelter if you remove the clear plastic covering (which makes it too hot inside) and replace it with another, solid covering. A deer blind or an ice-fishing shanty can serve as a chicken shelter for a few birds. We knew a woman who turned a two-hole outhouse into a two-nest chicken coop that was actually quite charming. Use your imagination, but remember the neighbors and any restrictions your neighborhood may have.

Reuse and recycle

Rob says: "One of the cleverest examples of alternative housing for chickens I ever saw was an old van converted to a small chicken coop. The front part of the van was removed, leaving just the back main body with rear axle for easy moving. The windows were painted black for hen privacy and were reversed on the van body so they could be opened from the outside. With a little ramp to a small, chicken-size door, some nest boxes, and roosts, the coop was good to go.

"There are tons of old vehicles waiting to be recycled out there. Just make sure your municipality will let you have one sitting on your property, and make it look neat and shiny."

Building from scratch

Building a coop is more expensive in most cases than refurbishing an existing structure. It also typically takes more time and more building skills. But it may be the only way to get the coop you need or want.

Only you know whether you have the skills and time to build a chicken coop. Chicken housing can be simple or elaborate, but you don't need to be a master carpenter to create a coop — chickens are very forgiving about crooked walls and slanting floors. Just be sure to keep the needs of your chickens first and foremost as you plan and build.

You may be tempted to slap something together from some old wooden pallets and a roll of salvaged wire, and, admittedly, some people can do a good job with these types of basic materials. However, you probably want to build something that won't make your neighbors unhappy and will hold up to years of use. It makes more sense to initially build the right kind of housing than to build something shoddy that you'll need to replace in a couple years.

To build your own coop, at the very least, you should know how to hammer a nail, use a tape measure, and operate a power saw. Be sure to choose a simple design if your skills are few or rusty.

If you already have some building experience under your work belt, you can add more complex features to your coop, such as adjustable vents, manure pits, interior walls over the studs, insulation, exhaust fans, and heating systems. If you've done a lot of home renovation projects, these tasks should be simple for you.

With larger coops that require wiring and plumbing, you need to either have these skills already or learn them as you go. You may need to have your coop inspected by a building authority; in that case, hire out any tasks that you don't feel competent to handle.

If you're building a small coop or an A-frame, or if you're converting a shed or barn into a coop, even if you're a novice, you should be able to complete the project in a weekend if you already have all your supplies and you work diligently.

If you need help estimating the time you need for a coop project, take your plan to the local building supply store and tell them about your skills and the time you can devote to the project. They can probably guide you a bit about the timeline, but remember, this figure will be an estimate based on what you tell them. If you have an idea but no concrete plan, many larger building supply stores can help you draw up a plan and also make a supply list.

With some experience, a good work ethic, a little help from friends, and good weather, you can put together even an elaborate coop in a week. Most people need more time, however, because they work around their day job, their family, and weather interruptions. If you've done building or renovation projects in the past, you probably have an idea of how long your chosen project will take. Add a few extra days to your estimate, to be sure you'll be done before you need to move in the chickens.

Building housing takes time, so don't wait until the chicks are straining the sides of the brooder or your spouse is on his or her way home from purchasing the birds to begin construction. Give yourself time to factor in design plan changes, trips to the hardware store, bad weather, or any other unexpected interruptions in plans.

Designing and building a chicken coop can be a great family project. Whereas one family member may be good at the design part, another may be good at the construction part. Chicken coop construction is also a good way to hone those latent carpentry skills.

We can't cover everything about carpentry in a book about chickens, so we only touch on the basics here. You can get more information on building by reading home improvement books, checking out *Building Chicken Coops For Dummies* (Wiley), and perusing websites. You can find some good online information specifically on chicken coops at `www.backyardchickens.com`, `www.thecozynest.com`, and `www.buildeazy.com/chicken_coop_1.html`.

If you can afford it, you may be able to hire someone to build a coop for you. Then you can have all the fun of designing the coop while leaving all the work to someone else. Any good handyman can build a simple chicken coop. You may also want to ask a high school that has a building trades class whether the students would be interested in such a project.

Buying a chicken coop

As more people raise chickens on a small scale, more pre-fab (built) chicken housing has become available for purchase. Some of this housing is excellent, but some pre-built housing is too small and cramped for even two hens. Plus, the housing may be hard to clean or may have wire that's too flimsy to keep out predators. Be sure to carefully evaluate any ready-made choices you're considering.

Pre-fab coop requirements

If you decide to purchase housing, you'll probably be looking at cages, either wire one or small shelter-and-run combined units. We discuss these options in Chapter 5. You can also purchase a shed for loose indoor housing or as a shelter for free-range chickens, but here we discuss cage-style housing.

Try to physically examine the housing you want to purchase. Consider the following:

- **Does the location of the doors make it easy to feed and water the chickens or collect eggs?** If you're going to keep laying hens, or *layers*, you need either built-in nest boxes or room to place these boxes where you can easily access them. Be sure the feed and water dishes are located near a door that's wide enough to allow you to insert and remove the dishes.
- **Do chickens have a place to roost?** Chickens like to have a *roost*, a place to perch off the floor. Some pre-built shelters have a slide-out pan or door under the roost. Because most of the manure in the shelter collects under the roost area, this feature is desirable.
- **Is the flooring easy to clean?** The absence of a floor can be a good option because it allows you to move the coop around on grass or set it on another surface to clean under it. Wire floors with trays under them to collect waste are another good option. Solid floors are the least desirable because cleaning them in small coops is often difficult. Even if floors are removable, they require more frequent care. Solid floors must be sturdy enough that they don't sag under the weight of birds and bedding.
- **If the floor is wire, does it feel smooth, and are the spaces small enough to prevent a chicken's foot from slipping through?** These considerations protect your chickens from injury.
- **If predators such as dogs and other large animals are a problem, is the housing sturdy enough to protect the chickens?** The mesh on housing needs to be sturdy wire, not plastic or chicken wire. Doors need good latches.

If the housing you choose doesn't have a slide-out pan or door under the roost to facilitate cleaning, you may want to install a pit under the roosting area. Alternatively, you can lay a flat board under the roost that you can remove and scrape to clean.

If the housing will be outside, make sure it's suitable for the weather in your area. The shelter should always have a waterproof top. In cold areas, the shelter needs thick walls or some form of insulation.

It always helps to see what you're buying in person so you can try opening doors, sliding out floors, and so on. If a friend or relative has purchased housing for chickens, take a close look at it to see whether something similar may be right for you. Looking at a catalog or website may not give you a good enough idea of what you are thinking of buying. Make sure you look at the description, too — it should tell you the dimensions, the weight, and what materials the coop is made from. If you have questions, call sellers and ask them.

Pre-fab shopping

If you want to buy housing, check out farm magazines, poultry supply catalogs, garden magazines, and online sources. If you live in a rural area, a farm store may carry poultry housing or have it available for special order. If it doesn't, it may be able to refer you to a place that does. Don't forget to check your local newspaper and online options such as Craigslist, Amazon, and eBay for bargains on new or used chicken housing. When buying a used coop, thoroughly clean and sanitize it before use. (It's difficult to do this with wood structures.) Following are just a few of the online sources for housing:

- `www.mcmurrayhatchery.com` A wide selection of poultry equipment, housing, and baby chicks. You can request a printed catalog by calling 800-456-3280.
- `www.henspa.com` Poultry housing and equipment.
- `www.horizonstructures.com/storage-sheds-chicken-coop.asp` Pre-built housing. This company can also design and build your coop. You can get more information by calling 888-44-SHEDS (888-447-4337).
- `www.strombergschickens.com` A great resource for books, supplies, and equipment. Request a catalog by calling 800-720-1134.
- `www.backyardchickens.com` Forums that advertise used housing.
- `www.backyardfarming.com` Forums that advertise used housing.

Choosing the Right Location

After you've gotten all the information about building restrictions and applied for any necessary permits, it's time to choose the spot where you want to place your chicken housing. If you have a small lot, you may have only one obvious choice. But if you have more room, the ideal spot is one with these characteristics:

- **Close to the house:** You want the chicken coop close to the house so that it's easy to service and you can keep an eye out for predators or other unwanted visitors. However, if you have more than a few chickens, you want them far enough away from the house that you can't smell them on warm, wet days.

- **Close to utilities:** Running electricity to the chicken coop is a big plus. Doing chores in the dark is no fun, and lighting keeps chickens warm and more content throughout the winter months. Plus, you and your chickens will both be happier if the coop can be located near a water source, to make cleaning and watering less cumbersome.

  If you can use the wall of another structure (such as a garage or barn) as part of the coop, you may have an easier time running water and electricity to the coop. You can save money and materials, too.
- **Away from the neighbors:** Even if keeping chickens is legal and your neighbors say they love these birds, it's just not fair to put the birds under the neighbor's bedroom window or in view of the pool. Including the neighbors in the decision about where to locate the coop may make them happier with your chicken-keeping project.
- **In a well-drained spot:** Chickens don't like to get their feet wet, inside or out. Don't put your chicken housing where the ground is low or where water drains toward the area. If the area is sometimes damp, you can add gravel to the runs; however, that doesn't always work well because water can stand on top of or saturate the gravel spaces. Wet chicken manure doesn't smell nice.
- **Away from potential environmental issues:** Try not to put chicken housing where manure from the coop will wash into lakes, ponds, streams, or other water. If you have a well, keep your chicken housing at least 50 feet from it, if that point isn't regulated by law.

Oh, the importance of planning

Kim says: "Here's an example of poor planning. My husband and I wanted to try the chicken tractor/pastured poultry method of raising broilers and replacement layers. We built several 4-x-8-foot pens, 4 feet high. They were made out of plastic pipe and chicken wire with tarps for the ceilings, and they were easy to move around the pasture. However, they were a pain to service. At 4 feet tall, they were hard to lean over to retrieve feed and water containers, and we had to pull back the tarp to do so. If we lifted a side, we risked the chickens escaping, especially the pullets. We inserted some doors in the sides and wired the feeders to the walls; however, when it came time to catch the birds, another problem reared its head. If we lifted the whole thing, birds went everywhere. I had to crawl through the small door, crawl around on my knees to catch the chickens, and hand them outside to my husband. Needless to say, we rethought our housing for the next season. This may work for you, though, if you have small kids who aren't afraid to catch chickens!"

When your plan is finally on paper, go out to where you intend to place the coop and measure the spot. Make sure your design will actually fit in the space you have in mind for the coop. You may not think this step is necessary if you're building a small, moveable chicken coop, but consider the smallest location to which you may need to move the coop — will it fit? Make sure you check overhead, too. If the coop is 8 feet high, will you need to cut any tree limbs or move any wires?

Combining Form and Function: The Basic Coop Blueprint

Chicken coops come in all shapes and sizes. In Chapter 5, we discuss many types of chicken coops and weigh the pros and cons of many coop styles. Before you begin building a coop, take a look at Chapter 5 — or at least check out the coop of a friend or neighbor.

Whether you're doing this building project by yourself or someone is helping you, putting your plans on paper is always a good idea. When you write down ideas and dimensions, you can spot any mistakes in your planning and better judge how much lumber and other supplies you need. Erasing a line is easier than tearing down a wall, so take the time to draw out a plan.

Begin with the major planning issues. Grab a pen and a piece of paper, and make a list of all the things you *need* to have in your chicken housing. For example, include the amount of indoor and outdoor floor space needed per bird; lighting, heating, cooling, and ventilation appropriate for your area; and anything you must do to satisfy building codes, such as having a cement foundation on an indoor shelter.

Other items to think about are nest boxes for egg-laying and roosts for sleeping. These considerations are discussed fully later in the chapter, but be sure to factor them in when planning. You may also want to plan for a manure pit or a dropping board under the roosts (see the earlier section "Pre-fab coop requirements" for more on this).

Coop size and shape: Giving your birds some breathing room

You need to allow a minimum of 2 square feet of indoor shelter area per bird (actually, 3 to 4 feet is better), and you need 3 to 6 square feet of outdoor run area per bird. To compute the correct amount of indoor space, multiply the number of chickens you want to keep by 2 to get the minimum square footage of floor space: For example, 2 hens × 2 square feet = 4 square feet of indoor space. Your shelter could be 2 × 2 feet, which would be most practical, or 1 × 4 feet. If you had six hens, you would figure 6 × 2 = 12 square feet, so you'd need a 3-×4-foot or 2-×6-foot shelter.

Use the same basic formula for figuring outdoor square footage, but multiply the number of birds by either 3, 4, 5, or 6, depending on how much outdoor space you decide on for each bird.

Ventilation: Allowing fresh air to flow

Chicken shelter areas need good ventilation year round. Ventilation can be accomplished through windows, roof vents, exhaust fans, and other means. You may be able to salvage some of those components, like windows, doors, and even exhaust fans, from home remodeling projects or buy them used, to save money.

Good ventilation usually consists of some way for cooler air to enter near the floor, be warmed, and then exit near the top. If the coop is tall enough to walk in, you need high and low ventilation points. Screened vents are generally used at the bottom for incoming air, and space between rafters or at the roof peak allows air to exit. Windows that you can open increase ventilation in warm weather. Small coops you service from outside can use the entrance point for chickens as the bottom source of air and then use vents near the top of the shelter for stale air to exit.

The amount of ventilation a shelter needs depends on how many chickens are kept in what amount of space and the weather in your area. Exhaust fans speed the exchange of air if natural means aren't enough to keep the air free of ammonia fumes and excess moisture, both of which are bad for chickens and for you. Exhaust fans are located near the top of a structure — and sometimes additional fans are used near the bottom — to pull cooler air inside.

We cover ventilation in more detail, including its importance in chicken health, in Chapter 5.

Roost and relaxation

Roosts are an important part of chicken coop furnishings. Because chickens are vulnerable when they sleep, they prefer to *roost* (perch) as high off the ground as they can when sleeping. The more "street-savvy" birds also pick a spot with overhead protection from the weather and owls. Roosts or perches can accommodate these needs. A roost is a pole or board suspended off the floor. Chickens like to roost in the same spot every night, so once they're used to roosting in your chicken coop, they'll head back home at nightfall, even if they've managed to escape that day or they're allowed to roam.

Some of the heavy breeds, including the heavy broiler-type chickens, can't get themselves very high off the floor. Broilers don't need any roosts for their short life span, but you'll want to accommodate other heavy breeds with low, wide roosts.

The chicken must have enough room to sit upright between the roost and the top of the shelter, even though most of the time it sits in a crouched position on the roost. A rooster likes to sit on the roost to crow, and he likes to be able to extend his head and neck a bit higher than normal. If your shelter is small, you may not have room for roosts, and your chickens then will have to sleep on the floor.

The area under a roost accumulates more droppings than any other place in the shelter. Some people like to put a pan or board to collect droppings under a roost, to make cleanup easier. Figure 6-1 shows an example.

Figure 6-1: Place dropping pans underneath a roost for easy manure cleanup.

Illustration by Barbara Frake

Alternatively, you can place a manure pit or additional bedding under the roost to make cleaning the coop easier. If you plan to add a pit, make sure you can access it with tools for cleaning purposes.

Room to roost

You can look at roost size in two ways. Some people think the chicken should be able to span the roost with its toes and grasp the perch. A two-by-four on edge or a 2-inch-round wooden dowel is about right for full-size chickens. Bantam chickens, however, are more comfortable with a smaller-diameter roost.

Other people believe that the chicken prefers to sit flat-footed on the roost. If you subscribe to this theory, you may want to turn that two-by-four on its side or provide a 4- to 6-inch-wide board as a roost. This option may be a better one in cold climates because the belly feathers will help prevent frostbite by covering the toes while the bird sleeps.

The roost should provide about 1 foot of length for each bird. If you have six chickens, for example, you need 6 feet of roost length. If the coop isn't wide enough for a single length of roost, you can place several roosts 2 feet or more apart. Whatever type of roost you provide needs to be strong enough to support the weight of the birds without sagging, and you'll need to anchor it so that it doesn't wobble as birds get on and off it.

Roost placement

If you can, place the roosts at least 2 feet off the floor. Even heavy hens should be able to fly up to roosts 3 feet off the floor. Lightweight hens and roosters can go even higher: 4 or 5 feet if you have enough space in the coop. You can use a step type of roost system, with each roost being a foot or so higher than the next in a staggered arrangement; see Figure 6-2 for an example of a stair-step roost. Don't put one roost directly under another roost; the chickens on the lower levels will become coated with droppings from above. The highest-ranked birds in a flock generally take the highest roosts.

If you have more than one roost and they're all on the same level, allow about 2 feet of space between them. To avoid contamination, don't place roosts over feed and water dishes. If you have the space, don't locate roosts over nest boxes, for the same reason. Be sure that nest boxes are covered if you have to place them under roosts. And you may not want to place roosts directly over areas where you walk to service the coop.

Put your roosts on hinges, or construct them so that they sit into grooves. That way, you can remove or lift them when you're in the coop to clean or catch birds.

Figure 6-2: A stair-step roost may be in a corner, as shown, or up against a wall.

Illustration by Barbara Frake

Feathering their nests

Nest boxes are important if you're keeping chickens for their eggs. They provide a safe and comfortable spot where layers can lay their eggs and where you can easily gather them. They're also essential if you want your hens to hatch their own eggs.

Young pullets like to "play house" as their hormones begin to prepare them for laying. If the proper nest boxes are available, they will try them out: sitting in them, arranging nesting material, and practicing crooning lullabies. Dark, comfortable, secluded nest boxes attract them. Try to have nest boxes in place by the time pullets are 20 weeks old.

You may also want nest boxes for pet chickens and ones you keep for show, especially if you want to hatch your own eggs. Chickens will lay eggs without nest boxes (in fact, some birds will never lay their eggs in the nest box), but it's far easier to collect eggs from a box than from the floor.

You don't need nest boxes if you're keeping meat birds. Chickens don't need boxes to sleep in and prefer to roost instead. Meat birds are often too heavy to fly up to a roost, so they just sleep on the floor. Make sure they have some soft litter on the floor, and they'll be happy.

One nest box for every two hens is ideal. You can use fewer boxes if space is limited, but one box for more than four hens leads to fighting and laying eggs outside the box. Hens do share or wait their turn to use nest boxes. You can have one box for each hen if you have the room, but you may find that some boxes aren't used at all.

Size does matter

Chickens like nest boxes that are big enough to turn around in but not so large that they don't feel comfy. Nest boxes that are roofed need to be tall enough for the hen to stand up. For average-size hens, a 12-inch-square nest box is the minimum size to use, but a 12-×-18-inch box or a 16-×-16-inch box is usually better. Boxes can be any configuration — 14 × 14 inches, 14 × 16 inches, and so on — they can even be round. Bantam hens can use smaller boxes, and extra-large breeds may need boxes at least 16 inches square.

Having boxes that are too big encourages several hens to go into the same box at the same time, which often causes egg breakage. Boxes should have a lower side in the front that encourages the hen to step into the box rather than jump in over a side. If more than one hen uses the box, hens may jump in over the top and break eggs. Sides can be anywhere from 6 inches high (if not roofed) to a foot or more high (if roofed).

Making it a suite

Hens enjoy company while laying, so don't scatter the boxes around the coop. Place nest boxes in a sheltered area of the chicken coop, and try to keep them out of drafts. In cold-weather areas, keep nests in the warmest area of the coop, to make the hens more comfortable and keep eggs from freezing. Figure 6-3 shows a setup of four nest boxes.

If you have numerous nest boxes, place some on or near the floor and some higher. Coops often have nest boxes stacked above each other, up to about 4 feet off the ground. All nests that are more than a foot off the ground need a perch or board in front of them. Hens need a spot to fly up to and land on before entering the nest. However, it has been our experience that hens don't use the highest nest boxes often, and heavier hens prefer low nests, so we keep nest boxes lower than 3 feet off the ground. Hens willingly accept floor boxes if you don't have enough height in the shelter to raise boxes off the floor.

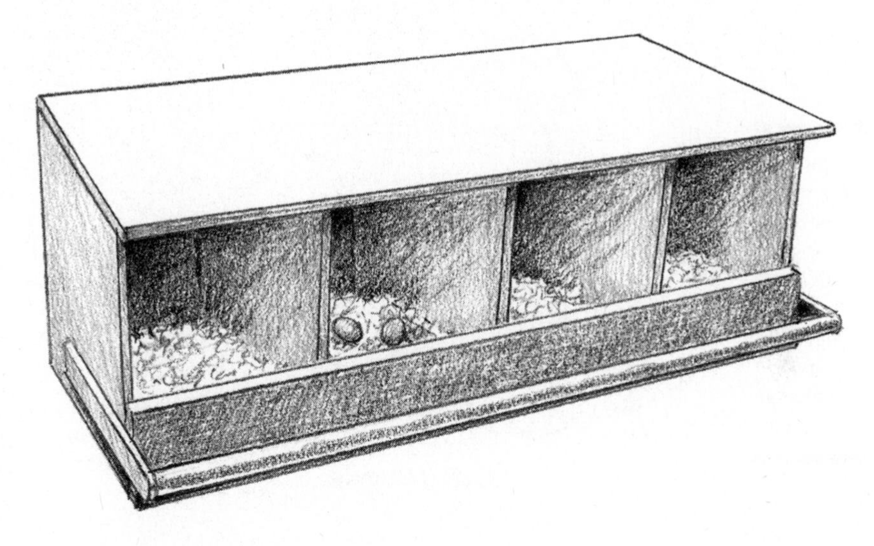

Figure 6-3: Side-by-side nest boxes are easy to make.

Illustration by Barbara Frake

To avoid a situation in which all the chickens prefer the same nest, make all the nests the same size and color, and provide them with the same degree of darkness (hens prefer nest boxes that are dimly lit and feel secluded). Even under these circumstances, however, hens often favor a certain nest box, and you can't do much about it. The hens will line up to use that box and may even squabble over it.

Provide extra nests if you intend to let some of your hens sit on eggs to hatch them, because a nest will be out of commission when a hen is sitting.

Buying or repurposing items for nest boxes

Many commercial nest boxes are available on the market. The old standard is a set of metal boxes; some of the newer nest boxes are made of plastic resin materials. The boxes have round openings and come in various configurations — two nests, four nests, and so on. Some have attached backs; others are open-backed, and you attach them to a wall.

Plastic and metal are easier to clean and disinfect, but if you choose to make your own nest boxes, you may want to build them using wood or to repurpose common household items as nest boxes. A covered kitty litter box makes an ideal nest box. Deep plastic tubs with a hole cut in the front can sit on the ground or on shelves. Even sturdy cardboard boxes work in a pinch.

A free-standing nest box is like a covered kitty litter box. See Figure 6-4. It has a handle on top so you can pick it up and collect eggs without stooping.

Figure 6-4: A portable nest box is good for easily collecting eggs and moving between coops.

Illustration by Barbara Frake

If you're using cages or small pens to house chickens, you can remove these boxes when hens aren't laying, to save room. Free-standing boxes also allow you to clean only the dirty nest boxes instead of having to clean a large unit.

Wiring, Fixtures, and Other Important Amenities

The availability of electricity in any kind of coop is a big plus. With electrical power running to your housing, you can provide light, heat, and fans much easier than if the coop isn't wired. Make sure you know how to safely install electricity, or hire someone for this part of the job. We cover the nuts and bolts (in a manner of speaking) in the following sections, but for detailed guidance, consult a registered electrician. Your building supply store may be able to help, but remember: Working with electricity can be a shocking experience.

Providing the hookups (electrical, that is)

Electric outlets and fixtures need to be out of the birds' reach or enclosed in metal boxes or wire cages. This setup keeps curious chickens from being "fried" or starting a fire.

If you're not familiar with electricity, it's best to leave it to a capable soul who understands how to figure out the load requirements for heating and cooling devices.

Lighting up

Along with natural lighting from windows, you need to plan for artificial lighting from light fixtures. Having more than one light fixture allows you to vary the amount of light and avoid a blackout if a bulb burns out. You need at least two fixtures, one for the strong light source and one for the dim one. You may need more fixtures, depending on the size of the coop and the wattage of the bulbs you intend to use. The inside of the coop needs to be bright enough that you can easily read in the coop. This arrangement usually requires one 100-watt or equivalent bulb per 200 square feet. Your dim light should give you just enough light to see the hens in the roost area (a 15- to 25-watt bulb or equivalent), which is enough light for them at night.

For optimum light, locate fixtures near the coop ceiling. If the chickens can reach the light bulbs, cover them with wire mesh or plastic covers designed for light bulbs so they don't get broken. Place a light switch near where you enter the coop, just inside or outside the door.

Chickens need 14 to16 hours of bright light and 8 to 10 hours of darkness or dim light to lay well year round. If you're not worried about egg production, light the shelter area at least 10 hours a day so birds can eat and drink normally, unless the shelter has good natural light through windows. Meat birds need to have the shelter lighted 22 hours a day so they can eat enough to keep up with their fast growth. For more about the light needs of laying hens, see Chapter 15.

When chickens sleep, they really sleep. Total darkness makes chickens go into a kind of stupor. They're an easy mark for predators at this point because they don't defend themselves or try to escape. A dim light in the shelter lets them defend themselves a bit and helps them find their way back to a roost if they're disturbed. It also prevents panicked flapping around and possible injury when strange noises wake them.

You don't have to go broke lighting the coop. If the shelter has windows, you can probably reduce or even eliminate the light during daylight hours, depending on how many bright, sunny days you get in your area in the winter months. Make sure, however, that your lights are on in the early morning and late at night to extend the light cycle to 14 hours. If you aren't able to turn lights off and on at regular times, you may want to put the lights on an inexpensive timer. You can also install a dusk-to-dawn switch on the dim light so it comes on when the coop darkens.

If you have a small pre-fab housing unit for two to four hens, a 100-watt light is too much. In those units, a 40-watt light is usually enough. If the hens in those units spend all their time in the outside run area except at night or when they're laying, you may want to hang the light over the outside run. You'll need a brighter light for that. The birds will still sense the natural nightfall and may retreat to the shelter, so lighting these small units, inside the shelter or out, may not be as effective as lighting larger shelters.

If electricity isn't available at the site of your chicken housing, consider one of the new battery-powered lights or use solar lighting.

Most light fixtures handle a variety of bulb types. To save money and help the environment, consider using compact fluorescent or LED bulbs instead of old-style incandescent bulbs for overhead lighting. Incandescent bulbs are being phased out of production. Both compact fluorescent (CFL) and LED bulbs cost more to begin with than the old bulbs, but they save you money on your electric bill and last a lot longer. Studies in commercial chicken farms have determined that chickens prefer LED light over CFL lights because the CFL bulbs have a flicker that birds can detect. However, either type of light bulb is better than incandescent. You can buy LED lights made specifically for poultry housing at farm supply stores.

Baby, it's cooold outside!

Chickens generally survive cold better than heat. Unless you live where winter regularly brings subzero temperatures for long periods of time heating a coop generally isn't necessary. Adult chickens can survive below-freezing temperatures if they stay dry and out of the wind, but they may not lay as well during long periods of subzero temperatures.

Some people like to heat a coop because it's more comfortable for the birds and for them when they're caring for the flock. Chickens can suffer from frostbite, and they consume more feed to keep their bodies fueled if the coop isn't heated. Heating the coop keeps the water from freezing so you don't have to bring water to the birds several times a day. And some fancy breeds, such as the Serema, don't do well in temperatures below freezing.

If you decide to heat the coop, you just need to keep it above freezing — 34 to 40 degrees Fahrenheit will do. Any more than that, and you may cause a moisture buildup that will impact the birds' health. Any type of heater works to heat coops. If you're unsure of what size heater to buy for the coop space you're heating, ask a heating salesperson at a building supply store to help you. A heater with a thermostat is preferable to one without.

Heat lamps and electric heaters are safer than heaters that use a flame. These types don't usually require special venting or chimneys. If the heat source has a flame, such as propane, natural gas, or wood heaters, make sure you follow the exact instructions for ventilation and chimney requirements. If you're unsure, consult a professional — better safe than sorry. Install a carbon monoxide detector with heaters that use a flame. Electric heaters and heat lamps don't require carbon monoxide detectors. Remember, birds are very sensitive to carbon monoxide poisoning.

Use any form of heat with great care. Make sure the ventilation is good, monitor temperatures to keep them from becoming too warm, and keep combustible materials (including birds!) far enough away from the heat source. Also make sure heat lamp bulbs won't be splashed with water, which may cause them to explode.

Fighting the heat

Heat can be more harmful to chickens than cold, but most home flock owners don't need to consider air-conditioning. Cooling the coop to below 85 degrees is desirable and can generally be accomplished by ventilation and fans. Heat stress can cause birds to slow their laying or result in soft-shelled eggs. Meat birds are prone to dying from heat stress, so if you can't cool the coop, you may want to raise chickens for meat at cooler times of the year.

If your summers are very hot, plan for more space per chicken in your coop and maintain a good cross-flow of air. Consider installing exhaust fans in the upper part of the coop; they require an opening near the bottom of the coop from which to draw cooler air. You also can place window fans in coops. Be sure to frequently clean dust buildup off electric fans, and cover all fans with small mesh wire so birds aren't injured.

Other ways to keep chickens cooler are to provide shade over the chicken coop, insulate the coop ceiling, and paint the roof with a reflective coating. Some people use sprinklers on the coop roof to provide evaporative cooling. Make sure cool water is always available to the birds, too.

Being Mindful of Materials

Building codes may restrict the types of materials you can use for building, so check first before you decide on what to use. You need materials for the framing and roofing. And then there's the flooring. And fencing. And posts. And don't forget the chickens! Oh, wait — that's another chapter.

Building materials are usually sold in specific sizes, so try to draw your plan with those sizes in mind. For example, most wood panels are cut to 8 feet. So instead of planning a 7-foot-high wall, make it 8 feet high to avoid a lot of excess cutting and waste.

When you've completed your paper plan and you've ensured that your planned coop fits the site, start making your supply list. Count the number of two-by-fours, fence posts, rolls of wire, wood or metal panels, and so on that you'll need by looking at your plan. Don't forget to add nails, screws, or other fasteners to your list, as well as windows, vents, paint, and roofing material. You may also want to add building materials to build the nest boxes and roosts. Most people build those items at the same time they construct the coop.

It's a good idea to do some comparison shopping to find the best prices on the materials you need. After doing a bit of price checking, you may want to alter your plans. If money is a factor, consider building a simple coop that you can add on to or modify as funds permit.

Getting to the bottom of flooring

Wood or cement flooring is preferable to dirt floors in shelters because those types of flooring prevent pests and predators and also make cleaning easier. Rubber and other types of roll-out, waterproof materials designed for barn floors make excellent coverings over cement and wood, if you can afford them. Materials such as roll vinyl made for human homes and metal sheeting may be easy to clean, but they can be slippery underfoot.

Dirt floors are acceptable for outdoor runs. Adding a layer of sand or fine gravel to the runs makes them more absorbent. In all but the biggest runs, the dirt floor will soon be hard packed.

Cages may have solid or wire floors, and slatted wood flooring is used occasionally. Wire flooring should be smooth and it needs to be strong enough to support the birds without sagging. It's easy to keep birds clean with wire floors, but wire prevents some of the natural behaviors of chickens, such as scratching at the ground. Toes can get caught in wire floors, too, causing injuries.

You should be able to easily clean solid floors in cages and small shelters. These floors may be pull-out tray floors or else floors that drop down on a hinge so you can remove droppings and litter.

Constructing the frame

You don't need top grades of framing materials and panels to build chicken housing, but you do need materials that are straight and sound enough to make your job relatively easy. Most chicken housing is built of some sort of wood because wood is probably the easiest building material to work with. You can also use lightweight aluminum panels like those used in pole barns. Plastic lumber works great for chicken coops, too, but it's quite expensive.

In areas where hawks and owls are a big problem, or when you must keep chickens inside the run, you may need a roof for your outside run in addition to the roof on your shelter. First, decide whether you want to be able to walk upright in the outside run. Then plan how to support the roof. If you have a wide pen, you may need additional posts inside it to hold supports that keep roof material from sagging in the middle.

Treated lumber is fine to use when building chicken housing because the chickens rarely chew on it. For your shelter, use exterior grades of wood unless you intend to cover the wood walls with siding. Paint or stain the walls with exterior-grade products for further protection. If you're unsure of what paint or stain is best, ask a building store salesperson for recommendations based on your area and what you can afford. Metal roofs don't need any kind of finishing, but wood roofs need shingles or roll roofing to protect them.

Wrapping your head around fencing

The kind of wire you need for any fencing exposed to the outdoors depends on the predators in your area. You need fencing strong enough to keep out predators, with openings small enough to prevent the chickens from squeezing through.

Inexpensive chicken wire is almost synonymous with chickens, but we have to say that it has its limitations. Chicken wire has hexagonal openings in ½-inch, 1-inch, and 2-inch sizes. It comes in several gauges, with the smallest-numbered gauge being the strongest. It keeps chickens in, but it doesn't keep out strong predators like dogs and coons. It rusts and becomes weaker after a few years, and it can be difficult to work with.

Many new types of nylon and plastic fencing are on the market. They can be good choices for chicken fencing if they're strong enough because chickens don't chew out of enclosures. But some predators may be able to chew their way in. Whereas the fine black nylon netting used to protect crops from birds and deer can work on the tops of pens to keep the escape artists home or protect them from hawks, we don't recommend it for the sides of enclosures.

Electrified netting is now available for pastured poultry setups. It requires the use of special posts, often included in a kit with the netting. The netting isn't very strong, but it keeps chickens in and predators out because they get an electrical shock if they touch it. It's lightweight and fairly easy to move, but it does have limitations. Most setups use solar power to charge a battery to power the fence. Long stretches of gloomy weather can cause a failure, as can wet vegetation or other objects that touch the fence and ground it.

One of the most common mistakes new chicken owners make is to underestimate the strength and cunning of predators. Predators lurk even in cities, and most of them love chickens and eggs. Even a raccoon can pull apart chicken wire. Raccoons also learn to open doors and climb over fences. Big dogs make short work of flimsy wire; skunks and opossums work on it until it rips. We highly recommend using strong welded wire on outside runs.

Welded wire comes with all kinds of opening sizes and in varying heights. A 1-× 2-inch, 2-× 3-inch, or 2- × 4-inch mesh is fine for adult chickens. Four- or 5-foot-high fencing works for most breeds.

Ice and snow can build up on any kind of netting or wire that's used on the top of an enclosure, causing the roof to sag or break if the load gets too heavy. You can stretch netting or wire tightly to keep ice or snow from clinging, but you may still need some roof supports.

Supporting fencing with posts

On the sides of the coop, chicken wire requires a top and bottom board to keep it stretched tightly; welded wire probably doesn't need this. All wire needs posts, however, and metal posts you pound in the ground are preferable to wood posts. Wood posts rot and break over time. However, many people prefer the look of wood posts, and they're easy to work with. Treated landscape timbers can make good posts but 4 × 4 treated fence posts will last longer.

In recent years, many people have taken to using PVC pipe (the kind designed for plumbing or electricity) to make supports for various types of chicken homes. This type of piping makes a lightweight, easy-to-move structure.

Some types of PVC piping become brittle in cold weather, and the types that don't become brittle are much more expensive. For permanent, year-round use in cold weather areas, it's better to use wood or metal frames and posts.

Place posts every 10 to 15 feet and at corners for outdoor runs. If the soil is sandy, you may need to use cement to keep the posts in place. You'll need about half a bag of ready-mix, quick-set cement per post.

Chapter 7

Coop, Sweet Coop: Furnishing and Housekeeping

In This Chapter

- Making cleanup easy with good bedding
- Luring hens to nests with comfy nest material
- Setting up food and water dishes
- Keeping your coop spick-and-span

Some chicken-keepers barely have time to keep their own homes clean, yet they have to find a way to keep their chickens' homes clean, too. No matter where you live, cleanliness is important to the health of your flock, and it may be important to your health, too. If you're an urban chicken-keeper or a chicken lover with close neighbors, keeping things neat and clean is even more important. If you become frustrated and feel overworked caring for your chickens, or if your neighbors complain of smells and sights, your chicken-keeping days will be numbered.

In this chapter, we talk about some creature comforts every chicken coop should have. You can add personal decorating touches like curtains for the windows and pictures on the walls, but the chickens certainly don't expect it. In this chapter, we also lay out a plan for you to keep the coop clean enough to meet your needs.

Bedding Down

Most people prefer to have some type of bedding material on the floor of chicken shelters to make cleanup easier. Bedding is also called *litter.* It's essentially floor covering, not something chickens need for sleeping. Bedding makes the floor of the coop easier to clean. It absorbs moisture and keeps down smells, making the coop a better place for you and the birds. Several types of bedding work, so you need to consider the cost of the various options and how they best suit both you and the chickens. Consider some examples of bedding:

- **Wood shavings:** This common material is available in pet stores as well as farm stores. Use pine shavings but not cedar or walnut shavings: Cedar and walnut shavings contain chemicals that can irritate your chickens' lungs and skin. Pine shavings also are more economical. You may be able to get a discount on the shavings if you buy several bales at one time. Remember that the bales are compressed and will cover a lot of area when opened.
- **Sawdust:** You may be able to get sawdust for free if you live near a sawmill. In rural areas, you can sometimes get truckloads of sawdust delivered. The one drawback to sawdust is the large amount of dust it produces, and some caretakers don't like breathing it in. Avoid sawdust from cedar or walnut trees because of the chemical irritants they contain.
- **Wood pellets sold as horse bedding:** Farm stores sell large bags of these compressed sawdust pellets. The pellets are very absorbent and less dusty than regular sawdust. Just pour them out and then lightly moisten to fluff them up. Make sure you moisten them before turning the chickens back into the shelter, however, because the birds may eat the pellets. They aren't poisonous, but they may swell in the crop and cause problems. Don't use wood pellets sold for fireplaces; they may contain chemicals.
- **Wood chips or shredded bark:** These materials are commonly used as garden mulch. If the cost is right, mulch with smaller pieces can work as bedding; however, it isn't very absorbent. Stay away from artificially colored mulch because it may harm the chickens or stain their feathers.
- **Sand:** Sand is a great flooring material, but it's heavy to move when cleaning. Sand is also heavy to carry and to shovel, so be sure to take its weight into consideration.
- **Straw or hay:** Straw isn't the best material for chicken housing. It's not very absorbent, and it can be dusty as well as messy. Chopped straw makes for better bedding, if you can find it or have some way to chop it yourself. Hay is also a poor choice, even though chickens love to pick through it. You can use hay or straw if you get it at a good price, but other materials work much better.
- **Other materials:** If you have access to products such as peat, coir, shredded sugar cane, shredded leaves, shredded paper, alfalfa pellets, fine gravel, ground oyster shell, and so on, you can evaluate how they work in your chicken housing and use them if they're more economical than other types of bedding.

Certain materials don't work well as bedding, so you want to avoid them:

- **Kitty litter:** Unscented, cheap clay litter is dusty and may swell in the birds' crops when they pick it up, thinking it's grit. Other types of kitty litter may be treated with chemicals for scent, which can harm chickens if they consume it.
- **Pine needles:** This material is too slippery and isn't absorbent.
- **Whole leaves:** Most leaves pack down when wet, and eventually they turn into a moldy mess.
- **Vermiculite or perlite:** Neither of these materials is good for use in chicken shelters. The dust from them can harm the lungs of both the chickens and their caretakers. In addition, these materials tend to blow around when dry and are relatively expensive.
- **Treated mulch:** Any garden mulches treated with weed killers, insecticides, or dyes can harm your chickens.
- **Cocoa shell:** Cocoa shell is generally expensive. It's sharp to those little chicken feet, is light enough to blow around, and may be toxic if consumed. (It's harmful to dogs, but no studies have been done specifically on chickens.) You may love the chocolate smell when you put down a fresh layer, but you may lose your liking for chocolate when you remove the cocoa shells loaded with chicken droppings.

Making Nests Comfy and Cozy

Nesting material helps cushion the eggs and keeps them from rolling around and breaking. It also makes hens feel more comfy — they love to rearrange and fluff the bedding as they sit in the nest. Unlike some birds, hens don't carry bedding material to a nest. However, they may scratch out a hollow in loose vegetation or soil and then sit in it, turning around and around to form the perfect nest. They then reach out with their beak and tuck bedding material in around them. In the wild, hens like to have the nest deep enough so that when they sit in it, the top of their back is the same height as the nest wall.

Straw or soft hay makes the ideal nest bedding. You also can use shredded paper or wood shavings. Put 3 to 4 inches of nesting material in each nest box. Some people simply use a piece of rubber matting in nests, which is easy to clean but not very natural for the hens. Hens will kick out, break up, and pack down nest bedding over time, so you'll need to replace it frequently. Also be sure to replace it if it gets dirty. (See the section on coop cleaning later in this chapter.) Clean nesting material keeps eggs clean.

Setting the Table and Crafting a "Pantry"

In the old days, you may have seen Grandma throwing chicken feed from a bucket onto the ground. In charming movie scenarios, Grandma even throws it from her turned-up apron. This may work for free-range, half-wild chickens, but your nicely confined birds need a better method.

Proper feed and water containers help you avoid wasted feed and save you money. You'll find no shortage of inexpensive chicken feeders and water containers on the market. (However, you're sure to find just as many expensive, elaborate feeders and waterers out there as well.) But feed and water containers are easy to create from materials you already have on hand. Your choice should suit your chicken-keeping setup, be easy to clean, and make feeding and watering convenient for both you and the chickens.

You can buy or build containers that hold a lot of feed or water and dispense it slowly to your chickens. This helps control waste because just a small amount of feed or water is available at one time, and what isn't needed at the moment isn't being wasted. These containers also save time because some are big enough to hold feed for several days. Automatic pet feeders and waterers can work for small groups of chickens. Farm supply stores and catalogs carry a wide variety of chicken feeders and waterers, ranging from small red plastic dishes that screw on a mason jar, to huge metal dispensers. The following sections show you options that require daily filling, as well as those that you fill less often because they dispense food and water as needed.

Self-feeding and watering systems save time and labor, but you need to check them every day to ensure that they're full and functioning properly. Chickens can suffer or die if valves become clogged or feeders are empty and you're unaware of it. Similarly, nipple self-watering systems may freeze up.

Feeding containers

Feeding equipment for chickens need not be elaborate, but it must accomplish these tasks:

- Hold at least a day's supply of feed for the chickens
- Allow all the chickens to eat at the same time
- Keep chickens from scratching out or otherwise wasting feed
- Be easy to keep clean and sanitary

Most chicken feeders today are made of galvanized steel, aluminum, or plastic so they can be easily cleaned. Occasionally, homemade feeders are made of wood. Figure 7-1 shows a few of the feeders that are available for purchase; if you prefer to make your own feeder, you can build a small wooden trough, use a small piece of metal gutter, or use a narrow pan such as a loaf pan.

If you live in a cold weather area, remember that plastic becomes brittle and cracks easily in freezing weather; you may want to avoid plastic dishes.

Some feeders hold more than a day's worth of feed. These are called *self-feeders.* They usually work by gravity — when the dish is empty, feed flows out of a container attached to the dish to fill it.

If you choose a self-feeder, you still need to check every day to see whether the feeder needs filling and is properly dispensing the feed. If the dish is full of scratched-up bedding, the feed may not flow out.

Feed dishes need to remain clean. Be sure you don't place feeders under perches or roosts where manure will fall into them. For the same reason, avoid feeders that have tops the birds can perch on.

Sizing things up

Each adult chicken needs about 6 inches of feeding space at the feeder. To maximize space, most feeders either are round or allow access to both sides of the dish. It's best to have enough feeder space so that all chickens can eat at the same time. Chickens eat according to their pecking order: If feeding space is limited, the highest-ranked chickens eat first, and the lowest-ranked chickens may not get the feed they need.

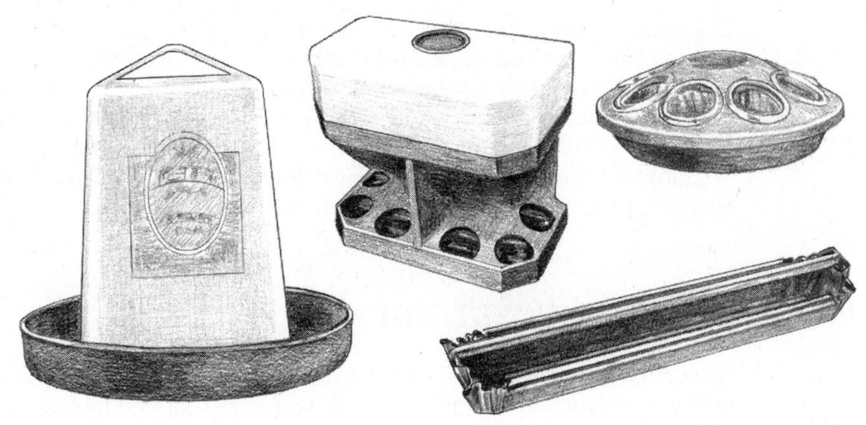

Figure 7-1: Some feeders, both automatic and manual filling, that you can purchase for chickens.

Illustration by Barbara Frake

Of course, dishes must be sized appropriately to the chickens. Baby chicks require smaller, shallower dishes than adults. Each chick in a brooder should have about 4 inches of feeding space at the feeder.

Provide more than one feeder if sufficient space isn't available at one feeder. This is especially important when the feeders aren't self-filling or when an assortment of feeds is offered that the birds can choose from. The top-ranked birds will get all the choice feeds or pick up the bigger particles of feed if feeding space isn't sufficient for all birds to eat at the same time.

Keeping feed in and chickens out

Feed dishes should be designed to keep chickens from getting in them and scratching out the feed, so they're generally long and narrow. Most feeders for chickens are in the form of either a circle or a long, rectangular, narrow trough. The narrow trough should be deep enough to keep chickens from easily throwing out feed.

The dish can be covered, with holes to let the birds access the feed, or it can just be too narrow for them to climb inside. Some feeders have a bar over the center of a long dish, and the bar rotates if a chicken tries to perch on it. Other feeders are placed on the outside of cages or pens, and the birds stick their heads through the wire of the pen to reach the feed. Narrow edges discourage chickens from trying to perch on the dishes. The feeder needs be heavy enough or attached to something so that it can't be tipped over easily.

If chickens persist in tipping dishes or scratching out feed, you can suspend the feeders by a chain or wire so that they're about level with the birds' backs. If you hang the feeders on a wall, make sure they have steep slanted tops or covers to discourage perching.

Chickens not only scratch in open dishes if they can get inside, flinging feed out, but they also may sit in mash or crumbs and take a dust bath, showering feed everywhere. Chickens use their beaks to fling pieces around, too, especially if the diet contains whole grains and they're trying to find what they like. Some of this feed may be picked out of the bedding, but a lot is wasted. To minimize the loss of feed, you can put the feeders in large plastic or metal trays or use feeders with narrow openings that chickens can't fit into.

Watering containers

All sorts of watering devices are on the market for chickens, from simple dishes that you fill daily, to automatic waterers that you fill less often and nipple devices that chickens must push to dispense a drop of water (see Figure 7-2 for some examples). Whatever water container you choose, it must

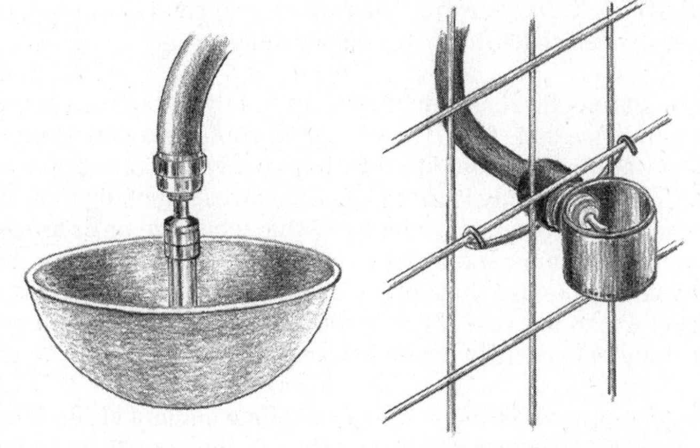

Figure 7-2: Some examples of automatic watering devices.

Illustration by Barbara Frake

hold enough water to satisfy all the birds for a reasonable period of time. However, unlike food dishes, water containers don't need space for all birds to drink at the same time.

Water containers must be clean at all times. The rims need to be narrow enough to discourage the chickens from perching on them. Also avoid placing water containers under roosts.

Choosing water dishes

Look for water containers made of a material that won't crack and break if ice forms in the container. Glass and many kinds of thin plastic, for example, are known to crack in cold weather. If you have electricity in the coop, you may want to purchase a heated water dish. Aluminum dog dishes work well to hold water for small groups of birds in cold weather. If ice forms in the dishes, simply turn them over and pour a small quantity of warm water on the bottom. You'll hear a pop as the ice is released from the pan.

For small groups of birds, a crock pot may be the answer to the winter water woes. Make sure that it's set to low heat and that the cord is protected from the birds. Check the water from time to time to ensure that it isn't getting too warm for the birds to drink.

Opting for automatic watering systems

To minimize your daily chicken-keeping chores, you may want to consider using an automatic watering system (refer to Figure 7-2 for examples), which continuously offers water to the birds. This type of system needs to be hooked to the household water system and works with a valve that shuts off water when it reaches a certain level in the dish that the birds drink from. You need to check every day to make sure the system is working properly, but you don't need to fill the water dishes daily.

For more than ten chickens, you may want to buy or build an automatic nipple watering system. This type of system contains a plastic pipe connected to a water source and fitted with nipple valves that drip water when touched. Chickens quickly learn to use the valves, touching their beaks to the nipple to get a drop of water. Remember that while the birds are getting used to this system, you must have dishes of water available to them. You may have to train them to use the nipples by frequently touching them and leaving beads of water. Because these systems ensure that fresh water is always available, they're especially useful in warm weather.

Even though nipple watering systems save time and are clean, they have some drawbacks. They need to be attached to the household water system. They won't work if they freeze or if hard water deposits or dirt clog the valves. And you may find it difficult at first to see whether all the chickens know how to use the system.

Other types of automatic waterers simply hold a large quantity of water in a reservoir and release it to a dish as birds drink. This kind of waterer also needs to be checked each day to see whether the reservoir needs more water. How often you need to fill the reservoir depends on how many chickens you have and how much water the reservoir holds, but this system may allow you to go a few days before refilling.

Self-watering containers in which water flows from a reservoir into a dish often become clogged with bedding. The birds then don't get water, even though the water reservoir looks full. Check these types of water containers often and clean out any junk from the dish area. Elevating the container off the floor or placing a plastic tray under it with no bedding in it also can help.

Proper feed storage

All types of feed are perishable and start losing nutrients the moment they're processed. The older feed gets, the more likely it is to become infested with insects, which further degrades the quality of the feed. So the first lesson is not to buy more feed than you can properly store and use within a reasonable time, generally about a month.

Storing feed can be as simple as purchasing a 20-gallon metal trash can, placing your bag or bags of feed in it, and keeping the lid closed. If you have only a few chickens and you buy one bag of feed at a time, a trash can is probably the best method. If a coon or other animal learns to take off the lid, you can drill a small hole in both sides of the can and run a bungee cord or chain through the handle of the lid to the holes on both sides.

If you buy larger quantities of feed to save money or because you have a lot of birds to feed, you need to come up with a secure way to store your feed. Metal containers work best because they're lighter than wood or resin-type plastic and rodents can't chew through them. But you can also use a storage unit of heavy wood if you don't mind the extra weight. Heavy-duty, resin-type plastic, like that used in stock tanks, is fine for storing feed; rats and other pests can chew through thinner plastic.

An excellent way to recycle an old freezer or refrigerator is to use it for feed storage. These appliances have close-fitting seals on the doors and may keep feed fresh longer than a can. If you have an upright model, place it on its back. Remove the door latch or jam it so that if children or pets climb inside and the door closes, they can push it open. The weight of the door should keep pests from lifting it, but if weight alone doesn't work, you can easily add some sort of lock.

Other feed storage options include these possibilities:

- Metal truck storage boxes
- Metal bins that are sold to hold trash cans
- Recycled metal drums with lids

Don't use barrels or tubs that held pesticides or other toxic substances. Some residue may remain and contaminate the feed.

Tight lids are important for all recycled containers. Also keep in mind that grain products can be dusty and bad for the lungs, so be careful when handling them. Always keep feed dry, too, and discard any feed that gets moldy.

You can pour the feed directly into the container or leave it in the bag. We prefer to leave it in the bag, to keep the fine particles on the bottom of the bag from building up in the storage unit. If you purchase more than one bag at a time, the unused bag will remain fresh longer sealed in its original bag.

If you must store feed in bags in the open, store it off the floor and about a foot away from any walls. You can stack many bags of feed on a pallet — the higher off the floor, the better. Cover the bottom of the pallet with a piece of metal or thick wood before laying the bags on it, to prevent mice and rats from going under the pallet and between the pallet slots to get to the feed. Keeping some clear space all around the pallet also discourages pests.

Feed bags stored near some pesticides or other chemicals can absorb dangerous products from a spill, or even from gasses in the air, so store feed away from all potential contaminants, including gasoline, antifreeze, oil, pesticides, and pool chemicals.

Cleaning House

Just as your teenagers are content to live in a room with empty pizza boxes under the bed and clothes all over the floor, chickens don't mind a little disorder and dirt. But we know that dirty, wet conditions smell and attract pests. Plus, for their health and ours, chickens need dry litter, clean nesting material in the nest box, clean dishes to eat and drink from, and fresh air to breathe.

Undoubtedly, most of us adjust our housekeeping methods as we gain experience and decide what works best in our unique situations. You want to accomplish the task of cleaning the coop as quickly and easily as you can. The level of cleanliness you maintain depends on your preferences or needs. (Luckily, chickens aren't too fussy about their accommodations — but then, they don't tip well, either!) Knowing your options in terms of how often and how thoroughly you need to clean is helpful for establishing a routine. As experienced chicken-keepers, we give you some insights in the following sections, and we try to keep you from making the same mistakes we did.

Gathering cleaning supplies

Before you start cleaning, you need to gather some tools, most of which you probably have lying around. At the very least, you need the following:

- A rake
- A shovel
- Something to carry manure and debris away in (most people want a large "muck" bucket for small coops or a wheelbarrow for larger ones)

Other good items you may like to have around include the following:

- **A dust mask:** The dust from chicken droppings and their feathers is bad for human lungs, and when you clean, you stir it up. We highly recommend wearing a dust mask when cleaning out the coop. Skip the painters' masks and get an N95-rated mask with two straps and an exhale valve. Wear the mask drawn tightly to your face; check the fit by placing your palm lightly over the mask and inhaling.

- **Work gloves:** A pair of gloves keeps your hands clean and protects them from splinters and blisters.
- **A fan:** A fan set on low speed can be helpful by blowing the dust away from you.
- **A broom and a stiff brush:** Unless your family members are happy with sharing, we don't recommend using brooms or other items from the human household in the chicken coop.
- **Window cleaner and some paper towels or rags:** These supplies are needed only for housing with windows; otherwise, no window cleaning is necessary.
- **A long-handled dust pan or a large, flat shovel:** These tools can help you scoop up messes.
- **A wet/dry vacuum cleaner:** Some people use these vacuums on small coops, but be aware that they raise a lot of dust.
- **A hose:** If you have cement or other solid-surface floors, you may be able to use a hose to wash them after you remove most of the litter and manure.

Don't use water to clean unless the floor drains well, the day is warm and sunny, and you can use ventilation to dry the coop quickly. It's best to avoid getting anything wet that won't dry before nightfall.

You don't need a lot of cleaning products for chicken housing. A general-purpose cleaner and a cleaner for windows (if you have them) are fine. Feed stores and animal supply catalogs sell several cleaning products that are safe for general use. If you have meat chickens or laying hens, read the label to see whether the cleaning product you're considering is safe for food-producing animals. Most of the common human household cleaners, such as Lysol and Pine-Sol, are safe to use. Check cleaner labels to see whether they say they're harmful to pets or other animals. Steer clear of ammonia, which isn't good for the lungs. You're looking for cleaning products, not necessarily disinfectants (although the preceding brands do claim to disinfect as well as clean).

Unless you've had a disease problem, don't worry about disinfecting the general quarters. If you've had a disease problem, ask a veterinarian for recommendations on products to use to clean the coop, and always read and follow the label directions exactly. Remember, though, to clean all surfaces before sanitizing. For sanitizing, a solution of one part common household bleach and three parts water makes a good general disinfectant if you feel you need it, but be sure to rinse surfaces.

If you've had a problem with lice or mites, don't spray the coop with kerosene or other old-time remedies. Animal supply catalogs and farm stores sell pesticides that are safer and more effective. Also don't use household pesticides or products for other pets until you've read the label completely. Some pesticides are extremely toxic to birds; the label should mention use on or around poultry.

Seeing what you need to do and when

What you do when cleaning the coop and how often you do it are fairly subjective decisions that vary based on how many chickens you have and how you're raising the birds. (Some circumstances — like free range — make coops easier to clean than other situations.)

Some people believe the right way to go is to be immaculate. They clean droppings at least once a week, hang flower boxes full of petunias under the windows of the coop, and install automatic air fresheners (we're not joking!). If you live in an upscale residential area with close neighbors, this may be the way to go.

Other people are a little more laid back. They clean once a month or so and make sure pests are well controlled, smells are minimal, and environs look neat and tidy. Many people fall into this category of chicken-keepers. If you live in an urban or suburban area, this method is the least you should consider.

A third method of chicken-keeping works well for many people, although it's really appropriate only for people who live in rural areas. It's often called the *deep litter method.* With this method of chicken-keeping, the housing is cleaned and litter is removed only once or twice a year, and minimal routine maintenance is kept up. This method can actually work pretty well if the bedding is kept dry and loose. In fact, several studies have shown that chickens raised this way may be more resistant to many diseases and parasites. As long as the chickens aren't too crowded and are kept dry — and the keyword is *dry* — smells will be minimal.

If you decide to go with the deep litter method, you need to add fresh litter when manure piles up, and you may need to remove and replace litter under the water container if it gets wet. You must keep the housing well ventilated so moisture and ammonia don't build up. You also don't want the litter to pack and crust over. If it does, the litter won't absorb moisture and smells, and the chickens won't be able to keep maggots and other insects under control, resulting in an insect problem. You may need to stir up the bedding if it gets compacted and hard. You can do this with a rake or even a small rototiller. Even throwing a little scratch grain down on the floor and letting the chickens dig for it can work.

Of course, many compromises and customizations go along with these general guidelines for cleaning. For example, many people put pits or dropping boards under the roosts because many droppings are concentrated there. They then clean the pits or boards frequently and clean the rest of the housing less often.

Regular, essential cleaning

At the very least, your chickens need a dry space, clean litter and nest boxes, and clean food and water dishes.

Keeping feed and water dishes clean may mean cleaning them more frequently than the rest of the coop. Brush out any caked feed, wash and rinse them, and then soak them in a solution of one part household bleach to three parts water. You can also use a commercial disinfectant, being sure to follow the directions on the packaging. Rinse and dry the dishes in the sun if possible. Make sure feed containers are totally dry before refilling.

Keep algae, slime, and scum from accumulating on water dishes. You may need a bottle brush to clean some of them. Check the nipples of automatic water devices for rust or hard-water scale buildup; if needed, soak them in a lime and scale remover liquid. An old toothbrush can help for cleaning nipples and other small surfaces.

Be sure nest boxes are clean at all times as well, and frequently replace any dirty or lost bedding. Clean nests make clean eggs, and clean eggs are healthier for both eating and hatching.

Don't clean the nest of a hen you have left to sit on eggs. If you notice smashed or leaking eggs, remove them along with any soiled nest material that you can. If the area around the nest gets filled with droppings, you may want to pick them up. After the eggs have hatched, immediately clean out that nest box completely.

Deep cleaning

Whether you're an urban or rural chicken-keeper, clean out everything once or twice a year, when the place smells or gets wet, or when the bedding gets so high that you can't walk in the coop without having to lean over.

Start your indoor cleaning by shooing out the chickens. Then take the following steps:

1. **Scrape off the roosts.**
2. **Dust out the cobwebs.**
3. **Brush down the walls.**
4. **Remove all the dirty litter.**

   Some people lightly dampen the litter to lessen the amount of dust that gets stirred up, but don't overdo the wetting — you can more easily remove dry litter than wet stuff.

5. **Sweep the floor with a damp broom.**
6. **Carefully wipe light bulbs after they've had a chance to cool.**

   The bulbs get coated with dust, which reduces light.
7. **Clean any windows.**
8. **Wipe the blades of any exhaust fans, and wipe off the screening that protects them.**
9. **Place fresh litter in the coop, if you use it.**

Disposing of manure and old bedding

Sometimes the hardest part of keeping animals, including chickens, is finding a way to dispose of their waste, especially in urban areas. Before you have your first wheelbarrow load of manure, you need a plan for where you will go with it. You may want to check with your city, township, or county officials to see whether any laws govern the disposal of animal waste.

You have a handful of options for getting rid of manure. Although your first thought may be to use it as fertilizer, be aware that doing so on a small urban lot may be more of a hassle than it's worth to you. Even if you have a good place to store it as it ages, a time may come when you simply have too much to use in your own yard and garden if you have a large flock. If you have more room, composting manure and bedding is an excellent green way of disposing of the waste. The following sections discuss your options in the order we recommend them.

Composting and using as fertilizer

Composting is an excellent way to use manure and chicken bedding, breaking down the materials into a soil-like substance. You need to have an area where a compost pile or bin won't offend neighbors or the rest of the family. If what you want to add to the pile is heavy on manure and light on bedding, you need to add coarse, carbon-filled material like straw, dry leaves, shredded paper, or coarse sawdust.

A *passive compost pile* is just a pile of manure and bedding in an out-of-the-way place that's allowed to break down slowly. Breakdown of the material occurs faster if the compost pile is well managed. Chicken manure and bedding are usually dry when they're removed from the housing, and compost needs to be reasonably wet to begin working — about as moist as a sponge that has been wetted and then wrung out — so you may need to moisten it. Too much moisture causes the pile to stop decomposing correctly, though. Anaerobic bacteria results, causing the pile to smell. Turning the pile and adding more dry material can help with the smell. Free-range chickens love to turn compost piles for you, but they can make a mess.

Homeowners should never compost dead chickens and butchering waste; composting such waste only attracts animals and insects and creates smells in a compost pile. Instead, bury this waste under at least 2 feet of soil.

You can buy compost tumblers and fancy bins to work with a small amount of materials, but all you really need is a circle or square of heavy wire to contain the pile. If the waste is well composted, your neighbors may be interested in taking some off your hands. You can apply compost directly to your lawn and garden. On bare soil, work in as much as you can; on planted areas, dispense lightly an inch or so at a time, to keep from smothering plants. You can put 3 to 4 inches of compost between established plants as mulch.

Chicken manure mixed with bedding can be a good fertilizer after it has aged, but it will burn plants if applied to them when fresh. You have to let it age first — from 4 to 6 weeks in warm weather, to several months in cold weather. After manure has aged, applying it to tree crops, flower beds, and lawns is fairly safe. Lawns where children play may be the exception because some disease organisms may remain. Use about a bushel per 500 square feet several times a year.

The problem with using manure as a fertilizer is that after you apply it to the surface of the soil, it can become a major water pollutant if its nutrients run off in heavy rain and pollute any surface water. For this reason, don't spread manure on slopes or within 25 feet of any water, including roadside ditches that may drain into streams. City storm sewers often drop untreated water into surface water like streams or lakes, and the excess nitrogen from manure can pollute that water. Thus, where you can spread manure may be regulated. Call your local municipality's environmental health department to find out whether it's legal to spread manure on your property. It's usually illegal to put any form of manure into storm sewers. Nitrogen, phosphorus, and other components of manure can also leach down through the soil and pollute groundwater if applied heavily to the surface. Never apply manure within 50 feet of your well.

The best way to use chicken manure as a soil amendment and fertilizer is to work it into the soil either by hand or by using a rototiller. This causes the waste products to bind with soil particles, making them less likely to pollute the water.

Don't use manure on food crops within several months of harvest. You can put manure on vegetable plots or on small fruit crops, such as strawberries and grapes, in the fall, after everything is harvested. This delay avoids food pathogens like salmonella, which can be present in manure.

Burying chicken coop rubbish and other disposal methods

People in urban areas may not have room to compost manure, or doing so may be prohibited. In such situations, burying waste is the best solution, if you have the room. Some places may allow you to bag the waste and set it out to be collected with lawn waste or with your regular trash pickup, especially if small amounts are involved.

If you run out of places to bury it or can't put it out with the trash, you need to get creative. Check with gardeners; some people may want the waste for composting and will pick it up from you. Otherwise, you may have to pay to take it to a landfill. Or perhaps Uncle Joe will let you dump it on the back 40 acres of his farm.

Part III

Caring for Your Flock: General Management

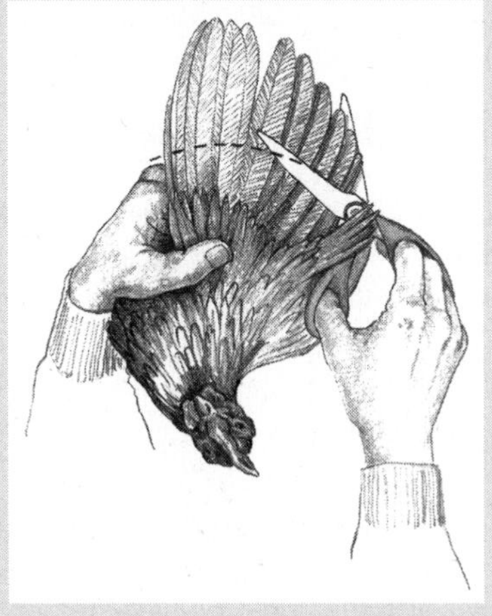

Illustration by Barbara Frake

Is organic the right choice for your chickens? Check out a helpful article at `www.dummies.com/extras/raisingchickens`.

In this part . . .

- Find the best diet plans for your chickens. Whether they're chicks or hens getting ready to lay eggs, giving your birds a balanced diet will help you maintain a healthy flock.
- Protect your chickens. Pest and predators are an unfortunate reality of keeping chickens, but staying informed on how to guard against these dangers will give you more time to enjoy your chickens.
- Know how to keep your chickens healthy and happy. We introduce the concept of biosecurity to keep unwanted agents out of the coop and prevent anything harmful from entering your home.
- Find out how to treat illness and disease. We also offer advice on warding off parasites and treating injuries.

Chapter 8

Feeding Your Flock

In This Chapter

- Understanding the nutritional needs of chickens
- Choosing the right chicken diet
- Using treats, scraps, and supplements
- Knowing what not to feed your chickens

Probably more controversy swirls over feeding chickens than any other part of chicken raising — *everybody* has an opinion. Some believe you don't need to feed chickens at all; they say you just turn the birds loose to find their own feed. Others devise elaborate diets that chickens love but that really aren't necessary. Chickens aren't fussy eaters; they'll eat just about anything. However, they're like small children: When faced with an abundance of choices, they don't always make the wise ones.

In this chapter, we explain why thoughtful diet planning is an important part of chicken-keeping, and we cover the spectrum of what nutrients chickens need, what options you have for providing those nutrients, and what special considerations you need to take into account depending on your chickens' ages and purposes (laying, providing meat, and so on). We also tell you when you need to supplement your flock's diet and with what, talk about treats as a means of keeping your chickens happy, and explain how much water your chickens need to stay properly hydrated.

Feeding Basics

Chickens aren't vegetarians by nature. They're *omnivorous*, meaning that they eat both animal protein and plant-produced foods like seeds, fruits, and leaves. Wild chickens get their protein in two ways:

- **They actively hunt for it.** They chase butterflies and grasshoppers and gobble up tiny frogs and baby mice.

Chicken digestive system overview

It takes only about 2½ hours for food to pass completely through a chicken's digestive system. The food a chicken picks up in its beak is first sent to the *crop,* which is a pouch-like area in the neck for storage. The crop is stretchy and allows the chicken to quickly grab sudden food finds and store them for a slower ride through the rest of the digestive system.

When the food leaves the crop, it's slowly moved along to the *proventriculus,* or true stomach, where digestive enzymes are added. Then it moves to the *gizzard,* an oval sac composed of two pairs of strong muscles. The gizzard stores bits of gravel and grit that the chicken picks up. With a squeezing action of the muscles, the grit helps grind down food particles much like your teeth break up food. The food moves next to the intestines, and the intestine walls absorb nutrients.

Both liquid and solid waste are combined in the end part of bird digestion and pass through the cloaca. The *cloaca* is a multipurpose organ through which waste passes, the chicken mates, and eggs are passed.

- **They scavenge it.** They have no qualms about picking apart a dead carcass, including the body of another chicken.

In a domesticated setting, chickens still love to roam the grounds looking for food, but the places we choose to keep them don't always support their nutritional needs. Most chickens need to be fed by you, the chicken-keeper.

The sections that follow spill the beans on why you need to feed your flock, what they need from their food, and what your food choices are.

Understanding why you need to manage your birds' diet

People often tell us that their grandparents raised chickens just fine by letting them run around the farmyard and throwing them a little corn every day. We'd like to remind chicken owners that matters are different now. Most yards don't have large animals close, by spilling grain and passing undigested grain in their manure for chickens to find. Modern chicken-keepers usually don't have large piles of manure breeding a tasty crop of maggot protein for their chickens, either. And when chickens are confined to a pen most of the time, they certainly can't find the food they need.

Grandma expected to get few eggs during the winter. She expected to raise chickens for meat for many months, unlike our broilers that are ready to eat in as little as 8 weeks. She anticipated that many birds would die over the year, many from poor nutrition. And good, nutritionally balanced chicken feed was seldom available, assuming she had the funds to buy it.

You can raise chickens like your grandparents or great-grandparents did. Outhouses are still legal in some places, too. Or you can read this chapter and see how modern chicken owners feed their birds.

The land is unreliable

Ideally, you could turn your chickens loose, and they would find all their food. This scenario works, however, only if the environment contains the right mixture of foodstuff in sufficient quantity during all seasons of the year. A suburban lot can't support many chickens very well.

Even a large rural area with room for chickens to roam may not have enough of the correct nutrients to maintain chickens year-round, especially in cold-weather areas. Your chickens need to be fed if any of the following apply to you:

- You don't want them to go too far or become dinner for something else.
- You want eggs all winter or tender meat raised quickly.
- You want your chickens to survive when winter brings snow-covered fields to your homesite.

If chickens have unrestricted access to a large amount of relatively wild land, they may be able to find enough of a properly balanced diet to satisfy both their hunger and their nutritional needs. But it takes a lot of productive land to fully support chickens, and the more chickens you have, the greater the chances are that your free-range birds aren't meeting their nutritional needs, even in the best weather.

The more exercise the birds get, the tougher their meat becomes. So, we especially recommend avoiding free-range conditions for meat birds. Use managed pasture if you feel strongly about raising your birds on grass. Free-range meat chickens vary tremendously in terms of how fast they grow and how tender they are when butchered. If you insist on raising meat birds with free-range diets and you want a good rate of growth, you need to provide some broiler feed.

Chickens don't make good choices

Many people think that chickens will choose the right diet if they're given many food choices. Not so, dear reader. Chickens are a lot like children. You need to guide their food choices. They may fill up on something they really like, even if it isn't the best nutritional choice. A chicken thats forced to eat a wide range of items to feel satisfied tends to balance its diet in the long run. But a chicken that has an abundance of a favorite food simply fills up on that.

If chickens come upon a bonanza of mulberries, for instance, they won't eat just a few and wander away looking for protein to supplement the meal. They'll stuff their little crops as fast as they can until they're bulging. Over time and territory, however, chickens tend to balance their nutritional needs in a wild setting.

Chickens have little sense of taste, and they often eat things that aren't really foods, like Styrofoam beads, lead paint flakes in the soil, and rubber bands. Some of these "foods" may harm them. Some may be caught in the gizzard, and some may pass through the body. You can't count on chickens knowing what's good for them to eat.

Chickens also can ingest pesticides from eating grass and other vegetation. Small urban and suburban lots are mowed and trimmed and treated with pesticides to reduce the insect population. Even rural fields can be treated with insecticides and weed killers. If you let your chickens roam for part of their dinner, make sure they aren't eating from "poisoned plates."

Knowing what nutrients chickens need

Because chickens, like children, don't always make the best choices in terms of food, you need to be in control of your birds' nutritional needs. The sections that follow show you the general nutritional needs of all chickens and mention the variations depending on the birds' age; we explain the specific dietary nutrient proportions various chickens need (called *rations*) later in this chapter. For more information, you can access the National Research Council's "Nutrient Requirements of Poultry" report at `www.nap.edu/openbook.php?isbn=0309048923`.

Poultry cannibalism

People are often disturbed when they find their chickens greedily picking the meat off the fast-food chicken bones they discarded. But chickens have no clue that they're cannibalizing one of their own, and they wouldn't care if they did.

Rob says: "Before I had chickens, I visited a friend who was raising a small flock. The friend said, 'Watch this!' as he tossed a recently laid egg onto the ground. Within the blink of an eye, the whole flock was fighting over the delicious treat smashed on the ground." (Of course, this isn't advisable for a plethora of reasons, but I remember being shocked and amazed that chickens love eggs so much. It seems most people are pretty amazed by this as well.)

Protein

Protein actually consists of a variety of amino acids, and each type of protein has a slightly different balance of these amino acids. In nature, chickens obtain their protein from insects such as bugs and worms. These bugs have a wide range of amino acids but are generally a small part of the diet in volume.

Protein is the most expensive part of the diet, so a little good-quality protein containing all the right amounts of amino acids is preferable to large amounts of poor-quality protein. Choosing the right sources of protein is important, as is balancing the protein content of chicken feed between too much and too little.

Chickens need 16 to 24 percent protein in their diet, depending on their age and intended use. Too much protein in a chicken's diet is a waste because it's secreted in the manure as nitrogen and ammonia. These elements are pollutants of both water and the air. Another downside to feeding too much protein is that a chicken's body has to work harder to metabolize protein and break it down into urea to be eliminated. The process creates body heat, and birds on high-protein diets may suffer more from heat stress. Many people think that chickens should have more protein in the winter because it creates body heat, but this assumption is a faulty one. Too much protein places stress on the bird's organs, and energy to break down the protein can't be used to produce meat or eggs.

Diets that are low in protein or in some amino acids cause problems such as slow growth, poor feathering, and fewer and smaller eggs. Molt is stressful to birds and requires good nutrition to see the birds through it. Feathers need protein to develop properly, and it doesn't hurt to increase the protein level at this time. A little supplementation with a high-protein feed, such as sunflower seeds, safflower seeds, pearl millet, cooked eggs, cooked peanuts, peas, or beans can be helpful at this time.

Laying hens need good-quality protein in their diet to produce quality eggs for long periods of time. If broiler-type birds don't get enough protein, their legs and wings may become deformed, and they may become unable to walk. They then need to be disposed of because they won't eat well.

Fat and carbohydrates

Chickens don't require a lot of fat. High-energy broiler feeds need 6 to 7 percent fat, and laying feed and chick starter need about 3.5 to 4 percent fat. Hens do need some fat for the production of egg yolk, but chickens can use carbohydrates for most of their energy requirements. As we all know, unneeded calories are converted to fat and stored. Excess fat in chickens is deposited around the abdominal organs and under the skin and can cause many problems.

Carbohydrates should furnish the bulk of the diet, and a commercial diet based on grain easily supplies them. Carbs are burned in the bodies of all animals as fuel for all life processes, such as breathing and growth. Diets based on pasture or vegetation need grain supplements to provide enough carbohydrates.

Vitamins and minerals

Chickens require all the vitamins and minerals we do except vitamin C, which their bodies can make. Table 8-1 gives you an idea of what each vitamin and mineral does and the benefits they provide. Chickens need many other vitamins and minerals, but the requirements for some still aren't fully known. So in the table, we cover the most important ones. If these vitamins and nutrients are adequate, the others are almost always available in healthy quantities. Chickens get these vitamins and minerals from a properly formulated feed.

Some types of chickens may need more of some nutrients. Layers, for example, need more calcium than other birds to continue laying well. This variation is why having the expertise of chicken nutrition specialists to formulate rations is so valuable.

Table 8-1 The Benefits of Vitamins and Minerals

Vitamin or Mineral	*Benefits It Provides*
Calcium	Promotes strong, smooth eggshells; prevents rickets; aids in hatching
Choline	Promotes growth, egg production, and a healthy liver
Cobalt	Promotes growth and general good health; helps prevent hatching problems
Folic acid	Promotes healthy feathers, growth, and egg production; prevents anemia
Iron	Prevents anemia
Magnesium	Helps prevent sudden death in broilers
Manganese	Helps prevent *perosis* (twisted legs) and hatching problems
Niacin	Helps prevent leg deformities; keeps mouth and tongue healthy
Pantothenic acid	Promotes healthy skin, mouth, and feet
Phosphorus	Promotes strong, smooth eggshells; aids in hatching; prevents rickets

Vitamin or Mineral	Benefits It Provides
Riboflavin (vitamin B_2)	Prevents curly toe disease; increases egg production; promotes growth
Thiamine (vitamin B_1)	Promotes a healthy appetite; sustains life
Vitamin A	Promotes growth, egg production, and good general health
Vitamin B_{12}	Promotes growth; helps prevent anemia and hatching problems
Vitamin D	Promotes growth, egg production, and strong eggshells; prevents rickets
Vitamin E	Decreases leg joint swelling; strengthens immune system; helps prevent mental problems
Vitamin K	Promotes good blood clotting; helps prevent bleeding in the muscles
Zinc	Helps prevent shortened bones and feather abnormalities

Comparing your feed options

When you know what constitutes a balanced diet for your chickens (see the preceding section), the next step is to figure out how to provide it. Should you head to the feed store for a bag of chicken chow, design your own custom blend, or send your chickens out to the pasture? Before you decide, it helps to weigh the pros and cons of the choices available to you. You have four main options for feeding your chickens:

- **Commercial feed, specially formulated for your chickens' needs:** If you have layers, you want to optimize their diet for egg production, so feeding a commercial ration that's properly balanced for laying chickens is highly recommended (instead of feeding them only whole grains or scraps). Commercial feeds have the proper ratios of calcium and other minerals added for egg production. Even pastured layers in pens should have access to a good laying ration.

  We prefer to use commercial feed for meat birds, too, because it blends all the ingredients and prevents birds from picking and choosing what they like and wasting the rest. Feeding whole grains and scraps isn't a good idea for meat birds. Pets and show chickens also do better on a balanced commercial diet that doesn't allow them to pick and choose.

- **Individually mixed feed for which you supply the recipe to a feed mill:** Some people have an abundance of homegrown grain or other products available to them and want to use those ingredients for feed. These folks often enjoy experimenting and tinkering with formulas and have a specific idea about what they want to achieve or avoid when feeding their birds.
- **Individually mixed feed you prepare in your own home:** Another way to make homemade mixes is to buy the ground ingredients separately (or purchase a small grain grinder) and mix small batches of feed at home. The problem is that most places require you to buy the ingredients in large quantities, such as in 50-pound bags. You don't use even amounts of the ingredients, and ground grains begin to lose nutrition quickly, so the grains may go bad before you get to them.
- **Pasture:** Chickens confined to tractors or other movable pens are sometimes referred to as "free range." They're really pastured poultry. These birds aren't able to get the right amount of feed with the right nutrients unless you supplement their pasture with other feed.

Yes, chickens love table scraps, and in the old days, chickens lived on them and whatever else they could find. But in those days, people didn't expect as much from their chickens and didn't know much about nutrition. Some people still collect stale bakery goods or restaurant garbage to feed chickens. If your chickens are pets and you don't worry about egg production or good meat, you can still feed your chickens this way. Just as it's difficult to balance a pet dog's diet with table scraps, however, it's also difficult to feed a pet chicken a balanced diet with table scraps. If you carefully choose human foods for your pet chicken, you may be able to compose a balanced diet. However, it's safer and easier just to feed pet chickens a good commercial feed and save scraps for treats.

Pet chickens shouldn't be fed a diet of dog or cat food. These pet foods are too high in protein and fat for regular chicken consumption, and they don't contain the right blend of vitamins and minerals for chickens.

Other than *grit* and *scratch*, which you need to add as supplements for whole-grain or pasture-fed poultry (we cover those topics later in the chapter), regular supplements of vitamin concoctions or other components advertised as "immune boosters" or "energy creators" are unnecessary if the feed is balanced. Most of these are just a waste of your money, but some can actually harm your chickens. For example, too much calcium and phosphorus can cause kidney stones, and too much vitamin K can cause bleeding problems. All kinds of problems can develop if a supplement goes too far:

- Alters the taste of the birds' water, causing them to avoid drinking it
- Makes the water taste salty, causing the birds to drink too much
- Gives the birds' feed a flavor they don't like

Don't waste your money buying supplements. Instead, spend it wisely on a good, balanced feed.

The virtues of commercial feed

Most feed mills or manufacturers try to use feedstuffs that are abundant locally. Corn and soybeans form the basis of most chicken feeds. Meat and bone meal are sometimes added to increase the protein content and add some vital amino acids. Meat and bone meal from ruminant animals, such as cows, sheep, and goats, isn't allowed in chicken feed because of disease concerns, but meat and bone meal from pork and fish are often used.

Some chicken feeds use only plant sources for protein and then supplement the amino acids that plant sources are low in with synthetic amino acids. Commercial poultry feed ranges in protein content from about 12 to 26 percent, with the higher-protein feeds made specifically for growing meat birds. To feed your meat birds, you can purchase commercial meat bird, game bird, turkey starter, or broiler feed. (It has different names in different places.)

Almost all commercial feeds use synthetic and natural vitamins and minerals to balance the feed, including all vitamins and minerals referred to in Table 8-1, plus some additional ones. Chickens on commercial diets rarely have vitamin or mineral deficiencies.

Poultry feed is generally based on vegetarian sources of protein, as noted earlier. But the vegetarian protein sources in commercial feed may not have the correct amount of some important amino acids, such as methionine and lysine, for optimal health in chickens. These amino acids are generally added to commercial rations as supplements.

Carbohydrate sources need to be chosen carefully, too. A diet high in fiber may make the birds feel full before they've consumed enough feed to get the protein, vitamins, and minerals they need. Birds on pasture that has become old and fibrous, and birds that don't have access to some high-calorie grains and protein typically lose weight, grow slowly, lay poorly, and suffer from vitamin and mineral deficiencies.

Chickens that are being given a lot of treats or additions to a commercial diet or that are fed a homemade diet may suffer from deficiencies from time to time. When birds get full on morsels they like, they may not eat the properly balanced diet in sufficient quantity to get the nutrients they need. And without some animal nutrition education, few chicken owners can devise properly formulated diets.

Because the nutrients available in grains and other products vary from batch to batch, these ingredients must be tested to see how much protein and other nutrients are in the feedstuff. Large companies employ animal nutritionists who test feeds and decide on the recipe or blend of ingredients necessary to meet the standards of the feed product to be made. These specially trained animal nutritionists also test the feed for mold and other toxic substances and make sure the feed is safe and nutritious after mixing.

Flavoring or color is rarely used in chicken feed. Some feed manufacturers add preservatives to keep the feed fresh longer. Growth hormones aren't used in chicken feed, and if a medicine is added to the feed, it must be on the label.

Medicated feeds

In most cases, medicated feed is starter feed, the feed you use on baby chicks. Whether you choose to use a medicated starter feed is up to you. Most feed stores carry both medicated and unmedicated starter feed. We believe that using medicated feed on baby chicks gets them off to the best start possible and results in fewer chick deaths during the early weeks of life.

The most common medicine added to chicken starter feed is a *coccidiostat*, a medicine that controls the disease *coccidiosis*, which can be quite deadly in young birds. Common medications used in feeds for controlling coccidiosis include Monensin sodium, Lasalocid, Amprolium (Amprol, Corid), Decoquinate (Deccox), Sulfadimethoxine, and Salinomycin. Medicated starter feeds are fed for the first 16 weeks for chickens meant to be layers or pets. Check the feed bag for when to stop feeding medicated feed to chickens being raised for meat. Most modern medicated feeds for meat birds can be fed right up to the day of slaughter, but some require a short 5- to 7-day period when you stop giving medicated feed before butchering.

Don't add medications for coccidiosis to the water if you're using a medicated feed. You will overdose the chicks. Also don't add antibiotics meant for other species to chicken feed without consulting a veterinarian, even if well-meaning fellow chicken-keepers urge you to do so. The FDA has special limitations for animals that produce food such as meat and eggs, regarding what antibiotics and other medications can be used on them. Some medications are actually illegal to use for food animals. With other medications, if you don't know the dosage for chickens, they may be harmful to the birds. Keep pets and livestock from eating medicated feeds because some medications that are safe for birds can harm or kill other animals.

The plain truth about homemade feed

Despite what most people assume, the homemade recipe approach has more drawbacks than positives. In this section, we give you some advice on making your own feed, in case you feel strongly about doing so. But first, consider these cold, hard facts:

- **Homemade *isn't* cheaper than store bought.** Some people think that making up their own recipe for chicken feed is cheaper than buying commercial feed. If you have only a small flock of chickens, it's extremely rare to get a cost savings from devising your own chicken feed if you have to buy all or most of the ingredients.
- **Making your own recipes requires a complex understanding of chicken nutrition.** Keep in mind that feed companies hire experts to formulate and test their feeds. If you intend to devise a recipe of your own, you need to thoroughly study chicken nutrition.
- **Most mills don't custom-mix less than 1,000 pounds of feed.** This amount of feed is difficult for the average person to transport and store — much less use before it goes stale. In many cases, a custom order by a person who isn't a co-op member or who's using relatively small amounts of grain (and 1,000 pounds is a small order) is charged a milling fee, too.

  If a group of chicken owners in an area band together and order a large amount of a custom feed, they may realize some cost savings.
- **Costs for each ingredient vary as the cost of grain goes up and down.**
- **You may have to pay for bags or furnish your own.**

If you *really* want to make your own feed, and if you have a feed mill in the area that grinds and mixes feeds on a custom basis, you may be able to devise your own feed formula using locally grown grains. Some mills have vitamin and mineral mixes and protein supplements on hand to mix complete feeds. They may even have an experienced animal nutritionist on staff. Most, however, expect you to furnish them with a recipe or use formulas they have on hand.

These mills generally just grind grains together to a certain size and then mix in a powdered protein mix and a vitamin-mineral supplement. If you want soybeans in the recipe, they often have processed soybean meal on hand. But usually these recipes aren't cooked or bound together, and the chickens may separate or pick through the ingredients. If the operation has a pelleting machine, you must pay extra to have the feed made into pellets.

Most of the time, the mix is a fine grind, and some of the separation of ingredients can be overcome by moistening the mix before feeding.

If you have a moderately large flock — 100 layers or 200 or more meat birds — and you want to devise your own rations, you can consult with a poultry expert from your county Extension service or local college, or you can use the services of an animal nutritionist employed at a large feed mill that caters to special mixing. These experts also may help you with smaller flocks, but they generally work with larger quantities of feed, and their

My homemade happening

Kim says: "One summer when we were raising three pigs for meat along with some broilers, I decided feed would be cheaper if I had my own recipe mixed. Pigs and chickens are fed a similar diet, and I thought I could customize one mix that would suit both. After a lot of researching formulas, consulting with the guy at the mill, and pricing the various feedstuffs, we ordered 1,500 pounds of feed. We loaded that feed in the back of an old van and started the 10 miles home."

"Each batch of the finely ground and dusty feed had to be mixed with water at every feeding in quantities small enough to be eaten in a few hours because it would mold quickly in hot weather. The huge stack of feed bags attracted mice by the dozens. It was messy and time consuming, and after paying for the feed bags and a milling fee, our wonderful recipe cost us at least as much as commercial feed would have. Our meat tasted wonderful, but I vowed never to go the homemade route again."

formulas may be difficult to reduce to small-flock quantities. The cost for mixing small quantities of feed usually outweighs any benefit. In fact, some mills won't even mix small quantities.

Letting birds feed on pasture

Chickens were not designed by nature to subsist on vegetation. Wild chickens eat a lot of insects and seeds to balance their diets. We also expect more eggs from domestic chickens than wild birds lay, and we expect our meat birds to grow heavier and faster than wild birds that subsist on a natural diet. Domestic chickens need some concentrated protein and calories to thrive. However, chickens that get some time every day to roam freely may get a third to a half of their food from foraging at certain times of the year.

Even chickens that have unrestricted access to large pieces of land may need feed at some times of the year. They may be able to find bugs and seeds to round out their diet in some seasons; however, in other seasons, the chickens may find those vital ingredients missing. Feeding your chickens well has the added advantage of keeping them closer to home, where you can find the eggs they lay and keep predators away.

If your pasture is properly managed and the weather keeps vegetation growing rapidly, you can meet more of the birds' dietary needs. Some pasture plants furnish more protein and carbs than others. Latino clover, alfalfa, and some other legumes, for example, raise the protein content of the pasture. If you want pasture or vegetation to be the primary feed for chickens, consult with your county Extension agent or a poultry specialist at a university about what to plant in your area. You also need the proper machinery to plant this pasture and maintain it.

If you have well-managed pasture and not just a grassy spot in the yard, pasture can furnish a large part of the diet of layers, breeding birds, show birds, and pet chickens. Managed pasture includes plants that furnish quality protein and carbs. The birds usually need some high-protein feed on the side, but the amount of feed is greatly reduced. You also need to supplement the pasture-fed poultry diet with whole-grain feed (called *scratch*) for a few extra carbs, as well as *grit* to aid digestion.

Vegetation alone can't keep the broiler-type (Rock X Cornish) chickens growing properly; their diets must be supplemented with concentrated carbs and proteins. The feed for these birds should be a commercial feed of about 16 percent protein or a mixture of whole grains and a high-protein supplement.

Getting pasture-raised birds to a good eating size may take a little longer. Because pasture-fed birds move around a lot more, you'll find more dark muscle meat and less fatty breast meat in their carcasses. Also, the skin of the birds may be a bit more yellow because of pigments in the grass.

Chickens must be moved to clean pastures frequently. How often depends on the weather, the rate of vegetation growth, and the number of birds. Move them before all the grass has been eaten to the roots or the pen is too dirty. However, keep in mind that you must take care when moving pastured chickens. You don't want to harm them or stress them too much. Water must be available at all times, and the birds should have a shaded place to go to when the sun is too hot.

Food to avoid feeding chickens at all costs

In this section, we list some items that you absolutely shouldn't fed to chickens. Because experts often disagree on what's poisonous to poultry, and because whether a given plant is poisonous may depend on the circumstances in which it's grown, we leave out many disputed plants. Some plants that are poisonous to other forms of livestock aren't toxic to birds, and some books don't differentiate between species of livestock when they list poisonous plants.

Some plants or plant parts, like apple seeds, may be poisonous to chickens if eaten in large quantities, but in all practical and normal uses, a chicken would never get enough to harm it. If you have a question about the safety of a plant, call your local Extension office or consult a veterinarian.

Never feed your chickens the following:

- **Alcohol:** Never give your chickens the stale beer that went with last night's chips. Birds' bodies can't tolerate much alcohol.

- **Anything moldy:** Some molds are harmless, but many have toxins that can damage organs or cause neurological damage. The average person can't tell which molds are harmful.
- **Avocado:** Birds shouldn't eat avocado because it contains a fatty acid called *persin* that can be fatal to birds. No guacamole dip for the chickens.
- **Chocolate:** Most animals can't tolerate large amounts of chocolate, especially dark chocolate. A little chocolate pudding or chocolate cake in scraps won't hurt, but baker's chocolate or semisweet chocolate chips may be harmful, even in small quantities.
- **Green areas of potatoes or potato sprouts:** Both are poisonous, and you shouldn't eat them, either.
- **Leaves from tomato, pepper, potato, eggplant, or nightshade plants:** Chickens normally don't eat these, but they may if they're really hungry.
- **Morning glory, sweet pea, and datura (angel's trumpet, jimson weed) seeds:** All of these are common garden plants.
- **Raw dry beans:** Fresh beans (except soybeans), like green beans, are safe. Some types of raw dried beans have toxins, so cook all beans. Bean plants are safe to feed.

  Don't feed whole raw soybeans to chickens. Soybeans need to be processed and cooked before they're fed to animals. They have enzymes that need to be removed, or they'll cause digestive system problems and poor health.
- **Raw peanuts:** Peanuts can have a fungus called *aflatoxin.* Cooking helps destroy the fungus but may not totally eliminate it. Peanuts for human consumption are generally safe if roasted or boiled.
- **Rhubarb leaves:** They contain oxalic acid, which is a poison.
- **Tobacco:** From the leaves to cigarette butts picked up off the sidewalk, avoid tobacco in all forms. The nicotine in tobacco is a poison, and birds' small bodies can't handle it.

Don't believe everything you hear. Some people associate a bird's death with eating a certain item, without any hard evidence to prove that the product actually caused the death.

Choosing the Right Commercial Feed

As with other animals' feeds, chicken feed runs the gamut from top-quality, name-brand feeds to poor-quality, cheap feed. If a feed is labeled "complete and balanced," it must contain the recommended amounts of nutrients set by the National Research Council for poultry of the listed type and age.

Many feed mills produce one feed that's labeled as several different brand names. The same feed may cost different amounts just because it has a different brand name! Some brand-name feeds may be made by different mills and may contain different ingredients in different areas of the country. So it pays to look at the ingredients and the guaranteed protein and other nutrient levels rather than purchase feed by cost or by brand name. Chickens don't have a high degree of taste, but at times we've seen chickens prefer one brand of feed over another.

To find the feed that's best suited to your chickens, you need to understand the purpose of different types of feed, the forms in which feed is available, and the info you can expect to find on package labels. The following sections address these topics.

Demystifying commercial rations

Feed labels indicate the age and/or type of chicken they're designed for, such as "meat bird starter," "layer," or "all adult poultry." Most feed stores sell a general-purpose poultry feed that you can give to birds not intended to be meat birds or layers, as well as a few specialized feeds.

When choosing commercial feed, first look for feed that suits your chickens' age and type. The following sections provide a rundown of what you're likely to encounter.

Starter rations (for chicks)

The ration for layer-breed chicks should be 20 percent protein. It's usually just called "starter feed." From the time they start eating, meat chicks need a high-protein feed of about 22 to 24 percent protein for the first 6 weeks. It's called "meat bird starter" or "broiler starter."

If you've just received some baby chicks that you intend to be laying hens in the future, look for "chick starter." Buy the same feed for pet or show birds. But if you have meat-type chicks, buy "meat bird or broiler starter." Both types of starter feed can be medicated or not. We discuss medicated feeds earlier in this chapter.

If you have different types of birds in a brooder, it's better to feed the higher-protein meat bird feed to all the chicks than use a lower-protein feed. However, separating your future layers from meat birds when they leave the brooder is strongly recommended. Meat birds need a protein level of about 22 to 24 percent until they're butchered, which is too high for layers.

Layer rations

When chickens that are being raised to be laying hens leave the brooder, or at 6 weeks, you need to change from a starter ration to the next level of feed. Taking care to feed young laying chickens correctly results in healthy birds that give you optimum egg production. Because they're around longer than meat birds, you should take an additional feeding step when raising them from chicks.

Here are the types of rations you should use:

- **Grower pullet ration:** The ration for growing pullets, from leaving the brooder at 6 weeks to about 14 weeks, should be about 18 percent protein. You want them to grow slowly enough to develop good strong bones and to reach a normal body weight before they begin producing eggs. High-protein diets tend to hurry the birds into production before their bodies are ready. Finding a feed labelled "grower" with this level of protein may be hard. If you can't find the 18 percent grower feed, use the 16 percent finishing ration listed in the next bullet. These feeds are usually called "grower-finisher" feeds.
- **"Developer or finishing" pullet ration:** From 15 weeks to 22 weeks, it's ideal to lower the protein level of the feed to 16 percent. The object is to get pullets well grown without too much fat.

  Your feed should have normal levels of calcium and other vitamins until the birds start laying. Feeding an adult layer diet high in calcium and phosphorus to birds that aren't laying yet can damage their kidneys. If young hens start laying early, switch them right over to a laying ration.
- **Adult layer rations:** After the hens reach the age of 22 weeks or begin laying, and throughout their laying careers, they need a protein level of 16 to 18 percent. The calcium and minerals should be formulated for laying hens. A rooster housed with a laying flock will be fine consuming laying rations. These feeds are appropriately called "layer feed."

  Don't force extra calcium and minerals on hens by adding things to a properly formulated feed. Too much calcium can cause kidney failure. If you're getting a lot of thin eggshells or soft-shelled eggs, give your hens some calcium in the form of crushed oyster shells in a feeder where they can choose the amount.

Broiler rations

Home flock owners may encounter two types of meat birds: the Cornish X Rock crosses that grow extremely quickly and require precise diets, and the excess males from heavy breeds of chickens, which require slightly different management. Here are the details:

- **Cornish X Rock broiler hybrids:** After the first 6 weeks, you can lower the protein percentage for these birds to 18 to 20 percent until they're butchered. "Meat bird" or "broiler grower-finisher" is generally used on the label for these feeds aimed at meat birds in their last weeks. Many people feed starter rations for broilers right up to butchering — that strategy is fine, although grower rations are generally cheaper. (Check the label, if the starter feed is medicated, to see if you need to stop feeding that brand a few days before butchering.) A slightly lower protein percentage in the last few weeks of these broiler hybrids' lives may keep them from "sudden death syndrome," a condition in which rapidly growing birds flip over on their backs and die, and prevent them from getting weak legs.
- **Heritage birds, dual-purpose males, and free-range meat birds:** These types of meat birds grow more slowly and add less muscle meat than the broiler hybrids. They take longer to reach a satisfactory butchering rate. After the first 6 weeks, you can lower the protein to 18 to 20 percent for the next 6 weeks; after that, protein content can be 16 percent.

Whole-grain mixtures or scratch feed

Chickens do love these whole-grain mixes, but they're almost impossible to accurately mix to provide all the nutrients a chicken needs. Many of the cheaper mixes are filled with pieces of corn cob and seeds that chickens don't seem to care for, like milo. Chickens pick through these mixes, and the dominant birds in the pecking order get first choice, often concentrating on the corn or another part of the mix and leaving the lower-pecking-order birds with little variety.

Scratch and whole-grain mixtures are best used to supplement the diets of chickens that have good free-range or pasture conditions or to relieve boredom in confined birds. A little scratch grain is thrown on the floor for the birds to search for, and in the process, they stir up the litter. However, remember that if you use only whole grains for confined chickens, some important amino acids found in protein of animal origins may be lacking or deficient in the diet.

Whole grains tend to attract mice and pest birds like sparrows more than processed feeds. They're also more likely to be infested with weevils and other insects. Relying on only one grain type, such as corn, doesn't meet all your chickens' needs, so don't use a single grain as the sole food for chickens.

Variations for breeders and show birds

If you're breeding birds, you may want to increase the protein content for roosters and make sure hens have a laying ration. If you keep any heritage or free-range birds for breeding, switch their diets to a layer diet.

The diets for show birds are often tinkered with based on elaborate formulas thought to grow feathers or improve color. Really, these birds just need a good, balanced diet that has a relatively low protein content (14 to 16 percent), unless they're molting, when protein can be increased slightly. Birds that are caged in small areas should have a more fibrous-based feed to make them feel full without getting fat. You can add small amounts of whole wheat or oats to the diet for this purpose.

Selecting a form of feed

Most chicken feed is ground, mixed, cooked, or steam-treated and then turned into mash, crumbles, or pellets. When products are mixed and bound together, each piece contains a balanced proportion of the mix. The vitamins and minerals don't sift out to the bottom, and chickens aren't able to pick out favorite pieces and waste others.

Feed comes in three forms:

- **Crumbles:** In research done on feeding chickens, the birds seemed to prefer medium-size pieces of feed, commonly known as crumbles, and they seemed to grow better and lay eggs better on this type of feed. Crumbles are actually broken-up pellets. Finisher rations for pullets or meat birds are commonly crumbles, but adult feeds also can be crumbles.
- **Pellets:** Pellets are the second-best method of feeding and the second most popular form of feed with chickens. They're long, narrow, cylinder-shaped pieces of compressed feed. Pellets are usually used for adult birds.
- **Mash:** Mash is the least preferred form of feed. It's finely ground feed; the texture is like cornmeal. Starter feeds for chicks are usually mash texture and are best for small chicks.

  If mash is the only type of feed available to you for older chickens, you can add a little warm water to the feed just before serving it, which gives it the consistency of thick oatmeal. Chickens generally gobble this down. Water from cooking potatoes or other vegetables or milk also can be used. Serving mash is a good way to use up the fine pieces of crumbles or pellets left in the bottom of a bag or the feed dish. However, don't let this wet mixture sit too long; it will spoil and become moldy, which may harm the chickens.

Some areas sell a pellet-and-whole-grain mix, usually under the name of "all stock" or "sweet feed." It's covered with molasses or another sweetener to hold it all together. While these rations sometimes list "poultry" on them (or, more often, include a picture of a chicken), they really aren't formulated for poultry. You can use these feeds on your other farm animals, and you don't need to worry if the chickens steal a bite. But definitely don't use them as your sole chicken feed.

Double-checking the label

To be sure your feed contains what your birds need, and to ensure that you don't eat eggs or meat that may be contaminated with medication from feed, be sure to check the label . . . twice!

Here's what the label tells you:

- The ingredients, including percentages of basic nutrients like protein and fat and the percentages of recommended dietary vitamins and minerals.
- The manufacturer's address, in case you have a question or complaint.
- Any medications used in the feed and how long the chicken must be off the feed before eggs or meat from the bird can be eaten.
- A "manufactured on" date or an expiration date. (Feed loses nutrients the older it gets, and after 6 months, most feed is stale. Stale feed may not harm the birds immediately, but long-time use of old feed can lead to vitamin and mineral deficiencies.)

Supplementing Diets with Grit

Because they have no teeth, chickens need grit in their *gizzard*, a muscular pouch that's part of the digestive system, to help break down food particles. Grit in the gizzard is especially important if the chickens are eating a fibrous diet, such as whole grains and pasture. Commercial diets are easy to digest and don't require grit to grind them. However, some chicken owners feel that chickens are happier when they have grit — even if they don't need it because of the diet they're on.

In nature, chickens pick up small rocks, pieces of bone, and shells and store them in the gizzard to help digest food. If you're feeding any kind of home-made diet, if you're giving whole grains, or if you have your birds on pasture, you need to supply them with some kind of grit. If you're feeding only a commercial mash, crumble, or pellet, your chickens won't require additional grit because these feeds are already quite easy for chickens to digest. Birds that range freely part of the day will pick up enough grit.

You can purchase grit in feed stores. It consists of crushed limestone and granite. Different sizes are available for chicks and adult birds. If you have just a few chickens, you can purchase canary or parakeet grit in pet stores. It's finely ground but is fine for chicks or, in a pinch, for older birds.

Make sure the grit you're purchasing is for birds. Feed stores sell a coarse salt-and-mineral mix for large livestock that some people mistake for grit, and it can cause serious harm to your chickens.

If your birds need grit, you can supply it in a small dish from about the fifth day of life. Chicks should be eating their regular feed well before you add grit, or they may fill up on it. Make sure the dish is covered or narrow so the birds don't dust-bathe in it. Discard it and add clean grit if it becomes contaminated with chicken droppings.

If you're feeding a whole-grain diet or pasture to laying hens, you may want to offer them a dish of crushed oyster shells or a calcium-and-mineral mix designed for hens. Both are available at feed stores. They supply the calcium and minerals that high-producing hens need. If you're feeding a commercial laying diet, you don't need to add these extras.

Some people feel that grit and oyster shells are important for all chickens because one bird's dietary needs are different from the next, and what works for one chicken on a commercial diet may leave another lacking some nutrient. Offering grit and oyster shells to all birds does no great harm, as long as they're free to eat it at will and their diet is such that they won't fill up on it.

Deciding When to Put Out Feed

Most people fill their chickens' feed dishes so food is available much of the day, or they use feeders that hold several days' worth of feed. You can use this feeding method for all types of chickens. It's the way chickens eat in nature; they eat small amounts frequently.

Other folks still feed their chickens at certain times of the day, generally morning and evening. That way, they can control the amount of feed that may attract pests. And if the chickens are too heavy, it restricts the amount they can eat. Usually, however, it's just a matter of preference; some people like to observe and tend to their chickens more often than others. This method works well for all but meat birds.

Chickens that are given free range may be more inclined to come to the coop to lay if they're fed there in the early morning. And if you need to lock the chickens up every night to protect them from predators, feeding them in the coop in the evening entices them there.

Because of their heavy rate of growth, the meat-type broiler chickens need to have food available to them for longer periods of time. Remember, chickens don't eat in the dark, so the lights must be on for them to eat. For the Rock-Cornish crosses, the lights need to be on 18 to 22 hours a day, and feed should be in the feed pans for those hours. A few hours of darkness helps prevent dying early from what is known as "sudden death." Use more dark time in hot weather and less in cold weather. You can regulate eating by putting your lights on timers so that lights are off for 2 to 6 hours at night. Laying hens, pets, and show birds are fine with more restricted times of feeding and don't need feed at night.

Birds on pasture are probably going to be subjected to natural daylight and darkness, which is one reason they grow a little slower. Since chickens don't eat in the dark, this can make a big difference when you're raising meat chickens on pasture in early spring and late fall, when nights are longer.

Determining How Much to Feed

Determining how much feed your chickens are going to eat per day or week is difficult because so many variables are involved. The type of chickens, whether they're growing or laying, and how active they are, affect the amount of feed each bird needs. How neat you are, the type of feeders you have, and the number of free-loading pests you support also change the amount of feed you need. The weather is a factor, too: Chickens eat more in cold weather and less in hot weather.

Our modern, high-production egg breeds convert feed to eggs very efficiently, especially if they're fed a ration formulated for laying hens. After they're laying well, it takes about 4 pounds of a quality feed of 16 to 18 percent protein to produce a dozen eggs. The breeds kept for dual purposes (eggs and meat) generally have heavier body masses to support and need more feed to produce a dozen eggs than a lighter-production breed.

About 2 pounds of feed are needed to produce 1 pound of body weight on a growing meat-type bird. So if a broiler weighs about 6 pounds at 10 weeks, it will have eaten about 12 pounds of feed. Remember that it ate less when it was small, and the amount of feed consumed increased each week. A medium-weight laying hen will eat about ¼ pound of feed per day when she begins producing. These figures are rough estimates, but they give you some idea of what to expect.

When you pick up your birds, they should feel well fleshed but without rolls of fat. If the breastbone feels very sharp and prominent, they're probably too thin. Chickens that don't get enough feed will stop laying. The thinner birds in the flock and the ones in which vitamin and mineral deficiencies show up first are the ones that are lowest in the pecking order; dominant birds tend to eat first and eat the best of what's available.

If you're using a lot more feed than you think you should, pests like rats may be eating it at night. You may want to empty feeders at night or put them inside a pest-proof container for all birds other than the broiler-type meat birds.

Keeping the Diet Interesting by Offering Treats

The first part of this chapter emphasized that a diet should be well balanced, but an occasional treat can be good for the birds. Treats can help relieve boredom in confined chickens, including birds that are being kept inside for the winter. They may reduce instances of chickens pecking at each other or eating materials they shouldn't, like the paint off the walls.

If you do feed treats to your flock, try to keep them nutritious and give them only a small amount — usually less than a cup per bird per week, divided up over several days. Make sure you feed only as many treats as the chickens can eat in a small amount of time. Treats left out may attract pests or smell.

Chickens don't care much for sweet foods, so avoid foods that consist primarily of sugar and fat. No treats should be moldy, either. Moldy food can cause a wide range of problems in animals. Too much of some foods, such as cabbage, onions, garlic, flaxseed, and fish, may cause your chickens' eggs — and even the meat — to have an off taste if these foods are fed for long periods of time.

Following are some good, safe treats for chickens. Remember that these are treats to be fed in small quantities. Clean up any treats the chickens don't eat right away.

- **Cooked potatoes and potato peels:** Don't feed raw potatoes or peels to chickens. The sprouts and green areas of skin can be poisonous. Remove very green peels and sprouts, and put the peels in a microwave for about 5 minutes; then cool, and they're safe for feeding.
- **Dark, leafy greens:** Avoid iceberg or head lettuce, which is basically just green-tinged water.

- **Eggs and eggshells:** You can cook and chop your cracked or old eggs and return them to your chickens. Be sure to crush eggshells into small pieces. Cooking eggs and crushing shells keeps chickens from developing the habit of eating their own eggs in the nest.
- **Fruits:** Apples, pears, and other fruit raked up off the ground make excellent treats, especially if they're wormy. Wash the fruit first if it has been sprayed with pesticides. You can feed most fruits to chickens, although they probably won't eat citrus fruit. Fruit can be soft or damaged but shouldn't be moldy.
- **Green, orange, and red vegetables:** Leftover veggies from dinner are fine, even in casseroles and sauces. Don't overdo cabbage, broccoli, Brussels sprouts, cauliflower, and onions; however, a small amount of these veggies is fine.
- **Meat and bone scraps:** Cook all meat and bones first. Chickens adore picking the meat off bones, even chicken bones, but you need to remove the bones from the coop after a day or two. If large bones are cracked, chickens will eat the marrow inside. If you don't have dogs or cats to give the leftover meat and fat scraps to, chickens will eat them. However, don't feed lots of grease, such as bacon grease.
- **Milk and other dairy:** These products are fine in moderation. Chickens will drink liquid milk, and sour milk is fine to give them as well. Cheese and yogurt also are fine.
- **Pumpkins and squash:** The "guts" from a jack-o'-lantern are quite popular with chickens. Even the rind can be fed after Halloween if it isn't moldy. Chickens also adore those monstrous zucchini no one else wants.
- **Spaghetti, other pasta, and rice:** Cook all these items first. Leftovers that aren't moldy are fine, even with sauces.
- **Stale bread, cookies, cake, cereal, and so forth:** Chickens adore these treats, and they're good to feed unless they're moldy. Don't feed too much and too often. It may be a good idea to scrape off sugary frostings before feeding. Don't feed a lot of very salty treats, such as chips, cheese puffs, and so on.
- **Weeds from the lawn and garden:** Most weeds are quite nutritious — just make sure they haven't been sprayed with pesticides. A little cut grass is okay, but don't overdo it with this snack. Every area has weeds that are poisonous, so consult a book or authority before you feed your birds unfamiliar weeds. Never feed yew trimmings (a soft-needled evergreen common in landscapes) to any animal, and don't include any mushrooms or fungi in your offerings. Dandelions, crabgrass, chickweed, and thistles are all safe.
- **Miscellaneous:** Cooked nuts are fine, as are raw crushed acorns, walnuts, hickory nuts, and pecans. Wild bird seed and sunflower seeds are fine, and it's okay to leave the hulls on. A little dry pet food and a few pet treats occasionally are okay, but don't feed too often or too much. Rabbit pellets can be an occasional treat as well.

Hydrating Your Hens (And Roosters)

Having a source of clean water is vitally important to your chickens. Chickens whose water intake is restricted don't eat as well as chickens with unrestricted access, and they don't grow as fast or lay as well, either. People often don't realize how important water is to their chickens until they go from pouring water in a dish once a day to a system that allows birds to always have fresh water available. The birds with unrestricted access to clean water grow better, are healthier, and lay more eggs.

In moderate weather, a hen may drink a pint of water a day. In hot weather, that amount nearly doubles. Broilers may drink even more as their metabolism works much harder, producing more heat and using more water. Birds roaming freely may drink more or less than confined birds, depending on the moisture content of the food they consume and how active they are.

Drinking can be restricted because water isn't available or because the water available is unappealing. Chickens don't like water that's too warm. In hot weather, providing an unlimited quantity of cool, clean water may mean the difference between life and death for your birds. Move water containers away from brooder lamps and out of sunny areas. You may want to change water or flush the pipes of automatic systems more frequently so the water is cooler. For more information on water containers and systems, see Chapter 7.

Chickens also drink less if the water has an off taste from medications or additives, such as vinegar, that people feel they need to add to drinking water. Make sure any medications are truly needed in hot weather, and avoid all those fancy additives so chickens will drink enough to avoid heat stress. Chickens also avoid dirty water full of algae, litter, dirt, and droppings, so scrub out those water containers.

In winter, if temps are below freezing, offer water at least twice a day in sufficient quantity that all birds can drink until they're full. Alternately, use a heated water container.

Any time birds stop eating as much as usual, check the water supply. Be sure to manually check nipple systems frequently to make sure they're working. We've heard horror stories of nipple valve systems becoming clogged by mineral deposits or other materials and failing to work. These devices freeze up easily in cold weather, too. If the chicken caretaker fails to notice that the chickens can't get water, the birds die of thirst.

Chapter 9

Controlling Pests and Predators

In This Chapter

- Identifying pests and predators
- Protecting the flock from predators
- Controlling pests

In this chapter, we talk about controlling the bad guys — enemies that want to hurt your chickens or steal their feed or eggs. Most chicken owners have to deal with pests or predators at some point in their chicken-rearing. Having to endure losses from predators can be extremely frustrating and heartbreaking. As a keeper of domestic animals, one of your responsibilities is to guard your chickens and keep them safe. The commercial poultry industry moved its flocks indoors decades ago because of the 4 P's: predators, pests, parasites, and pathogens. In this chapter, we discuss the pests and predators; in the next chapter, we discuss the parasites and pathogens, (diseases) that can attack your flock.

A predator in the hen house can ruin your day and end your chickens' days, so it's important to understand how to deal with the predators that roam your area. Pests may not kill chickens and are usually more of a problem to the chicken keeper than the chickens. They can spread disease however, and they often cost you a lot of money and time. Knowing how to manage pests is an important part of chicken keeping. Hopefully the information in this chapter will guide you in keeping your chickens safe and lessen your chances of being bothered by pests.

Keeping Pests from Infesting the Coop

Pests are creatures that don't directly kill and eat chickens or feed off them like parasites. Instead, they eat chicken feed or eggs. Too many scary disruptions and too many feet in the feed can be the start of major problems.

Pests cost you money and may make your neighbors unappreciative (to say the least!) of your chicken habit. Mice and rats, for example, can be found anywhere humans and domestic animals reside. When only a few are hanging around, you may not even know they're there. However, they multiply rapidly and become a serious problem seemingly overnight. The same goes for most other pests.

Preventing pests

Preventing pests is always better than dealing with an established population. Be diligent, clean, and tidy, and pest problems may never bother you or your birds. The following list shows you several actions you can take to prevent pest problems:

- **Store feed properly.** Part of dealing with pest problems is storing feed products correctly. Be sure to store feed in insect- and mouse-proof containers. Metal trash cans with tight-fitting lids work best. Insects destroy the nutritional value of feed, and mice can eat through the bottom of a feed bag in seconds. After people discover how to store feed correctly and keep freeloaders from eating it, they're often amazed at how little their chickens actually eat. Turn to Chapter 7 to find out about all things feed related, including information on dishes and storage.
- **Keep feeding areas clean and dry.** Clean up any spilled feed. Wet areas are most likely to support maggot growth, so keep the place dry as well.
- **Don't give rodents a place to hide.** Keep trash picked up, and remove piles of junk that can shelter rats and mice. Keep grass and weeds trimmed around buildings. Rats are more likely than mice to come from a neighbor's buildings or yard and to go from feeding at your place to sleeping at theirs, so you may have to enlist the help of the neighbor to control these pests.
- **Cover what you can.** Covering feeding stations and water containers outside can prevent wild bird droppings from getting into them. You also may need to cover your pasture pens with fine bird netting (like that used to protect fruit from birds).
- **Don't allow wild birds to nest inside buildings where you keep chickens.** This advice applies for all types of birds, from sparrows to swallows. Besides disease, wild birds can bring parasites like lice to your chickens. Excluding the birds from the building and removing any nests you find promptly are the best ways to handle this.

Identifying and eliminating common culprits

The most common pests are insects and rodents. They occur in both rural and urban areas. Clean conditions, proper storage of feed, and an action plan (for when you start seeing signs of pests) are your best defenses. In the following sections, we discuss various pests and the ways to control them.

Mice

Chickens actually eat mice, so mice seldom set up housekeeping right in the chickens' home. Instead they work on the fringe, getting into stored feed, chewing up building insulation and wires, and running under your feet when you least expect them. Mice don't eat eggs, but they can eat and soil a lot of feed, cause allergy problems for some people, and spread certain diseases. It's best to try to get rid of them or at least control their numbers.

Mice have small territories. They build nests close to a food source, and they don't travel far. You may have a mouse problem if you notice these telltale signs:

- Small oval droppings.
- Little round, ball-shape nests in concealed places.
- Tiny holes in walls, floors, and feed bags. (Mice can squeeze through openings as small as a ½ inch. They may make small entrance holes, but they don't chew large holes.)
- Shallow surface tunnels in loose soil, litter, or snow.

Mice and rats rarely exist in large numbers together, so the one good thing you can say about a mouse problem in the chicken coop is that you probably don't have rats.

Mice are curious about their environment and can be trapped rather easily. Numerous styles of mousetraps are on the market, and some work better than others. (Remember the old saying about building a better mousetrap?) The best ones are traps that you can squeeze to open and don't require you to touch the dead mouse. Other good traps electrocute mice so you can then dump them from the box trap. Trapping is an option when populations are high or you don't want to use poisons. Emptying traps and disposing of dead mice can be disgusting, though.

If you're kindhearted and choose a trap that catches mice alive, you then have to do something with them. Don't just take them outside the door and turn them loose; they'll be back inside before you know it. You can kill them, but why set a live trap to do that? You can feed them to the cats or chickens, but that's not such a wise or humane choice, either. The best bet is to take them to the woods or a field far from other homes where they can feed some animal you're trying to keep away from your chickens. Mice are designed by nature to feed something else.

Mice are controlled fairly easily with poison bait. However, place the bait where children, pets, and curious chickens can't get to it. Hardware stores and feed stores sell bait stations that hold bait safely. Replace bait as soon as it's eaten.

Don't let chickens or pets eat mice killed by poison. Some poisons can remain potent in the dead mice and may harm the chicken or pet.

Rats

Rats are larger, meaner, and more secretive than mice. They can eat eggs, and they've been known to eat baby chicks and even feed on larger birds as they sit on roosts or nests in the dark. Chickens and even most cats leave rats alone. Rats eat a lot of feed and destroy more by soiling it. They also do considerable damage by chewing on the structure and its parts, like wiring and plastic pipes.

The chance of rats actually eating live chickens is low, but it can happen when other food is scarce, the shelter is unlit at night, and the chickens don't roost off the floor. Chickens go into a kind of stupor in the dark and don't defend themselves very well. Rats come up under them where they sit on a nest or the floor and begin eating them alive. Just having a small nightlight on after dark allows the chickens to move around and defend themselves.

Rats travel farther than mice to get food, and their nest may be outside the chicken coop. You may have rats if you notice the following:

- **Tunnels through the soil, often with large mounds of soil around an opening:** These tunnels and holes are often large enough to make people think they have a woodchuck or some other animal. Woodchucks aren't active in the winter or at night. If new tunnels appear overnight or in cold months, you probably have rats.
- **Large holes chewed through heavy wood, plastic, cinderblock, or even cement:** Rat holes are larger than mouse holes and often have greasy, dirty smears around them.
- **Droppings that are much larger than mouse droppings but are shaped similarly.**

Rats need liquid water, unlike mice, and are sometimes found drowned in water buckets with steep sides. They can swim pretty well, however, and are adept at climbing, too. Rats are suspicious of new things, so it may take days to get them to eat poison bait or fall victim to a trap. Traps must be placed close to pathways that rats habitually use.

Traps aren't as effective at controlling rat populations as poison, however. Buy a poison specifically for rats, and change the type of poison you use from time to time so resistant populations don't build up. Read the label to find out if the bait works with one feeding or multiple feedings. Remember to use bait stations to protect pets, children, and chickens, or place the bait where they can't find it. You can place chunks of bait in tunnels if the chickens don't have access to the area. Otherwise, they might scratch up the bait digging out a dirt bath.

Don't throw any dead rats killed by poison into open fields or woods. Bury them instead, or wrap them tightly and dispose of them in the trash. Otherwise, birds of prey or other animals may eat them, and the poison left in the rats then may kill those animals.

Weevils, grain moths, meal worms, and other insects

Weevils, grain moths, meal worms, and other insects attack stored feed. They may provide a little extra protein for the chickens, but their feeding strips the nutritional value of the feed, leaving only husks behind. Grain mills and feed stores attempt to control these thieves with pesticides, but some level of infestation is almost always present. You know you've got grain pests if you notice the following in your feed:

- Fine webbing
- Tiny worms

These pests are more likely to attack whole grains than processed feed, but they can be found in either. Buy only the amount of feed you can use in about 2 months. Keep your feed in tightly closed containers, not open bags. Even with these precautions, however, know that insect eggs may already be present in the feed when you buy it, so you can still develop a problem. You can buy strips that have pesticides in them to hang in grain bins, or you can try sticky traps that lure grain insects with pheromones. Read and follow all label directions carefully.

If you have a way to freeze feed for a few days, doing so will effectively kill most grain pests. In the winter, just leave the feed in your car for a few days when the temperature is below freezing. If you still have insects in the feed, use that feed promptly if it isn't too badly infested, or discard it if it is. Then thoroughly clean the feed container with hot, soapy water and allow it to dry in the sun. Metal and plastic containers are better to use than wood because they don't absorb moisture and insects and other pests can't chew through them as easily. Be sure to clean up any spilled feed, and empty and clean the feed dishes.

Flies are another insect that may become a problem. Chickens love maggots, which are baby flies, and if they can reach them, few will make it to adulthood. But if the maggots are outside the chickens' reach, such as in a manure pile outside the pen, you may get large quantities of flies. To keep flies in check, be sure to compost manure in such a manner that you keep it hot and cooking. If you have a passive compost pile that you just let sit, keep it far away from your house and the neighbors.

You can use sticky paper to catch flies in a chicken coop, but hang it where the chickens can't grab it and where your hair won't get stuck to it. Don't use pesticide sprays unless they say they're safe for use in poultry housing, and follow the label directions exactly if you do.

Never treat feed with pesticide sprays or powders intended for garden or home use. Some of these products are highly toxic to birds. Animal supply catalogs and farm stores sell pesticides to use in feed areas that are safe and effective.

Nuisance birds

You've got good reason to keep wild birds away from your chickens. Wild birds are carriers of many diseases, including some that are harmful to human health, such as avian flu and West Nile virus.

Wild birds also can make your feed bill jump in a hurry. Feed your chickens inside a building, if possible. Cover windows to inside shelters with screens. Close doors and plug holes in eaves and under rafters. Most wild birds won't enter buildings from openings close to the ground (like most chicken shelters have for the chickens), so that's seldom a concern.

In pasture situations, keeping wild birds away is difficult to do. They're less attracted to concentrated feeds than whole grains, but occasionally they become a problem even with this kind of feed.

It's difficult — and illegal in many places — to poison pest birds like sparrows without killing unintended birds or your chickens. Live traps are sometimes available, but then you must dispose of the birds you catch in them. The best defense against nuisance birds is a good offense: Exclude birds and chase them away.

Ultrasonic devices said to scare off birds without you hearing them are useless. If they worked, they would seriously offend your chickens. But they don't work, so don't waste your money. Fake owls or hawks may work temporarily, but they won't be popular with your chickens, either. Birds are quick to learn that something isn't real, and any benefit is then over.

Fending Off Predators

Man isn't the only species that likes chicken nuggets. A predator in the hen house can be a huge frustration — and worse for the chickens it kills. *Predators* are animals that eat other animals or kill them just for fun. Chickens are high on the preferred food list for many predators. Some predators destroy all the birds they can when they gain access to a coop; others take one every so often when they're hungry.

Most chickens do little else but squawk and run to defend themselves, so knowing how to deal with predators is important. Case in point: it takes one loose dog just a few minutes to completely destroy a prized flock of chickens. And though chickens don't seem to be bothered much by cats and most farm animals like goats, sheep, and horses, it's not necessarily safe for them to be hanging around. Pigs, especially, will catch and eat chickens.

Predators are more likely to begin attacking poultry when the birds are raising young in the late spring or early summer, or in winter when food is scarce. And when predators find a good source of food, they often return. Remember that predators live in both urban and rural areas. In fact, some urban areas may have higher numbers of predators than rural areas.

The following sections help you keep your flock safe from harm and introduce you to the most common predators that chicken-keepers need to beware of. In case you do find yourself with a predator on your hands, we also tell you how to figure out which one you're facing and then deter it.

Providing safe surroundings

A little planning when building chicken housing goes a long way toward keeping your chickens healthy and safe. You may be surprised by what may be after your chickens, even in an urban area. Of course, free-range chickens are at the greatest risk for running into a predator, but predators can be remarkably resourceful about breaking into chicken housing. Build the strongest, safest housing and pens you can afford. Chapter 6 tells you all about constructing your chicken's home.

The following tips can help you further protect your chickens from predators:

- **Be careful about letting your chickens roam.** Chickens that range freely may disappear without a trace. In areas with heavy predator presence, letting chickens roam freely may be impractical. Keeping chickens penned until later in the morning and bringing them in early in the evening can help.
- **Prevent nighttime attacks.** Good, predator-proof shelters can be closed up at night. A light left in the coop at night further ensures the safety of your chickens.
- **Cover coop windows.** Put strong wire over open shelter windows.
- **Keep an eye out for dogs.** Keep your own dogs from chasing chickens, even in play. If neighbors let their dogs run loose all the time, your chickens are probably going to need to be penned up.
- **Fence your chickens in.** Pens of sturdy wire are the best protection your chickens can have. If raccoons are around, you need strong, welded wire to keep them out. To keep out larger predators, a foot of heavy wire bent outward at the bottom of the run fencing and either buried or weighted down with large rocks is advisable.

  Electric fencing is effective at keeping raccoons, foxes, dogs, and coyotes out of chicken pens. A single strand of electric wire near the top of outside runs and near the bottom of any flexible fencing should keep them out.
- **Steer clear of trees.** Make sure predators can't jump off nearby trees into the pen or get on the roof of the coop from nearby trees.
- **Beware of aerial attacks.** If you let your chickens roam, choose dark-colored birds, which are harder for birds of prey to spot. White birds are easy pickings if they roost outside at night, so make sure they roost inside. If hawks and owls take chickens out of fenced runs, you may need to cover them with nylon netting or fencing.

Recognizing common chicken predators

The most common predator of chickens is the dog; this animal may be man's best friend, but it certainly isn't a chicken's. However, many other predators occur in both urban and rural areas as well. We discuss the most common ones in the following sections.

Luckily, you don't need to worry about some animals. Domestic cats and even semi-feral cats seldom attack adult chickens, and they don't eat unbroken eggs. Groundhogs don't eat chickens or eggs, but they may occasionally eat chicken feed. Bats don't feed on chickens in the United States. Bobcats and cougars are rare or nonexistent in many parts of the country, so don't think of big cats first unless you actually see them.

Domestic dogs

Even the tamest dog may enjoy chasing chickens, and even if he doesn't intend to kill them, he may chase them into harm's way, cause them to pile on each other in pens, or just run them to death.

When domestic dogs get into a chicken pen or find free-roaming chickens, they generally kill all they can catch. The birds are killed in a variety of ways but not eaten. Dogs usually leave behind a mess, with blood and feathers everywhere. Birds left alive may have deep puncture wounds or large pieces of skin pulled off and generally need to be destroyed. Dogs that kill frequently may get more efficient and cause less mess. After dogs kill chickens once, they generally kill again and must always be controlled. Even small dogs can be deadly chicken killers.

Dogs are a special problem because dealing with them often causes problems with the dog's owners, usually your neighbors. We cover this issue in more detail at the end of this chapter.

Opossums

Opossums live in almost every part of the country — in cities and rural areas alike — except for the Western plains and mountains. An opossum is about the size of a cat and has grayish hair; a long, pointed snout; beady eyes; and a long, naked tail.

You may see an opossum in the evening if you make a surprise visit to the chicken coop because opossums are generally active at night. You may also uncover them in their big, messy nests tucked away somewhere in the daytime. They sleep so soundly that you could probably touch them — if you wanted to. Some actually snore, and you often can follow the sound to where they're sleeping.

Opossums may hang around chickens for years and never bother them or just eat an occasional egg. Many farmers used to let an opossum hang around a barn to eat rats and mice. However, we now know that opossums can pass many diseases to domestic animals, so they shouldn't be allowed to live with livestock and poultry. Sometimes an opossum discovers that eating chicken is great. If they kill, opossums usually kill one chicken at a time and consume part of it on the spot. They can carry eggs away, but they usually crack and eat them in the nest.

Opossums seem slow and dull-witted to most people, but don't let the act fool you: They're actually pretty intelligent. They can move quite fast, climb trees, and fight ferociously when cornered. If their quick movements and fighting don't work, they emit a foul discharge and play dead. Opossums eat just about anything, but they prefer lazy ways of getting food, resorting to killing only occasionally.

To keep opossums out of the coop, close or screen all openings at night. Possums climb very well, so keep a roof on your runs.

Raccoons

Raccoons are cute, but the cuteness wears thin when you know their habits. They're formidable predators and can easily kill an adult chicken. In the wild, their favorite foods include baby birds and eggs, and a chicken coop can be like heaven to them. Raccoons are found nearly everywhere in the United States, in cities and in the country, and they don't need to live by water.

Raccoons are active at night. When seen out in the daytime, they may be sick. They carry a number of diseases and parasites, including rabies, which can be a threat to humans and other animals. They shouldn't be allowed to nest in barns and chicken coops — they're so destructive that few people tolerate them for long anyway.

When raccoons kill chickens, they usually kill several or all birds at one time. They bite the heads off the birds and may carry the heads around before dropping them. They usually tear flesh out of the breast area of one or more birds to eat and may carry the meat pieces to the water container, leaving pieces of flesh and blood around it.

Raccoons may take eggs left in nests overnight, even reaching under sitting hens and carrying them away. Or they may eat them on the spot, crushing the egg pretty well and, once again, carrying pieces to the water dish. Raccoons don't, however, need water to eat. Sometimes they concentrate on eggs and chicken feed and never bother the chickens, but don't count on it.

Raccoons can make quite a mess of stored feed, ripping apart bags, tipping over bins, and just generally playing in the food, soiling and wasting it. Make sure you clean up any raccoon feces left behind with hot water and soap — wear gloves as you do so, to avoid raccoon roundworm, which can be quite harmful to humans.

Raccoons can rip apart chicken wire. They also can reach through wire to grab chickens and then pull the birds against it to kill them. They may pull the feet of birds through wire floors and chew them off. They open brooders and eat chicks with gusto.

Raccoons climb very well and are adept at opening doors and latches. They don't tunnel under structures, but they may run along an existing tunnel, like a drain culvert or sewer line.

To limit raccoon entry to your coop, don't use hook-and-eye-type latches or slide bolts on doors. Use dog-leash-type clip latches or even padlocks. Have good latches on brooders and cages. Cover windows with heavy welded wire, not chicken wire. Close all doors that enter the shelter at night. A strand of electric wire at the top and bottom of the run that you turn on at night is an effective barrier.

Foxes

Foxes generally live in rural areas. They sometimes become predators of chickens, especially of free-ranging birds or ones pastured out some distance from humans. They're active in the early morning and late evening.

Foxes generally carry away their prey to eat it. They take one bird at a time, but they may return time after time if not stopped. You sometimes find the partially buried carcass of a bird where they hid it. Foxes prefer to go under or through fences but can climb if they need to.

Use wire roofs on runs or install electric wires at the top and bottom of runs to keep out foxes. Close any doors they may enter at night. If foxes are a problem in your area, don't allow your chickens to roam freely.

Coyotes

Coyotes have become problems even in some large urban areas. Large populations may live quite close to humans, but they're seldom seen. They're skillful and opportunistic, and they eat a wide variety of animals, including chickens and small pets. They're nocturnal, but they may hunt during the day in cold weather if they notice little human activity. They look somewhat like small German Shepherds. Like many other predators, they grab a chicken and go off with it. Because they often travel in pairs or small packs, they may take several birds at once. Coyotes also eat eggs but seldom enter chicken housing to get them.

Strong, high fences keep coyotes out of chicken runs. Bury the fence a foot in the ground, to discourage digging under the fence.

Hawks, eagles, and owls

Chickens seem to have an instinctive fear of large birds overhead. When they spot one, they squawk loudly and distinctly and either run for cover or flatten themselves against the ground. Hawks and eagles are active only in the daytime. Owls, on the other hand, generally hunt at night, but when times are lean, as in the winter, some owls also hunt in daylight.

It takes a pretty large hawk to prey on chickens. Many small hawks don't bother adult chickens, but they may pick off baby chicks. Hawks are found pretty much throughout the United States, but they're more likely to prey on chickens in rural areas where they roam freely. They tend to avoid taking birds out of smaller fenced enclosures, but it sometimes happens. Bald Eagles and Golden Eagles are more common than they once were. They can easily grab a chicken and fly off. They hunt during the day.

Some hawks and eagles carry away their prey; others pick it apart where they nail it. A chicken killed by a hawk or eagle won't be found inside the shelter. These birds pull the meat and feathers off the chicken in clumps, often starting at the breast and pulling out organs. If left alone, they can pick apart a carcass in a short time.

Owls have been known to fly through windows or down from lofts to pluck chickens off roosts while they sleep. They usually take the chicken with them from a building, but outside they may sit on a post or even the ground to consume part of the bird. If chickens roost outside, they're a prime meal for owls. However, like some hawks, not all owls are large enough to kill chickens.

Owls that get into a pen or a house with lots of chickens may get a little kill-happy and kill several birds, usually by pulling off the heads. They may take a single bite out of each breast, or they may eat just one bird.

Hawks, eagles, and owls are federally protected birds across the United States, so humans can't kill or harm them. The best way to avoid these predators is to prevent them from getting your birds.

Minks and weasels

Minks and weasels aren't as common as they once were. They're generally found in rural areas near some source of water. They're active day and night but are very secretive, so human activity generally keeps them away during the day.

When minks or weasels enter a chicken coop, they're not looking for chicken feed; they eat only meat and eggs. The meat may come from rats, which are a favorite food. But if they find chickens while they're pursuing rats, they're quite happy. And if your chickens range freely by waterways, they're fair game.

Weasels typically kill one or two chickens by biting the heads, and then they primarily drink the blood, without eating a lot of flesh. Minks may kill many more chickens than they need, also by biting the heads, and then they pile them up. They may eat small amounts of flesh from one or more birds. Minks seldom eat eggs when chickens are available. Minks often enter chicken coops through the opening a rat has made or by following a rat tunnel. They can squeeze through quite small holes, so to prevent their entry, close all openings in housing that measure more than 1 inch. Keep chickens out of brushy areas near water, and keep a clear area around pens and shelters. Noise, such as that from a radio, and bright lights are more likely to discourage minks and weasels than some other predators.

Other bad characters

Skunks occasionally prey on chickens or eat eggs, and they love chicken feed. Skunks are more likely to be a problem when they take up residence under a chicken shelter. Skunks are nocturnal, and if you pick up eggs before nightfall, you won't have a problem. When skunks do eat eggs, they open one end neatly and eat the inside. In rare cases, a skunk may kill a chicken. Skunks don't climb, so fencing keeps them out.

In the far Southeastern states, alligators sometimes eat free-ranging chickens near water. Some large exotic snakes also may kill chickens. In the North, some large snakes swallow eggs whole, but they don't bother chickens.

Bobcats are rare and generally take only free-ranging birds. Cougars are becoming more common in many areas and are a big danger for chickens, pets, and humans. If you suspect that a cougar is in your area, seek professional help from your Department of Natural Resources (DNR).

Figuring out who's causing trouble

Determining which predator killed or injured your chickens can sometimes be difficult. If a chicken simply disappears, it's probably impossible to find out what happened, but in some cases, you may find clues. An inexpensive trail camera, usually purchased in a sporting goods department, can be invaluable in determining who's visiting your coop. These cameras run on batteries, work inside or out, and take photos at night as well as in the day. You can also set up security cameras or even use a baby monitor with a camera. If don't have camera evidence, examine the victims and chicken housing closely and consider the following questions (Table 9-1 can help you answer the questions):

- Can you see an obvious way the predator entered, such as through a large hole in the fence?
- Has more than one bird been killed?

- Were the birds eaten or just killed?
- What part of the body did the animal feed on, and where are the wounds?
- Were eggs smashed or carefully cracked? If cracked, how were they opened?
- Do you see animal tracks in snow or mud?
- Do you live in an urban or rural area and/or by water?

Table 9-1 Figuring Out Whodunit

Predator	*What It Goes For*	*Hunting Hours*	*Signs*
Dog	Chickens	Day or night	Chickens killed but not eaten; survivors may have deep puncture wounds or large pieces of skin pulled off; scattered blood and feathers
Opossum	Usually eggs; occasionally chickens	Night	Droppings look somewhat like cat droppings; tracks similar to tiny human handprints, with five toes and one toe pointing off to the side or backward on the back feet; one chicken killed at a time and partly consumed; eggs usually cracked and eaten in the nest
Raccoon	Baby birds and eggs; capable of killing adult birds, though	Night	Several or all birds killed; heads bitten off; flesh torn from breast area; pieces of flesh and blood around water container; feed bags and bins trashed
Fox	Chickens	Early morning and late evening	Chickens free-ranging or pastured away from humans in a rural area; may find partially buried carcass of a bird
Coyote	Chickens and sometimes eggs; seldom enter chicken housing	Mostly night; may hunt during day in cold weather if there's little human activity	Several birds missing at once

Predator	*What It Goes For*	*Hunting Hours*	*Signs*
Hawk, eagle, or owl	Baby chicks or adults; only large hawks, eagles, and large owls can kill adult chickens	Hawks and eagles during day; owls mostly nights, though some-time's days during winter	Hawks, eagles: Birds carried away or picked apart on the spot where killed Owls: Birds plucked from roost and carried outside; in coop, heads may be pulled off, with a single bite taken out of each breast
Mink or weasel	Chickens, eggs, and rats	Day or night, but they shy away from human activity	Chickens kept in rural areas near water sources; heads bitten off; small amounts of flesh eaten; birds may be piled up

Even if you've seen certain animals in the vicinity of your chickens, don't assume they're the killers. And even if certain animals have been around a long time without bothering the chickens, don't assume they *weren't* the killers. Determine the killers based on the facts, not assumptions, so you can better protect your chickens.

Catching the troublemaker

Most predators are relatively easy to catch, and you generally use the same methods for most of them. Before killing or trapping predators, check with your state's DNR about applicable game laws. The DNR may give you permission to kill or trap certain predators that are killing your chickens. Hawks, eagles, and owls are federally protected birds, though, and you aren't allowed to kill or trap them.

When we say "trap," we mean live trap. Other types of traps can kill or maim animals you didn't intend to kill, such as your chickens or the neighbor's cat. But before you set a trap, make sure you're willing to deal with whatever you catch, including a skunk. You must check live traps every day and promptly deal with what you trap.

We offer information on trapping because many people feel it's the only option they have. In reality, excluding predators with the proper fencing and shelters is preferable to trapping or killing them. We provide some suggestions for excluding each predator in their respective sections earlier in the chapter.

Live trapping isn't as humane as many people think. If you want to release the animal, you need to take it miles from your home, or it will come right back. Get permission before releasing any animal on property you don't own, including parks and nature preserves. Don't leave your problem on someone else's doorstep. Studies show that most animals released into strange areas die shortly after the release, from starvation or the current animals in that area.

Unfortunately, most live trapped predators will need to be killed. In fact, some places require live trapped wild animals to be killed because moving them spreads disease. Shooting is the most humane way to dispose of a live trapped animal — it's over and done with quickly. If you don't want to shoot an animal you trap, ask local animal control if it can handle the disposal for you. If you check with the DNR, as we suggest earlier, before you consider trapping, you may get suggestions on what you can do with the animal. Know what is legal to do and decide what you *will* do before you set the trap.

You can purchase various kinds of live traps, and they're all set in different ways, so follow the directions that come with the trap. Practice setting and opening the trap to release an animal before actually using it. Make sure the trap you pick says it's suitable for the predator you want to catch.

For all four-legged predators of chickens, a can of cat food or tuna makes great bait. If you have cats, though, you may want to bait the trap with whole raw eggs, which don't interest cats. Place the trap outside your coop so you won't catch chickens, but keep it close to the coop (or close to the place you store feed, if coons or other animals are getting into that). It may take a few days for the predator to enter the trap, so be patient.

Handling an angry animal in a trap can be quite an experience. If you get bit, you may need rabies prevention shots. Cover traps with tarps or blankets if you do move animals in traps: it calms them and protects you. Wear heavy leather gloves to protect your hands. If you catch a skunk in your trap, it won't spray you unless it can get its tail up. If you speak softly, move slowly, and keep dogs away, most skunks will let you release them without spraying. If you shoot a skunk in a trap, it may release scent as it dies. You can sometimes hire licensed professionals to trap predators and remove them.

Dealing with the neighbor dogs

When a neighbor's dog kills your chickens, you may have mixed feelings about what to do. You may not want to offend your neighbor, or you may even like the dogs that did the killing and don't want them to be punished or destroyed. And sometimes you may not be absolutely sure whose dogs did the killing. But from years of hearing about and helping chicken owners with just this type of problem, we can give you some pretty good advice.

First, if your chickens are roaming off your property into neighbors' yards or along public roads, you are just as responsible for their deaths as the dog owner is, and you probably won't get any damage award. The dog's owner may be ticketed if the animal control officer sees it off the property or if it doesn't have a license, but that's about the extent of it.

Making a claim

What can you do if the neighbor's dogs kill your chickens? Well, you can take them to Judge Judy's courtroom (or another courtroom) to recover damages. You'll need pretty good proof of whose dogs did the damage, such as pictures of the dogs in action (not just pictures of dead birds) or eye witnesses to recover damages.

If you're sure, or pretty confident, that a dog killed your chickens, report the incident to your local animal control agency or to local law enforcement. If you know whose dogs did it, pass along the information. You can also talk to the owner of the dog(s), but make an official report. Here's why. Almost all dogs that kill chickens will do so again if they get the chance. The owner may promise you that it will never happen again, but in so many cases, it does happen again. Most places excuse a first offense by issuing a warning to the owner. If you don't report the incident the first time, you may have to suffer a third raid on your chickens before the owner is ticketed or required to give up the dog.

If you don't know the neighbor well or you suspect the neighbor might overreact if you accuse his dog of killing your birds, have an animal control or law enforcement officer go to the neighbor's house with you to handle the situation.

Most dog owners are sorry for what the dog did and will pay whatever damages you ask. If you have an unapologetic neighbor and you have good proof that his dog did it, the local law enforcement agency may request that he pay you. If that doesn't work, you can take your neighbor to small claims court.

Collecting damages

How do you determine what the chickens are worth? In this kind of case, a court won't grant you money for pain and suffering (unless the dog(s) also physically hurt you) or sentimental value. Most courts won't award damages for future egg production or sale of offspring, either. You can expect to be paid what a chicken of that breed, age, and quality normally sells for. Come prepared with ads from breeders, hatcheries, and so on to show what the birds sell for. If you sought veterinary care for any chickens, you can ask for reimbursement of those bills. If the dog did damage to property as well, such as a fence or coop, you can expect to get reimbursed for that, too.

If after the first dog attack on your chickens you didn't have good proof of whose dog did the crime, try to make your chickens dog-safe behind a tall, strong fence. Then monitor your chicken coop or yard with a security camera or a trail camera. A picture, as they say, is worth a thousand words.

If you live in a rural area your township or county may have protocol for dog damage to livestock. The county or township often pays you and then attempts to collect from the dog's owner, if the owner is known. Dog license fees may fund this program. In this case, you usually don't need as much proof to show whose dog did the deed. A designated person will examine the dead birds and determine whether it's dog damage. You may be paid a set fee per bird, or you may be allowed to tell the examiner what you feel the birds are worth. To find out if your area has such a program, call your local animal control or the county or township government.

Chapter 10

Keeping Your Flock Happy and Healthy

In This Chapter

- Providing biosecurity for your flock
- Keeping your birds calm and collected
- Grooming your chickens
- Managing temperature extremes

The art of handling and managing animals is called *animal husbandry*. Caring for animals goes beyond knowing how to house them and feed them — it also involves knowing how to manage them so that they're healthy and content.

In this chapter, we discuss how to keep your flock healthy with preventive care and how to humanely handle chickens. In the next chapter, we discuss what to do if your chickens do get sick.

Providing Biosecurity for Your Flock

If you ever have the chance to visit a large poultry operation or a poultry research station, you'll be amazed by the precautions it takes to prevent disease or parasites from entering the flock. You'll have to wear special coveralls, plastic booties, and maybe a face mask just to walk inside a facility — and maybe even to get on the grounds. *Biosecurity*, as defined by the Environmental Protection Agency, is the protection of agricultural animals from any type of infectious agent — viral, bacterial, fungal, or parasitic.

As the owner of a small flock or a pet chicken, you don't have to take all the extreme biosecurity measures large operations do, but if you want healthy chickens and you want to keep your family healthy, some biosecurity measures are necessary.

Maintaining biosecurity

Home flock owners often wonder how their chickens became ill, especially if they don't take them to shows and sales. Studies have shown that 90 percent of diseases are carried to chickens by human caretakers. Disease organisms can survive on shoes, clothing, unwashed hands, car tires, and equipment. Keep that info in mind when you're visiting a poultry show or your friend's flock.

Follow some basic biosecurity guidelines for your flock's safety:

- If you've been visiting other flocks, at the very least, wash your hands before caring for your own birds. To go the extra mile, also change your shoes and clothing.
- Be very careful about who you allow to handle your birds. Sometimes you may need to ask someone to help you with your birds, but do this selectively and infrequently. Home flock owners like to visit and handle the chickens, but sharing birds is an excellent way to spread disease. Share photos instead.
- If poultry diseases break out in your area, avoid going places where poultry reside. If you do come in contact with them, take a shower and put on clean clothes and shoes before you take care of your flock.
- If you lend poultry equipment, cages, carriers, and so on, or if you buy used equipment, be sure to thoroughly disinfect it before you use it on your own birds or even store it on your premises.
- If at all possible, maintain a closed flock — don't add new birds to existing flocks. (Chicks that come from a reputable hatchery and that you raise in a brooder are an exception.) If you do add new birds, quarantine them for at least 30 days far away from your flock. Any bird that leaves your premises and is around other birds needs to be quarantined if it returns to the flock. We talk more about quarantines later in this chapter.
- Try to minimize your flock's contact with wild birds. Wild birds can carry many diseases, including strains of avian flu, to your flock. Don't let wild birds nest in or on shelters, and keep them out of chicken feed dishes. Don't mingle wild orphans such as baby ducks with your flock.
- Eliminate mice, rats, and insect pests like cockroaches from your coop because they can spread diseases. Try to control flies and mosquitoes.

- If many birds die within a 48-hour time period or less and you don't see any signs of a predator attack, contact your state veterinarian. (To find your state veterinarian, contact your state Department of Agriculture.) Describe any symptoms and follow their directions.

Knowing when to quarantine chickens

Every time you add birds to your flock, you increase the risk of introducing disease and parasites. One of the best ways to protect your flock is to quarantine all new birds or birds that have been to shows or have been off the property for some reason. Small flock owners are the least likely to do this, but if you frequently show birds or you add new birds often, sooner or later, you'll have a problem if you don't use a quarantine system.

To quarantine, you need a large cage or other area away from your flock where you can keep all new birds (or birds that have been to shows) for 30 days. The quarantine area needs to be at least 30 feet from your other birds, but more distance is even better. Feed and care for the birds in quarantine after you have fed and cared for the regular flock. Watch the birds for signs of illness, and examine them for lice and other parasites. If all is well in 30 days, you can add them to the flock.

It's a good idea to have a dedicated quarantine space available regardless of whether you plan on showing your birds. Illness, new birds added to the flock, and various other reasons require a quarantine space. If you have one at the ready before an issue arises, you'll be that much more prepared.

You also want to put quarantine in effect whenever you suspect a bird is ill or whenever a chicken is injured. Quarantine prevents diseases from spreading and protects chickens who aren't feeling well from bullying and even cannibalism by flock mates. It also allows you to alter the environment, such as by adding heat, and allows you to monitor whether the bird is eating and drinking and see what its poop and/or eggs look like.

If more than one bird seems to be ill with the same symptoms, it's fine to quarantine them together. Feed and care for sick chickens after you've cared for healthy ones, and wash your hands after caring for sick birds. Since some chicken diseases can sicken people, too, don't eat, drink, or smoke when caring for sick chickens. It might be best to keep children and anyone with a compromised immune system away from sick chickens. It's not also a good idea to bring sick chickens into your home.

If sick chickens seem to recover, give them several days more in quarantine before you return them to the flock. When you return them, watch to make sure they aren't picked on. The absence of one or more birds may have changed the pecking order, and fighting may begin. Read the section "Introducing new birds carefully," later in this chapter, for tips. To your other chickens, a bird that has been absent for a while is a new bird.

Sometimes chickens recover from a serious disease but remain carriers. In the next chapter, we discuss which diseases can leave some birds as carriers. Carrier chickens infect any chickens that aren't carriers themselves and may keep disease outbreaks and deaths happening in the flock for years to come. You have the choice of keeping these carrier birds in quarantine forever or killing them. Birds that are carriers of serious disease are rarely good egg producers and often get sick again when stressed. We highly recommend that you humanely dispose of birds diagnosed or suspected of having a disease that makes them carriers. Never sell birds that recover from diseases but remain carriers. It's unethical and unkind to the new owner and his flock.

Keeping Disease and Parasites Away

Following some basic biosecurity measures as outlined earlier in the chapter helps keep your flock healthy. Vaccination also can prevent some chicken diseases. Other diseases result from parasites, which owners can take steps to control as well. In this section, we discuss these preventive measures.

Diseases can also spread to home flocks from wild birds and from pests like raccoons, opossums, insects, and rats. See Chapter 9 for info on controlling pests and predators. See Chapter 11 for more details on disease symptoms and treatments.

Giving vaccinations

Just as you have your pets vaccinated for rabies and other diseases, you need to give your chickens some vaccinations, too. Vaccines aren't available to prevent every disease that can affect chickens, so you still must practice good management techniques to stave off disease problems. But preventing diseases with vaccination when you can is a good idea.

Some diseases are more prevalent in one area than another. Ask a local veterinarian or a poultry expert from your county Extension office (go to `www.csrees.usda.gov/Extension/`) about which vaccinations chickens in your area need. Home flock owners may decide to do without vaccinating their chickens and live with the risk. Only you, as the owner, can make that decision. If you never show chickens and you don't frequently buy or sell chickens, you may never have a problem. If you intend to sell live chickens, breed them, or show them, however, you need to get the vaccines recommended for your area.

Some vaccinations are for chicks, and some work only for older birds. If you don't mind giving vaccinations, you can usually buy the vaccines from poultry supply places. Follow the label directions exactly for administering and storing the vaccines. If you'd rather not tackle the task yourself, contact a veterinarian or ask an experienced chicken-keeper to help.

Following is a list of common diseases that chickens can be vaccinated against:

- **Fowl pox** causes sores and respiratory problems, and some birds die as a result. In areas where fowl pox has been a problem, chickens need to be vaccinated. The vaccine is given at hatching and again at 8 weeks or at 10 to 12 weeks. The vaccine is usually administered with a special device that's dipped in the vaccine and then stuck into the wing web.
- **Infectious coryza, avian encephalomyelitis (AE), Newcastle disease, Mycoplasma gallisepticum, avian flu**, and **fowl cholera** are serious diseases that cause poultry death for which vaccines are available. However, the vaccines need to be given only when outbreaks arise in your area. Your vet or county Extension office will be able to guide you if an outbreak develops in your area. Sometimes the state agriculture department coordinates vaccinations when outbreaks occur.
- **Infectious laryngotracheitis** is a viral disease similar to pneumonia in humans. The death rate is high in infected chickens, and birds that recover are carriers for life and can infect healthy chickens. If you're showing chickens or you regularly buy and sell chickens, you need to vaccinate your flock for infectious laryngotracheitis once a year. Use only tissue culture origin (TCO) vaccines in home flocks. The vaccine is given as an eye drop, so it's easy to give.
- **Marek's disease** causes tumors and death in young birds. Almost all hatcheries offer to vaccinate chicks for Marek's disease for a small fee, and it's a good idea for small flock owners. We highly recommend that every chicken get this vaccination. If you hatch birds, you can vaccinate them yourself. The vaccination is given by needle just under the skin. It's only given once. After 16 weeks of age, the vaccine isn't needed.

Chicken vaccines often come in large, multidose vials. Although even a thousand-dose vial isn't too expensive for most chicken vaccines, be sure to store the excess correctly and use it before the expiration date. You may want to consider purchasing the vaccines with several other chicken owners. You'll then have to pay only a fraction of the cost, and you won't have to worry about long-term storage.

Putting up barriers against parasites

Worms and other internal parasites are common in chickens, and at some point, most chicken owners have to deal with them. In the next chapter, we go into more detail about diagnosing and treating both internal and external parasites. However, preventing parasites is always preferable to trying to treat them after the chickens get them, so in this chapter, we discuss prevention.

A parasite uses your chicken for food. Two basic types of parasites prey on them: internal and external. Internal parasites, which include worms, coccidia, and a few other tiny creatures, feed inside the chicken in various places. External parasites feed on the outside of your chickens. They may live right on the chicken or just feed on them and hop off to hide. The most common external parasites are lice and mites, but others exist.

Preventing internal parasites

If you allow your chickens access to the ground, you will almost certainly have some internal parasites at some point. But that's not a good reason to prevent your chickens from enjoying the outside. Most flocks that are well fed and healthy don't show any signs of a mild parasite infection or have any effects from one.

There are two types of internal parasites to be concerned about in chickens: worms and coccidia. Chickens get several types of worms, but three are the most common: large roundworms, tapeworms, and gape worms. Different than a worm, coccidia are small parasites that infect chickens' intestinal tracts and can cause death in young chicks. We discuss how worms and coccidia affect the health of chickens in more detail in Chapter 11.

Internal parasites are picked up from the feces of other chickens or wild birds and from the soil. They can also come from chickens eating earthworms and other bugs as they roam your yard. These larger critters may host one stage of the worm or other parasite, and then the parasite finishes its life cycle in your chicken.

To reduce worm populations, try to move pastured poultry or chickens in small tractor-type cages every few days to new ground that has been without chickens for several months. This rotation helps prevent worm eggs from building up in the soil.

It's also a good idea to worm all new birds during their quarantine if they're at least 18 weeks old or when they reach 18 weeks of age. For existing flocks, we don't recommend regular preventative worming. Give worm medications only when worms are diagnosed and seem to be causing health problems.

Giving all chicks a good start by treating for coccidia early is highly recommended. *Coccidiostats*, medications that kill coccidia, can be given as medicated starter feed or in the water and don't require a prescription. They include medications such as Amprolium (Amprol, Corid) and Decoquinate (Deccox), and you can find them in feed stores and poultry supply catalogs. As chickens get older, they may still carry the parasite, but most develop immunity to its effects and generally don't need treatment.

For more information on internal parasites and their treatment, see Chapter 11.

Keeping external parasites away

External parasites, including mites, lice, ticks, and other creepy crawlies, are also fairly common in chickens. It's important to keep both external pests and internal parasites from getting to your chickens. They make your chickens uncomfortable and also make people very unsettled if they know the chickens have them, even if most of these pests can't infect humans. (We get itching just thinking about them.)

External parasites get to your flock from new birds added to the flock, infested equipment and housing, wild birds, or rodent pests.

To prevent external parasites from entering your flock through new chickens, closely examine all new chickens you get. Part the feathers and look for crawling creatures. The bugs may actually crawl on you, making spotting them easy. Check around the vent (under the tail). If the feathers around the vent look like they have dirt deposits near the skin, the birds may have mites. A good practice is to treat all new chickens with a product to control lice and mites while they're in quarantine. See Chapter 11 for suggestions on those products.

If you purchase used housing, cages, carriers, and other equipment, they may be carrying unwanted guests. Some chicken mites, ticks, and bedbugs don't stay on the birds all the time, but hide in the cracks of housing and equipment. Examine and then thoroughly clean and disinfect all used housing and equipment. Some external parasites can remain in used housing or equipment for more than a year and still be ready to attack your birds. You'll want to treat used housing with pesticides before you add your chickens, and be sure to thoroughly disinfect used equipment as well. Use a pesticide approved for poultry, not a garden or household pesticide or one for pets. Read Chapter 11 for pesticide recommendations.

To prevent external parasites, also remove wild bird nests, both old and new, from chicken housing and try to restrict wild birds from eating and roosting in your coop. Put screens on chicken house windows and fix holes where they can enter. If you free-range your chickens, you can't stop them from occasionally coming into contact with wild birds, but don't encourage contact by feeding birds like wild turkeys and ducks close to chicken areas. Also keep your chickens out of wild song bird feeders.

To prevent external parasites from hitching a ride into your coop on rodents, you will need to control the rodent population in and around your coop. Read Chapter 9 to learn ways to prevent and control rodent pests.

Learning about chickens and human health

In addition to diseases that affect your chickens' health, some bird diseases regularly turn up in the news as being harmful to humans. Unfortunately, these news stories scare people into believing that keeping chickens is dangerous. We're here to assure you that your risk of catching any disease from chickens is minimal. In fact, you put yourself at greater risk of contracting a disease by going to work than by keeping chickens.

However, in rare cases, humans can get some diseases and parasites from contact with chickens. We want to mention them here so you can take the proper precautions to keep you and your family healthy.

What diseases can people get from chickens?

Campylobacteriosis, salmonellosis, and E. coli are the most common diseases people can get from coming into contact with chicken poop or eating undercooked eggs and poultry meat. In most cases, these diseases cause an intestinal flu–like illness and most people recover. But small children, the elderly, and people with compromised immune systems can get very sick and even die from these bacteria. In rare instances, humans can also get psittacosis, a bacterial disease, from chickens. The viruses, avian flu, avian tuberculosis, Newcastle disease, and ersipeloid, can occasionally infect humans. These diseases are no longer common in domestic birds and require close contact with poultry in order for people to get the disease.

People can also get histoplasmosis and farmer's lung, which are caused by irritation to the lungs from feather dander, dust, and contaminated soil, complicated by an allergic reaction of the body's immune system.

Can the same parasites that affect my chickens affect me?

Some parasites that live off chickens can move on to humans. External parasites people can get from chickens include the chicken mite *Dermanyssus gallinae*, chiggers, bedbugs, bat bugs, and ringworm. Lice from chickens may get on humans and even bite them, but they don't live on humans long. Bed bugs and bat bugs are similar, but they don't live on either people or chickens; they just come out at night and feed on them.

People don't get the same internal parasites as chickens, so that's one problem you don't need to worry about.

Sensible safety rules

Of course, you always want to follow safe, sensible methods of handling and caring for chickens. These methods include the following:

- **Don't keep chickens in the house, even with diapers.** Although chickens make excellent pets when kept outside, your chances of getting sick from your chickens increases greatly if you share your living space. Chicken feet carry bacteria like E. coli, salmonella and Campylobacteria and deposit them around the house. Avian Tuberculosis and avian flu are easier to acquire when there is close contact on a frequent basis with an infected chicken.
- **Wash your hands carefully after handling and caring for chickens.**
- **Keep chickens away from your face — no kissing or nuzzling.**
- **Wear disposable gloves when handling sick or dead chickens.** Also wear gloves when you come in contact with chicken poop.
- **Wear a face mask.** Keep from inhaling dust when cleaning coops.
- **Cook chicken meat and eggs thoroughly and store them properly.**

- **If you let your chickens roam freely through your yard and garden, thoroughly wash vegetables and fruits that may have become contaminated by chicken droppings.** Also clean patio and picnic tables and railings chickens have soiled before you use them.
- **Quarantine sick chickens.** See the section "Knowing when to quarantine chickens" earlier in this chapter for more info.

Supervise children when they're handling chickens, eggs, or poultry equipment, and make them wash their hands immediately and thoroughly — not just splash them with water — when they're done. Don't let toddlers teethe on the shelter door handle while you collect eggs or kiss chicks for a cute photo.

Controlling Environmental Conditions

Many more chickens die each year from poor environmental conditions than from disease. In the housing chapters, we talk about designing housing to protect chickens from environmental problems such as dampness, heat, and cold. Disease problems often begin when poor conditions stress the birds' immune systems, so good environmental conditions are important in maintaining the health of your flock.

Dealing with heat, cold, and dampness

The temperature and the amount of moisture in your chickens' environment have a significant impact on their wellness. Chickens need protection from extreme heat and cold, and you must keep their environment dry. The following sections address these concerns.

Handling the heat

When the temperature rises above 85 degrees Fahrenheit, especially when the humidity is high, it's time to check on the chickens. If your birds are subjected to heat stress for too long, they'll die. For meat birds, death can occur within a few hours in heat above 90 degrees, especially with high humidity. Losses can start occurring rapidly — within hours — if temperatures rise above 100 degrees Fahrenheit for any type of chickens.

Chickens that are too hot are inactive and breathe with their mouths open. If your birds appear heat stressed, work slowly and calmly with them. Any more stress from fright may cause death.

Free-ranging chickens survive heat much better than penned birds because they instinctively find the coolest places. Heavy birds like the broiler hybrids suffer greatly in the heat. They may not eat enough and thus won't gain weight as quickly as cooler birds.

You can improve the conditions of penned birds by shading them, putting a fan in the shelter, or putting a sprinkler on the shelter roof. The wet roof cools the coop when water evaporates from it. Chickens don't like to drink warm water, so move their water dishes to the coolest areas, change the water frequently, and make sure water is always available.

If you see chickens that appear very stressed from heat and you can easily pick them up, immerse them in cool — but not cold — water for a few minutes.

Fighting off the cold

Chickens actually handle cold better than heat, as long as their shelter is dry and out of the wind. But frostbite of the combs, wattles, and toes may occur when temperatures get down near 0 degrees Fahrenheit. Frostbite causes blackened areas that may eventually fall off. If you live in a cold-weather area, try to pick breeds of chickens that have small combs and wattles because those parts lose less body heat and don't get frostbitten as easily.

Having wide roosts is a good idea because they allow chickens to sit with their feet flat instead of curling their toes around the roost. When chickens sit with their feet flat, their feathers cover the toes, which makes the toes less susceptible to frostbite.

In addition to causing frostbite, cold temperatures can cause hens to stop laying, or if they continue to lay, cold weather may contribute to egg binding (see Chapter 11 for more info on egg binding and other health problems).

Chickens need more feed in cold weather to keep their bodies functioning properly. Water is also important. You must either provide a heated water dish or furnish water twice a day. If chickens don't drink, they won't eat as much, and if they don't eat as much, they're more susceptible to dying from the cold.

To make chickens more comfortable, you may decide to heat the coop. If you do, don't heat it to more than 40 degrees Fahrenheit. More than that usually causes problems with dampness. For more about coop heating, see Chapter 6.

Doing away with dampness

Environmental conditions that are damp or wet are big contributors to illness in chickens. Damp conditions can come from moisture produced by birds' respiration and droppings in poorly ventilated housing, or from rain or snow. The warmer the air becomes, the more moisture it holds. Heating a coop too much in cold weather often results in the formation of condensation, which causes wet, unhealthy conditions.

Wet areas favor the growth of molds and fungus and cause additional discomfort in hot weather. Molds on feed and bedding cause a wide variety of illnesses, and they aren't good for the human caretaker, either. High humidity makes ideal conditions for many respiratory diseases. Always keep chicken bedding dry, and never give your birds moldy feed. Make sure housing is well ventilated. Keep outside runs well drained to prevent chickens from tracking in a lot of moisture on their feet.

Keeping your chickens from eating poisons

As with all livestock, the best approach with chickens is to anticipate problems before they occur and prevent them from happening. Chickens aren't particularly discriminating about what they eat, and although you wouldn't purposely feed your chickens poisonous substances, you may not realize what chickens view as *food*. Following is a list of some commonly overlooked dangers:

- **Garden seeds that are treated with fungicides or other pesticides:** If you use these types of seeds, make sure your chickens can't scratch them up and eat them or get to them in storage.
- **Granular fertilizers:** If a chicken happens upon some widely scattered bits of fertilizer, it may sample one or two and then move on. But if chickens see you scattering fertilizer, they may come running and, in the spirit of competition, gobble up quite a bit before either they quit or you chase them off.

  Keep all fertilizers and pesticides in places where chickens can't get to them and where children can't find them and feed them to your chickens. Lock up the chickens before you spread fertilizer, or make sure they aren't nearby to watch you do it.
- **Paint:** Watch out if you're scraping paint off old houses or buildings and the flakes fall into the soil. If the paint is lead based and the chickens eat the pieces — either purposefully or when eating food — you may have a problem. Lead poisoning is generally slow, and sometimes it's hard to connect the illness with the lead paint chips.

  Speaking of paint, when the time comes to apply a new coat, don't go away and leave that can of paint open next to the chicken shelter.
- **Pesticides:** Keep chickens out of any areas to which you've applied pesticides to kill weeds or bugs. Just getting some pesticides on their feet may be enough to harm your birds. Snail and slug baits are often in pellet form, and they're extremely toxic.
- **Rodent poisons:** Mouse and rat poisons are deadly to chickens if they ingest them. Chickens shouldn't eat mice and rats that have been poisoned, either. Be very careful in placing mouse and rat baits. Chickens are curious birds, and they may reach their heads under things and in holes to sample what they see.

Free-ranging chickens generally have a good sense of what's poisonous to eat as far as berries and vegetation go. Penned-up chickens are more likely to eat "food" that's not safe because they're hungrier for variety and fresh food.

Safely Handling Your Flock

Knowing how to correctly handle chickens is important because it reduces stress on your flock and on you. Handling involves catching chickens, carrying and holding chickens, and, to some extent, taming them. To a new chicken-keeper, handling chickens can seem frustrating or even frightening at times, but don't worry — you'll improve with experience.

Although you need to know how to handle your chickens, you also need to know when *not* to handle them. If they're suffering from the heat, don't handle them at all, if possible. The extra stress may kill them. Immediately after a traumatic event such as a predator attack, try to give the birds 30 minutes or so to settle down before you examine them, move them, or otherwise disturb them. Unless a bird is bleeding badly, you can put off first aid just a bit. This quiet period may help prevent shock, and it lets you and the birds assess the situation.

Don't handle chickens unnecessarily. We cover this more when we cover taming chickens. Chickens are prey animals, so every time they're caught and restrained, they experience some stress. Chickens under stress get sick more often and don't lay as well as birds without a lot of stress. Handle your chickens only when you have to examine, treat, or move them.

Catching chickens

To care for your chickens, sometimes you have to catch and hold them, but how do you catch a bird in your pen at home? If you can, wait until night and then walk in and take it off the roost. If you normally leave a night light on in the shelter, mark in your mind where the bird is, turn off the light for 10 minutes, and then go in with a flashlight to get it. Try not to disturb the other birds, especially if you need to catch more than one.

If you need to catch a bird and you don't have time to wait until night, lure it into the smallest area possible (usually the shelter) and block off any exits. Then go in calmly and slowly, and try to grab for the legs. Birds expect to be grabbed from overhead, so going beneath them to get the legs surprises them. Don't grab birds by the neck unless you're going to be butchering them anyway.

You can also use a net or a catching stick, but in our experience, nets spook birds and don't work much better than hands in close quarters. Nets are better for open areas. A catching stick has a hook at one end that you push beyond the bird's feet and then slide back, hopefully snagging the feet and pulling the bird toward you. You may need a little practice to figure out how to use one.

To catch a chick that's in a brooder, try putting one hand in front of the chick and using the other hand to sweep the chick into it. Don't pick up a little chick by its legs; scoop up the whole body. Putting your hand down over the back scares chicks because the motion resembles a predator swooping down, but sometimes doing so is the only way to catch them.

When a chicken actually flies the coop, new chicken owners often panic. Although chickens may take every opportunity to escape, they usually don't go far unless they're scared and in a strange place. If a chicken gets out of a pen at your home, don't be alarmed. Unless it's chased, it will generally hang around close to any remaining penned birds.

Even if the whole flock gets out, there's no reason to panic. First, go look at the pen and see how they got out. Fix the problem before you try to get them back inside.

Chasing chickens is your last resort — they're faster than you, so try other options first. If you have a special bucket you use to feed them or bring them treats, get it out, add some feed or treats, and show it to them. If they seem interested, lure them back to their pen, throw the treats inside an open door, and shut it when they're inside. If they won't go in the pen, you may be able to trick them into a garage, shed, or fenced yard that you can close up after them.

Most chickens want to go back to the familiar place where they roost at night when it starts getting dark. After the other chickens have settled down on the roosts, open the door or gate, and they'll often pop right in. If you can't catch a chicken by nightfall, watch from a distance to see where it goes to roost for the night. When it's completely dark, you can take a flashlight and simply go pick it up — if you can reach it.

What happens if the chicken escapes in an unfamiliar place, such as at a show? If you can, let it settle down a bit. Don't chase it out of the area. People who travel with chickens to shows or sales need to have a long-handled net or a catching stick with them.

Before you try to use a net or a catching stick, however, throw down some choice grain like corn to divert the chicken's attention. Or wait until it's attracted to someone else's caged birds, is crowing, or is otherwise engaged before you attempt to catch it. If conditions are crowded and catching the bird seems impossible, try coming back after the crowd has left.

If you have a really hard-to-catch chicken running loose, the old trap with a box and string or a live trap may work. Fortunately, chickens seem to get easier to catch the older they get, especially if their caretakers are kind to them.

Carrying and holding chickens

Carrying birds by the legs with the head hanging down is okay for birds you're taking to slaughter, but don't do it with your prize layers or pet birds. You can dislocate the legs or otherwise hurt them. Instead, tuck the bird under one arm with your other hand holding both feet, or cradle the bird in your arms with the wings under an arm. A firm squeeze and soothing talk soon calms most birds. The head can face forward or back, whichever works best for you.

If you need to restrain a bird for treatment and you don't have anyone to help you, you may need to tie the bird's legs together and lay it on a table while restraining the wings with one hand or loosely wrapping the bird in a towel to restrain the wings. Some people learn to hold the chicken with their knees while seated, with the feet up and the head facing away from them.

Don't squeeze birds too hard when carrying them or restraining them for treatment. A chicken needs to move its ribs to breathe well, and even if the mouth and nose are uncovered, you can suffocate a bird if you hold it too tightly. Suffocation often happens when children hold baby chicks too tightly. A chick needs to be held loosely in your closed hand, with its head peeking through your fingers.

If a bird is really wild and fighting, you can cover its head loosely with a hood or a piece of soft cloth, and it will probably settle right down. Don't try to carry too many birds at a time — take your time to do it right and humanely. Beware of roosters with long spurs on their legs. When they're struggling, the spurs can scratch your arms quite badly. Wear long sleeves to handle these birds.

Taming chickens

Chickens have become popular as backyard pets as well as food producers. But as a pet, chickens do have some drawbacks. With a few exceptions, chickens never grow to like cuddling or being held. As mentioned earlier, chickens are prey animals, and anytime something bigger than them catches and holds them, it's stressful. If the kids want a pet they can dress up or sleep with, chickens aren't a good choice.

Many chickens can learn to eat out of your hand or even jump on your knee if you're sitting quietly. You can even train chickens to do tricks by using treats. They may let you gently stroke them occasionally. But they don't like being grabbed or touched, for the most part. They may follow a kind keeper around the yard or garden and provide hours of enjoyment as you watch their antics. But chicken chasing isn't not fun for them — it's sheer terror.

Some breeds are calmer than others, and individual chickens have different personalities. The tamest breeds are generally production-type birds that lay brown eggs: Isa Browns, Rhode Island Reds, Wellsummers, and so on. These birds were selected for calm behavior over the years because calm birds are better layers. The brown egg color also seems to be linked genetically to calmer, quieter dispositions. Polish, Silkies, Brahmas, and Cochins are also considered to be calm breeds. Some strains of Ameraucana and Easter Eggers are quite tame; others are not. Chickens of other breeds also may be calm and prone to taming as well.

Some people think they should handle baby chicks frequently to make the chickens more tame when they're older, but this generally doesn't work well. Chicks that are handled frequently often don't grow as well as ones that aren't handled and are more likely to get sick. And in most cases, they won't be any tamer when they're out of the brooder and have more room to escape from you than ones that you didn't handle.

Routines do much for taming chickens, even older ones. Feed and water at the same times each day. People who are calm and quiet, who don't rush around, and who seem predictable feel safe to the chickens, so they respond more calmly. Providing small amounts of treats also makes chickens happy to see you. Chickens do recognize the person who feeds and cares for them, and they also remember people who scare or harm them.

We've found that chickens mellow as they age, as long as they've been treated kindly. Be patient. Young hens are more distrustful and "flighty," but experience will teach them that you can be trusted — or that they can easily outrun or fool you. As they age, they'll become much less frightened of you — and maybe too friendly, in some cases.

The trick to taming chickens is to provide a safe and predictable environment, give them a little enticement with yummy treats, and allow them to come to you instead of forcing yourself on them. Some hens will never be really tame, and that's their personality — just respect it.

Just part of the family, er, flock

Kim says: "I don't really attempt to tame my hens. I don't handle chicks in the brooder any more than absolutely needed. But I have hens right under my face as I work in the garden, hens that sit at my feet when I sit on the deck, and hens that try to land on my head when I show up with treats. They eat out of my hand — or rather, grab food out of my hands — like the bag of onion sets I was carrying to the garden. They often treat me as part of the flock. I think I'm even in the pecking order!"

Diffusing Stress

Stress affects all animals. Stress lowers the immune system response to disease and causes many undesirable behaviors. In this section, we discuss factors that may cause stress to your chickens and some ways to manage chickens to reduce that stress.

Managing the molt

Molting is the gradual replacement of feathers over a 4- to 8-week period. It's a natural process, not a disease. Chickens generally act normally during molt. If they seem sick or don't eat or drink well, something else is wrong. Molt is also a time for the hen's reproductive system to rest. It usually happens in the fall as the days get shorter, but lack of food or normal lighting can also cause it. Even when hens are kept under artificial lights, molting eventually happens, usually after about a year of laying. Both hens and roosters molt.

Chickens begin to lose the large primary wing feathers and the head feathers first. From there, the process spreads backward gradually. Chickens may look a little scruffy during this time, but they shouldn't look bald. You may notice a lot of tightly rolled *pin feathers*, which look like quills sticking out of the chicken, and a lot of feathers on the floor. The pin feathers gradually open into new, shiny, clean feathers.

Some breeds, particularly high-producing egg strains, have a quick, barely noticeable molt because they're selected to get through molt quickly. Individual variability also comes up in the length of molt. Hens that complete molt quickly are generally the most productive and healthiest hens in the flock.

The process of molting and growing new feathers is energy intensive. Chickens with a proper diet and few parasites have little trouble with molt. However, molting is a time when you need to make sure your birds' diet includes good-quality protein. Some people switch from layer ration to meat bird ration for a few weeks during molt because it has a higher protein ratio. You don't need to add medications, vitamins, or anything else if you're feeding a good commercial diet.

Because the immune system may be less active during molt, it's best to avoid bringing in new birds or taking birds to shows during this time. Hens generally quit laying during molt, although some lay sporadically. Both of these situations are normal. The hen's system needs to produce new feathers, not eggs. It's an egg vacation. The first eggs she lays when returning to production may be smaller than normal, but she likely will quickly return to laying normal-size eggs.

Introducing new birds carefully

You want to limit new introductions to your chicken flock for two good reasons. First, you risk introducing disease to your existing flock (see the earlier section "Knowing when to quarantine your chickens" for advice on how to avoid that calamity). Second, you introduce stress because all chicken flocks have a ranking system, or *pecking order.* (We discuss basic chicken behavior in more detail in Chapter 2.) Every time you add or remove a bird, the order changes, causing fighting and disorder in the flock.

Unless a chicken has been alone and is pining for friends, it generally doesn't like new birds and may viciously attack newcomers. Before a new ranking order is established and the new birds become part of the flock, you may witness bloodshed or even death. However, you can keep order with some strategies.

First, never introduce a new rooster into a flock with an established rooster. One or both of the birds may die in this case. If you want to breed a new rooster with your hens, you need to either divide the housing, runs, and ladies or remove the old rooster. A very young rooster that grows up with the flock sometimes is tolerated, but don't add a *cockerel* (young rooster) older than 6 weeks of age.

A rooster introduced to a flock of all hens typically has few problems. Sometimes a hen or two will challenge him, but he will quickly become lord and master.

Adding some new hens to old hens is the most frequent type of introduction in small flocks. Don't just toss the new birds in and hope for the best, though.

If you've ever watched females of any species fight, you know how vicious they can get. Try to introduce more than one bird at a time, to divide up the bullying a bit.

If you can move all the birds to new quarters, old and new, less fighting will ensue. Another way to introduce new birds is to put them in a cage or enclosure next to the old flock members for a few days. The old birds can let them know who's boss without actually harming them.

Usually, young hens allow older birds to dominate them if the older ones are active and healthy. But sometimes the tables get turned and an older bird gets the worst of it. Different breeds are sometimes more assertive or aggressive. Silkies, Polish, and other breeds with topknots or crests, along with Cochins and some of the smaller bantam breeds, may be bullied by younger birds of large active breeds.

If you can't pen newcomers nearby for a few days, put them in the pen at night, after the regular flock has gone to bed. Then keep a close eye on the flock in the morning. Or release new birds into the shelter area of a coop after the old birds are let out to do a little free roaming.

Expect some fighting, and don't interfere unless a bird is injured and bleeding. The flock is establishing a new order, and when they all know their places, the fighting will cease. Remove any bleeding birds because they may be quickly pecked to death. Often a rooster will interfere in the fighting. Roosters are attracted to the new girl or girls and sometimes protect them. Also, roosters don't like too much strife in their households and may punish the offending birds.

Keep an eye on newcomers for a week or so and make sure they're getting to the food and water. If they stay huddled in a corner, you may have to remove them. Never spray the birds with water or concoctions of scented products to try to confuse them about who is new. It doesn't work, and it just brings more stress to the flock.

Discouraging bullying behaviors

By nature, chickens pick on weak members of the flock. If they draw blood, chickens keep picking at the wound, often until they kill the injured bird. A dead bird may be pecked at and eaten if you leave it in the pen. This act doesn't mean chickens are vicious; it's just how nature designed them — to be opportunistic feeders. Immediately separate any dead, injured, or ill chicken from the others. Maintaining crowded conditions and frequently moving birds in and out of the flock causes more fighting, which increases the chance for wounds.

We talk about the ranking system that all chicken flocks establish in the preceding section and in more detail in Chapter 2. Chickens that are at the bottom of the ranking system are more likely to be picked on. Whenever you add or remove birds in a flock, the chickens have to establish a new ranking system, and some birds may be injured.

Chickens are conscious of colors and patterns, and they often pick on a bird that has different coloring, color patterns, or feathering than the majority breed, especially in brooder housing. Chickens with topknots are frequently picked on by other types of chickens. Watch these birds carefully and remove them if they're bullied.

Cannibalism and feather picking are signs of poor management. In a home flock, you likely won't have many problems with these issues, as long as you feed the chickens a proper diet and maintain housing that isn't crowded. In nature, chickens spread out and spend most of the day searching for food and keeping busy. Plus, they have enough space to avoid higher-ranked flock members. Home flocks that get at least some free-range time each day rarely experience problems.

Confined chickens need to have food available at all times and may benefit from some busywork, such as pecking at a head of cabbage, a squash, or a pumpkin, or looking for scratch grain scattered in the bedding. Raw vegetables, fruit, and whole grains provide roughage. Research has shown that chickens on diets without animal protein and/or low roughage are more prone to feather picking, which often leads to wounds and cannibalism. Properly balanced commercial diets supplement vegetable protein with the needed amino acids found in meat-based diets.

Very bright lights in brooders, stress caused by predators, too much handling, excessive noise, and other conditions sometimes cause feather picking. Changing brooder bulbs to infrared bulbs or bulbs with lower wattages may help. Changing conditions to reduce stress is always desirable. Sometimes only one or two chickens seem overly aggressive, and you may need to sell or destroy them.

Employing Optional Grooming Procedures

Marking your birds so that you can identify them has some benefits. If you're going to show birds, identify them in a permanent manner. In a flock of similarly colored birds, you may want to identify birds for medication, for planned mating, or for other reasons.

In addition to marking chickens for identification, other practices are designed either for the welfare of the chicken or for showing purposes.

Dubbing, for example, is the old practice of removing a chicken's healthy comb, usually in a rooster. The *wattles*, the fleshy lobes under the beak, are also removed. Dubbing was traditionally done with fighting breeds to keep them from being torn during a fight, and it remains a practice for showing Old English and Modern Game breeds in the U.S. Other countries have banned this practice. Cockfighting is illegal. If you're not showing birds, you have no need for this painful practice.

Marking birds for easy identification

You can purchase leg and wing bands from most poultry supply catalogs and at many poultry shows. Some poultry clubs and organizations also sell bands for their members. The bands may be made of metal or plastic, and they may be flat or round in shape. They come in different sizes to accommodate different breeds of chickens.

Bands are either temporary or permanent:

- **Permanent bands** are numbered bands that are slipped on the legs of chicks and remain on them for life. Metal or plastic numbered bands can also be clamped through the wing web for permanent marking. Some supply houses let you choose a combination of letters and/or numbers for bands, especially if you order a lot of them, but most supply houses sell prenumbered bands. Permanent bands are usually required at shows, except for meat birds.
- **Temporary bands** may be either numbered or colored. A temporary band usually consists of a coil that's spread apart and clasped on the leg for temporary identification purposes. Plastic write-on bands, similar to hospital bands, are also available. They can be slipped over the top of the chicken's wings at the shoulder. Temporary bands are useful for keeping track of certain birds for mating, for compiling egg production records, and for medicating.

Valuable chickens can also be microchipped like other pets. To microchip your birds, you must purchase the chips, something to insert the chips (usually a large syringe and needles), and a service to serve as a repository for the information contained on the microchip. A chip is about the size of a grain of rice. Microchips need to be read with a machine, and the biggest drawback with their use in chickens is that few people think to use a scanning machine to find an owner or prove ownership.

Trimming long, curled nails

Chickens kept in cages or on soft flooring all the time may develop long, curled nails that can get caught in flooring or make walking difficult. Be sure to trim these long nails. Any pet nail-trimming device can do the job. You'll also need someone to hold the bird while you trim the nails.

Don't trim too much off at one time because a vein that runs about ¾ of the way up the nail will bleed profusely if you cut it. It grows as the nail grows, so taking off only about ¼ of the nail at one time is all that's safe. If you look at the nail in good light, you can probably see the small red vein running through it. Cut just before this vein.

If you nick the vein, put pressure on the cut end with a cloth or paper towel to stop the bleeding. Styptic powders, sold in pet and livestock catalogs and stores, are designed to stop bleeding. Just wipe off the blood and quickly apply the powder. In a pinch, a bit of flour may stop the bleeding, or you can press the bleeding nail into a potato.

Make sure the bleeding has stopped before you return the bird to its pen. Other birds will peck at the bloody area.

Trimming wings and other feathers

Wing trimming is sometimes done to keep chickens from flying over the pen walls, although some lightweight birds may get enough lift to escape even with their wings trimmed. Trim only the large flight feathers, and understand that you'll need to do it again when the feathers grow back after the next molt. Sometimes trimming the feathers on just one wing is enough. Keep in mind that chickens can't be shown if they have trimmed wings.

To trim the wing feathers, have someone restrain the chicken for you. Pull the wing out away from the body and, using sharp, strong scissors, cut across the middle of all the long flight feathers (see Figure 10-1). If you hit an immature feather and it begins to bleed, grasp the feather close to where it joins the wing and pull it out. Removing the feather should stop the bleeding.

Pulling out feathers is somewhat painful for the chicken, so cutting them is better than pulling them out. Another advantage to cutting versus pulling is that feathers that are pulled out often regrow before the molt, whereas cut feathers don't regrow until they're shed during molt.

With breeds that have topknots, it's often helpful to trim the topknot feathers over the eyes so the chickens can see. When they can see better, they're

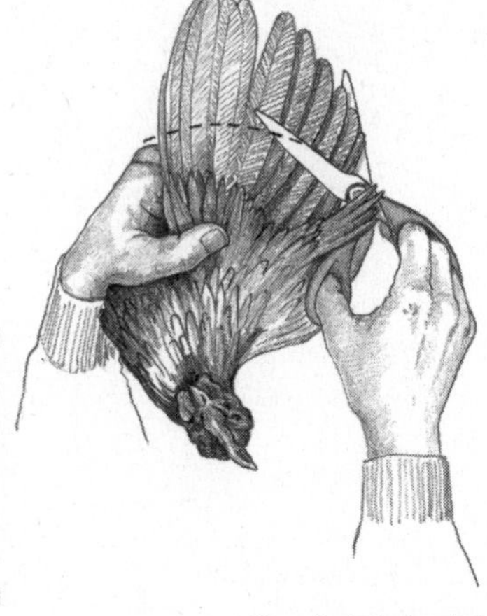

Figure 10-1: How to clip the wing feathers of a chicken.

Illustration by Barbara Frake

more active, and other breeds are less likely to pick on them. However, don't trim the topknot feathers with show birds, or they'll be disqualified.

In some heavily feathered breeds, feathers around the rear end may become matted. Even if they're not matted, they can interfere with mating. You can carefully trim these feathers.

Cutting feathers makes the trimming job last longer than plucking out the feathers. If a cut feather bleeds, however, you must pluck it out.

Chapter 11

Handling Health Problems

In This Chapter

- Deciding about treatment
- Handling injuries and first aid
- Dealing with parasites
- Treating common diseases
- Informing authorities on diseases and deaths

Many people worry about what to do in the case of illness or disease in the home chicken flock. We encourage you not to let your imagination run wild or become discouraged thinking about what terrible things may happen to your flock. The chances are pretty good that you'll never see any of the diseases mentioned. Injuries and parasites are much more common in home flocks, though.

Chickens are pretty hardy animals, but when they do get sick, they often die before you can do anything about it anyway. Often the first sign that something is wrong is a dead bird. Birds are adept at hiding illness. This behavior helps the species survive: Weak birds are easier prey for predators. And members of their own flock often pick on sick birds, combining illness with injury. It's helpful to be a close observer of your flock so you can catch illness and injury early.

Sometimes chickens die naturally without being diseased or attacked by predators. Old birds may have just lived out their life span. Because they're so large and heavy, broiler-type birds often die of sudden death syndrome, which is a combination of heart failure and other stressors. Genetic faults sometimes cause death as well.

In this chapter, we give you information on what to do if you encounter injuries, parasites, disease, or even death in your flock. We also explain how to administer medication and what to do when a bird dies. This book cannot go into great detail on poultry illnesses and injuries because of space limitations; however, we recommend *Chicken Health For Dummies*, by Julie Gauthier (Wiley), for more detailed reading on the subject.

Making Decisions about Treatment

When faced with sick or injured birds, you have to decide whether treating them is worth the time and money. Many small flock owners develop a sentimental attachment to their chickens and want to try to save them at all costs, and that's fine. Others know that they have limits to their time and resources, and that's understandable, too. Every situation is different, of course, but you need to think about what you're willing to do in case of injury or illness in your flock even before it happens.

In all cases, you need to consider what's best for you and for the rest of your flock. The next sections explore whether it's better to destroy one or two ill birds than risk the health of the remaining ones. They also explore your options if the disease is a treatable one and you choose to keep and treat the ill bird. We also provide some information on finding a vet to help with treatments and guide you in your decisions.

Choosing to treat ill chickens

When chickens are pets, the decision to treat a diseased chicken may seem easy to make: do whatever possible to keep the chicken alive. One option is to choose to treat the chicken and then keep it quarantined. This way, you can still keep the bird without it posing a risk to the rest of the flock.

The problem with many chicken diseases is that no effective treatment exists (or no product is registered to treat chickens). And in some cases, if the chickens are treated and recover, they will be carriers of that disease for life and will infect any new chickens they contact. These carrier birds often don't lay well and are never very healthy birds.

The choice to get rid of a chicken, especially one you consider a pet, is difficult, but often it is the best for the future of the flock.

Deciding to eliminate chickens

Humane euthanasia is preferable to letting a chicken suffer through an injury or illness that you can't afford to treat or that no treatment can address. For ways to humanely dispose of a chicken, see Chapter 16. If you can't do so, maybe a chicken-owning friend or a vet can do it for you.

In some cases, you may feel it's best to dispose of all the birds and then thoroughly clean the premises or wait the required time until it's safe to replace the flock. If you have sick birds in your flock but none has died, you may need to kill a living but ill bird to find out what's ailing the flock. Then you can submit it for a necropsy (the animal equivalent of an autopsy). Your vet (see the next section on finding a vet) or a university poultry specialist can guide you in all the decisions on cleaning, disposing of dead birds, and submitting birds for testing.

Sometimes the decision to treat or exterminate a chicken is taken out of your hands. If your birds are diagnosed with a serious disease, a health regulatory agency may order them destroyed. Or if disease is sweeping the area, your state vet may send people to inspect all chicken flocks. If that's the case, they'll inform you of what to do, how to clean the premises, and how to dispose of the chickens. In many cases, you will be compensated for the loss of your birds if you're ordered to destroy them, but you won't have a choice of whether to treat them.

Finding a vet to treat chickens

In a chapter that talks about chicken illness and injuries, we need to address a problem chicken owners often have: finding a vet who's willing to treat chickens. Finding a general practitioner who treats chickens is easier than finding a vet who specializes in birds or poultry. But, even suburban vets are seeing more pet chickens than they used to, thanks to the popularity of chickens as pets.

Vets who specialize in poultry usually go to work for a big commercial chicken business and won't treat the birds of small flock owners because of fears that they may transmit a disease to the commercial operation. A vet who is a general practitioner can give you prescriptions for poultry medications and handle minor trauma cases, and that's all most small flock owners need.

Start looking for a vet by asking chicken-owning friends if they can recommend a vet who treats chickens. You can also call local vets and ask. Do this before you actually need help, to save valuable time. Your county Extension office may be able to help. To find yours try this website: `http://www.csrees.usda.gov/Extension/`. A college that has a veterinary medical school also often runs a clinic that may accept poultry patients.

A vet who treats chickens usually charges you about the same amount she does to treat any other pet. Also just as for any other pet, the vet probably won't give you extensive advice over the phone without examining the chicken, nor will the office give you a prescription unless the chicken is a patient.

Treating Injuries

A chicken can be wounded in many ways. The most common injuries result from being bitten by predators or pecked by other chickens. Just like kids, chickens can get themselves into some interesting predicaments. If the grass is greener on the other side of the fence, they may get their heads stuck in the fence trying to reach it, injuring themselves in the process. Unfortunately, chickens, like many birds, may die from shock and stress when wounded instead of dying from the wound itself.

In the following sections, we tell you how you can identify and tend to the most common types of chicken injuries that home-keepers face — as well as how to keep your injured birds from going into shock.

How to give your bird the once-over

If a chicken has been wounded or it was in a situation that may have inflicted wounds, you need to check it over carefully. Feathers can hide some wounds that don't bleed much, including deep puncture wounds.

If the chicken is still up and walking, try to catch it without too much more stress. Especially try not to chase it. Sometimes chickens run to their shelters when scared, so try to shoo the injured bird into a safe area where you can close it in. Instead of chasing it, which can hasten bleeding of wounds, let the chicken settle down first. Then catch it as calmly as possible.

When you have the bird in hand, examine it carefully, handling it gently. Part the feathers, remove loose clumps of feathers, and sponge off bloody areas so you know the depth and extent of the wounds. Use gloves when handling injured birds, in case they have a disease.

If injuries are extensive and dirty, as from a predator attack, it's probably kinder to euthanize the chicken than try to cure it. The easiest and most humane method of killing a chicken is to dislocate the neck, which Chapter 16 tells you how to do. If you do euthanize the bird, be sure to bury the carcass at least 3 feet deep in the ground; see Chapter 16 for more info on discarding dead birds and waste.

Ways to keep an injured bird safe

If you decide that you're going to try to treat the bird, begin treatment quickly or proceed to a veterinarian. Do everything you can to minimize shock when transporting a bird to a vet or when treating at home. To soothe an injured bird and keep it from going into shock at home, create a comfortable isolation environment:

- **Put the bird in a safe, darkened place to calm it after treatment.** Injured chickens are subject to abuse by their flock mates. Bleeding wounds are attractive to other chickens, so be sure to isolate a bleeding bird to prevent further injury.
- **Unless the weather is hot, provide an overhead source of heat, such as a heat lamp.** Chickens need room to move away from the heat source if they get too warm. If they can't move well, check frequently to make sure they aren't too warm. If it's more than 90 degrees Fahrenheit you don't need additional heat. If the chicken is panting try to move it somewhere a little cooler.
- **Offer water at once.** You can dip the beak of a chicken into shallow water, but don't do any more than that to get the bird to drink.
- **Don't feed the bird for a few hours after injury.** Don't try to force food down the bird.

Skin injuries, cuts, and puncture wounds

Chicken skin is thin and tears easily. Scratches and small cuts from being caught on something, fighting, or being attacked by a predator may occur. Broiler-type birds may develop blisters on their breast area.

If the wound is shallow, wash the area with hot water and soap, and gently pat dry or clean the wound with hydrogen peroxide. If the wound continues to bleed, use *styptic*, or blood-stopping, powder or pressure to stop it. It's a good idea to keep some styptic powder around the homestead. This inexpensive powder is readily found in pet stores, farm stores, and animal supply catalogs. Bandages rarely work on chickens because of their feathers and the fact that the birds pull them off. If the cut is on the bottom of the foot, keep the bird on a clean surface for a few days.

Animal bites are very dangerous to chickens. Even if they don't look deep or dangerous, there's a good chance that infection will set in. If the puncture went through the flesh and into the body or brain cavity, the bird needs to be destroyed.

Thoroughly clean around puncture wounds or deep wounds with warm water and soap. Use hydrogen peroxide, iodine, or betadine to flush the wounds. Also try to get any dirt or other foreign matter out of the wound. Place the chicken in a clean area, and check the wounds for infection several times a day. You'll need to clean the wounds twice a day for several days.

The outlook isn't good with deep punctures caused by animals. Keep the chicken warm and quiet to prevent shock. If the chicken is very valuable to you, take it to a veterinarian as soon as possible.

Foot sores (also known as bumblefoot)

Bumblefoot is a bacterial infection or abscess of the foot. It generally occurs in heavy roosters, but it can affect other birds as well. It's caused by a cut or even a small scrape to the bird's foot that gets contaminated by bacteria, usually staph. Rough perches and wire cage floors are common causes of these cuts and scratches. Large, heavy birds that jump down from high perches also can injure their feet.

Bumblefoot causes a large swelling on the bottom of the foot or on a toe. It may feel soft in early stages and hard later. It looks red and inflamed and may feel hot to the touch. A black scab usually forms over the sore. The bird may limp and refuse to do much walking. If it isn't treated, bumblefoot can cause death.

Use gloves to examine or treat birds you suspect of having bumblefoot because the staph or other bacteria that cause the abscess can infect humans. Put the bird in a cage with clean, soft litter such as pine shavings. Isolate the bird from other chickens — the bacteria can infect them, too.

In early stages, you may simply need to administer antibiotics. Several registered antibiotics are available for chickens: Lincomycin and amoxicillin are two common ones. You need a veterinary prescription to use antibiotics on chickens. Read and follow the label directions to determine the correct dosage and find out how to administer the antibiotic to birds. You must give the antibiotics for the full time the label directs.

Soaking the foot also helps, especially if the injury has progressed to the hard stage. Put a cup of Epsom salts in a dish pan of hot water — water that feels hot but doesn't burn your hand. Then hold the chicken's foot in the pan until the water cools, about 10 to 15 minutes. Don't let the bird drink any of the water.

The soaking should soften the abscess. Gently remove the scab and try to open the wound by pulling it apart at the wound edges instead of squeezing it. Rinse the wound with hydrogen peroxide and try to gently clean out any pus.

Then apply an antibiotic ointment that's safe for birds (ask a vet for a recommendation). Pad the wound with a clean gauze pad and wrap it with first-aid tape or *vet wrap*, a bandage that sticks to itself. Clean, flush, and rewrap the wound once a day until it looks like it's healing. All dressings and soaking fluids will be loaded with bacteria, so dispose of them carefully.

If soaking and pulling apart the wound edges doesn't open the wound so it can drain, you can bring the bird to the vet, who can cut open the abscess. Watch birds being treated with antibiotics for diarrhea, which is caused by good bacteria that's also being destroyed; you can add some "digestive health yogurt" to the chicken's diet to help restore it.

Head injuries

The most common areas for head injuries are the combs, wattles, eyes, and beak. The following sections explain each in detail.

Torn or infected combs and wattles

Sometimes a bird tears its comb or wattle. These parts can get caught on something or get damaged in a fight. Generally, torn combs and wattles can't be repaired and may become infected. In such cases, you need to trim them off. You can either do it yourself or take the bird to a vet.

To do the trimming yourself, follow these steps:

1. **Wash the area with hot water and soap.**
2. **Clean the area with rubbing alcohol.**
3. **With sterilized, sharp scissors or knife, trim off the torn area of the comb or wattle.**
4. **Apply an antibiotic cream to the edges.**

   Ask a vet or poultry expert to suggest a good antibiotic ointment, or look for a pet ointment that says it's safe for birds.
5. **Isolate the chicken and keep it separated from other birds until the wound has healed.**

Because the comb and wattles have blood vessels, trimming them causes some bleeding, but it's minor. Some pain is involved; however, anesthetics usually aren't given for pain because the pain is quickly over and birds are tricky to anesthetize. If you take the bird to a vet for the trimming, he may administer local painkillers. Most birds recover quickly and completely.

Eye injuries

Chickens' eyes are sometimes damaged during fights or predator attacks. A little pirate patch on a chicken looks silly, so cleaning the eye and keeping the chicken separated from the flock is about the best you can do. Clean eyes with a nonmedicated eyewash for pets or humans. A chicken will be fine if it's blind in one eye, but if it's blind in both eyes, you need to cage it if you decide to keep it.

Beak injuries

Chickens' beaks sometimes get broken or cut off in freak accidents. Missing beak portions don't grow back. Depending on how much beak is left, the bird may or may not be able to eat normally. If a small amount is missing from the end of one or both beak halves, the bird will be fine. However, if large portions of either the top or bottom beak are gone, the bird will have great difficulty eating. It needs its beak to pick up feed (not to chew or crush it). If the bird isn't able to maintain a decent weight on its own, it's probably more humane to destroy it (see Chapter 16).

Try different sizes and shapes of feed to see what the bird is best able to eat. You can hand-feed the chicken by putting small amounts of feed at a time in its mouth or into the crop. The chicken should still be able to drink. We've even heard of people gluing portions of beaks back onto birds, and if the bird is a pet, it doesn't hurt to try.

Broken legs or wings

Broken wings are fairly easy for the chicken to live with; broken legs are not. If a bone is protruding through the skin, infection is extremely likely, and the chances of the bird making it are poor. Wings can be amputated, but even though chickens can exist with one leg, the quality of life is poor. A vet needs to handle any amputation.

A broken wing may drag on the ground or appear twisted. You can help heal it by folding the wing into a natural position against the bird and then wrapping the bird with gauze strips or vet wrap to hold the wing in place. Unless it's a show bird, a wing that heals crooked or droops is no big deal. Keeping the wing wrapped for 2 weeks is usually enough. The bird must be separated from other birds during this time, but keep it near the others: The bird will feel better if it can see and hear the flock.

A broken leg may look crooked and swollen, and the bird won't walk on it. Broken legs can be splinted, but it's best to let a vet or someone experienced in bird rehabilitation do this. In a young bird, the bones heal quickly. Once again, you need to separate the bird from your flock until it has healed.

Frostbite

Frostbite causes blackened areas on the ends of combs, wattles, and sometimes toes. In most cases, these areas eventually dry up and fall off. Keep an eye on frostbitten areas. If they get infected, you may need to trim them off and treat them with an antibiotic cream. (See the section "Torn or infected combs and wattles" for instructions.) But don't trim off the blackened area unless it gets infected — the blackened area gives some protection to the area below it. When you remove that, the area beneath it may be frostbitten next.

Rubbing chicken combs and wattles with oil, petroleum jelly, and other substances doesn't prevent frostbite. If your weather regularly drops to near 0 degrees Fahrenheit, hanging some heat lamps over the roosting area or heating the shelter may help. Don't heat shelters too much above freezing; doing so causes moisture problems that may be worse than the cold.

Roosters with frostbitten combs may be temporarily infertile. But it's not the frostbite that causes the infertility; it's the amount of cold the chicken has been exposed to. Usually fertility is restored after conditions improve and the rooster's body recovers from the stress.

Egg binding

Egg binding occurs when a hen has an egg that she can't pass from the oviduct for some reason. Egg binding isn't a common occurrence in poultry flocks that are fed and handled properly. Often this diagnosis is mistakenly made when a hen looks droopy before she dies and an egg is found in the oviduct upon postmortem examination. Hens often continue to produce eggs after they get sick, but the birds may die before they lay the eggs.

True egg binding in domestic chickens isn't common. It occasionally happens when birds are injured or have a hereditary defect. Hens that are weakened by poor feed, heavy worm infestations, diseases, or severe cold are more susceptible. Old hens also are more likely to become egg bound. If a hen has been egg bound once, she's likely to have problems again in the future.

If a hen is handled roughly just before she lays an egg, the egg may break inside her. So be sure to handle hens carefully, especially early in the day. Egg binding is a dangerous situation for the hen and seldom resolves safely. In fact, you can't do much for an egg-bound hen. A hen who's egg bound will sit on the floor or ground. Her feathers will be fluffed, and she'll be drowsy and act sick. Sometimes you'll actually see her strain as if trying to produce the egg. More often, you'll notice her tail pumping up and down.

A hen that's truly egg bound will die if she doesn't pass the egg. Don't try to handle it by sticking items like syringes full of oil up her vent; you're likely to hurt her and cause infection. Trying to break the egg inside her and extract the pieces isn't usually effective, either; it's likely to result in infection and death.

Moist heat is considered the safest remedy for egg binding. Place the hen in a cage with a clean towel moistened in hot water and wrung out on the bottom. Use a heat lamp to warm the cage to 85 to 90 degrees F. Check frequently to see that the cage isn't too hot and to moisten the towel again if needed. Make sure the hen has water to drink.

This treatment should have the hen passing the egg in a couple hours. If you see an egg, she hopefully will perk up and be ready to exit the cage. If no egg has passed but she seems more active and will eat, you probably misdiagnosed her; something else is wrong. If she continues to act droopy and ill, give her a few more hours of treatment. She will die unless she passes the egg, usually within 48 hours. You can't safely do much more to help her. However, a vet can give a hen an injection of calcium gluconate, which often causes her to pass the egg.

Getting Rid of Parasites

Parasites feed on a chicken's blood, other body secretions, or its feathers. They can be inside (internal) or outside (external) the chicken. Parasites are common in all kinds of chickens. A few hens in a city backyard may be less likely to pick up parasites, but chickens anywhere can get them. Chickens that range freely have less trouble with lice and some other external parasites. On the other hand, they may be more likely to pick up some kinds of internal parasites than penned chickens.

Completely preventing parasites is difficult. (We discuss preventing parasites in Chapter 10.) Wild birds carry many parasites to chickens. Your chickens also may pick up parasites from the ground, from shared carriers or equipment, from rodents and insects, and even from your shoes and clothing when you go to infected places.

The sections that follow describe all you need to know about internal and external parasites that commonly affect chickens.

A white silkie hen.

A Golden Sebright hen.

Japanese bantam rooster and hens.

Frizzle rooster. The frizzle is actually a feather mutation and not a breed. Many breeds can have frizzle feathers.

Minorcans come in black and white colors.

Pearl Leghorn rooster and hen.

Ameraucana hen with muffs and a beard.

Plymouth Rock hen.

A Turken rooster.

New Hampshire Red hens.
Notice the black tail feathers.

Friendly Isa Brown hens.

Big fluffy Brahmas come in many colors.

Silver Polish chicken with its colorful topknot.

Typical Cornish-Rock Cross meat chickens on pasture.

Temporary hoop housing for pastured meat birds.

A neat small coop for suburban yards.

Make your coop so it's easy to gather eggs.

A colorful selection of home-grown eggs. Eggs of all colors have the same taste and nutrition.

A hen in a nest box preparing to lay.

A rooster mating with a hen.

One of the easiest ways to raise chicks is to let a hen do it.

This is what a fertile egg looks like when candled.

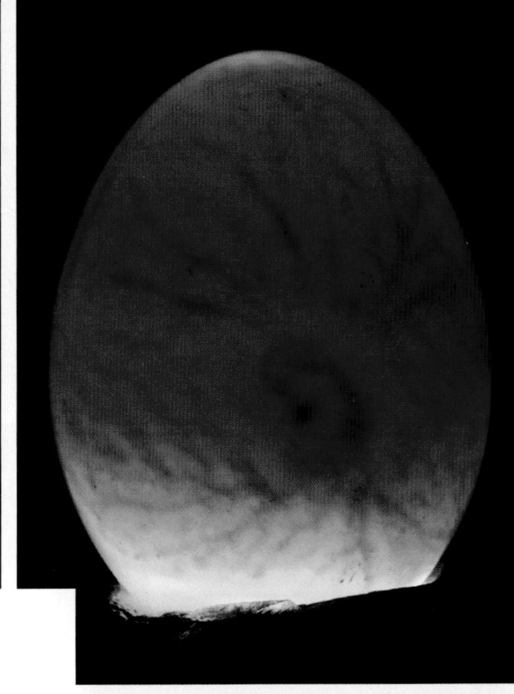

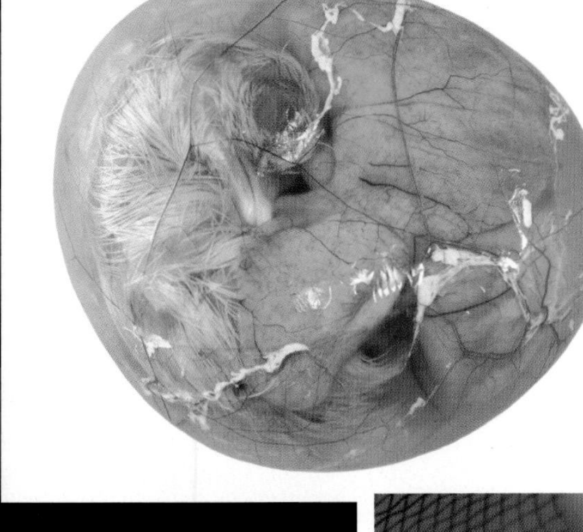

Late stage of a chick embryo.

chick beginning to *pip* through its shell.

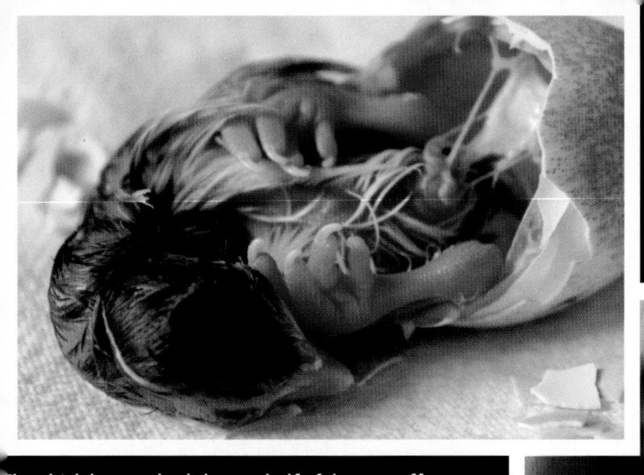

The chick has pushed the top half of the egg off.

A newly hatched and still wet chick.

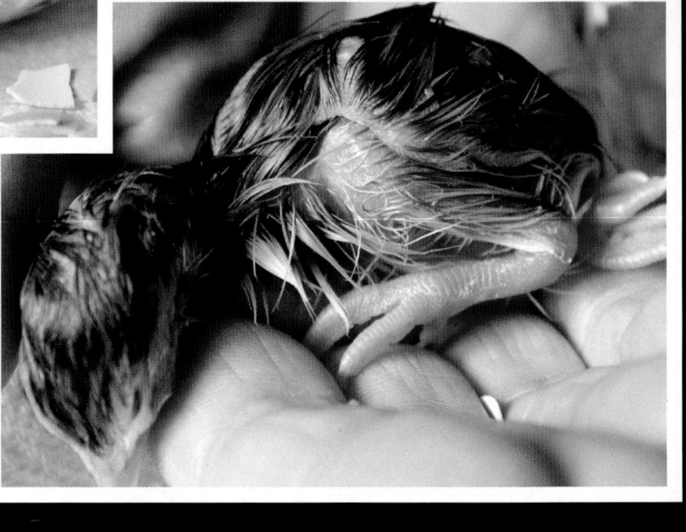

A chick after it's dried and ready for the brooder.

Chicks in a brooder made from a stock tank. These chicks are not directly under the light so they are at a comfortable temperature.

Internal parasites

Two common types of internal parasites that affect chickens are worms and coccidia.

Chickens with worms are more prone to cannibalism, retained eggs, and loss of egg color and size, and they may even die from blood loss and poor nutrition. Worms and other internal parasites become a greater problem when a chicken's system is stressed by disease or a poor environment. Heavy infestations may then contribute to death, poor growth, or poor egg production.

Another internal parasite, coccidia, isn't a worm; it's a microscopic protozoa. Coccidia infections are extremely common in chickens. *Coccidiosis* is the name for the disease caused by coccidia.

If internal parasites, like worms, are an issue, you have a choice of whether to treat them. If your chickens are acting healthy and producing as you want them to, you may decide not to treat them for internal parasites. Some evidence indicates that a mild case of internal parasites actually strengthens the immune system.

Worms

Chickens that have worms may look unhealthy and thin. They may gain weight slowly even though they eat more feed than chickens without worms, and they may lay fewer eggs. Sometimes, however, the owner notices no symptoms; many species of worms can live in chickens without causing any problems. If you aren't into the habit of checking out chicken droppings, you may never know your birds have worms. Even if you do examine chicken droppings (some people actually take pictures of them), you still may not know whether your chickens have worms because often they aren't visible in droppings.

A veterinarian can examine the droppings of a chicken in a lab and tell you whether your birds have worms in many cases. Vets look for worm eggs or actual worms. Sometimes these lab tests aren't successful because worm eggs weren't being produced when the sample was collected.

Chickens generally end up with one or more of the following types of worms:

- **Roundworms:** These worms are a common parasite of chickens. Chickens pick up roundworm eggs when foraging for food. The eggs go through the digestive system to the intestines, where they hatch. Roundworm eggshells are very tough, and most disinfectants don't kill them. They can remain alive in the environment for up to 2 years.

  Many times roundworms don't cause any symptoms in chickens. Young chicks are more likely to be harmed than older birds. Heavy infestations cause weakness and sometimes diarrhea. The worms are often passed in the feces and are easy to see. They're white and up to 3 inches long, with a diameter about the size of a piece of dry spaghetti. They may be coiled up or moving when spotted.

  Sometimes a worm comes out of the intestines and then works its way into the oviduct. It can then be encased in the chicken's eggshell and passed out in the egg. This occurrence is gross, but if you discard the egg, the problem is solved. However, people who see this often want to worm their chickens even if it's not necessary.

- **Tapeworms:** Tapeworms are fairly common in chickens, but they seldom cause serious symptoms in the birds — just an increase in eating because the birds are sharing food with the worms. Chickens pick up tapeworms when they eat earthworms, snails, slugs, grasshoppers, and some beetles and flies, which serve as *intermediate hosts*. The immature tapeworms, or *larvae*, that reside inside an intermediate host get into the intestines of the chicken when the chicken eats the host. There they mature and produce eggs, which are passed in the chicken's feces and consumed by another intermediate host. And so the cycle continues.

  Sometimes segments of tapeworms pass out in chicken feces and are seen as small, white, flat, rectangular pieces that move. However, most of the time people are unaware that their chickens have tapeworms.

- **Gapeworms:** Gapeworms are generally found only in birds that are free-ranging or in pasture pens. They're more serious than other worms because they attach themselves to the inside of the *trachea*, or windpipe, and interfere with the bird's breathing if many of them are present. Infected chickens breathe with their mouths open and may produce a strange grunting sound when trying to breathe.

  Like tapeworms, gapeworms require an intermediate host, which can be earthworms, snails, or slugs. When chickens eat these critters, they become infested.

Treating for worms

The only over-the-counter internal parasite medication for poultry that you can use is piperazine, which is effective against roundworms only. Some very effective worm medications exist, including albendazole, fenbendazole, ivermectin, and levamisole, but a veterinarian must prescribe them for chickens.

If you worm meat chicks or layers, you need to follow label directions about how long to wait after the last dose before you can butcher them for eating or before you can eat the eggs. These medications are passed to meat and eggs.

Well-meaning chicken-owning friends may suggest using worm medications that are meant for other species of animals and sold over the counter. If the label doesn't list poultry as safe for treatment, you're breaking the law when you use it on poultry. Some parasite medications can be harmful to birds, and even with safe ones, calculating the correct dose for your chickens can be hard if it's not listed on the label or prescribed by a vet.

Most herbal and "natural" remedies just don't work on internal parasites, including diatomaceous earth, vinegar, and garlic. Herbal and natural medications can actually be harmful to chickens. Old books often recommend tobacco as a worm remedy, for example, but that treatment is worse for the chickens than worms. And too much garlic given to laying hens can result in eggs with a bad taste.

Generally, treatment for worms consists of worming the entire flock. Some people prefer to worm chickens at least twice a year as a precaution, even if they don't see worms or symptoms. But when worm medications are used frequently, worms build up an immunity to them. You may want to alternate the types of worm medications if you worm frequently.

However, we don't believe that home flocks need to be wormed as a precaution if they appear healthy and are productive. If you notice worms and your chickens don't seem to be as healthy as they could be, it may be time to consult a veterinarian and treat as necessary.

Coccidiosis

Most types of animals are infected by species of coccidia, but each species seldom infects more than one type of animal. Chickens, on the other hand, can be infected by nine kinds of coccidia, but they all cause basically the same problems.

Coccidia line the intestinal walls. They interfere with the absorption of food and damage the intestinal walls, which causes bleeding. The severity of the disease is determined by how many coccidia are in the intestinal tract. Chickens get coccidia by ingesting *oocysts*, which are immature coccidia passed in fecal matter. The oocysts contaminate feed, litter, and soil and can last for a year in the environment. They can be spread by shoes, clothing, equipment, wild birds, pests like rats, and infected chickens.

Coccidia are most often a problem in young, growing birds, but occasionally they can cause problems with older birds, especially if the chickens get bacterial diseases such as ulcerative colitis. When birds become adults, they develop some resistance to coccidiosis, but they may still carry some coccidia. Birds under 3 weeks seldom show symptoms. Slightly older chicks from 3 weeks to 30 weeks may have bloody diarrhea, anemia, pale skin color, listlessness, poor appetite, or dehydration. Young birds with heavy infestations of coccidia often die.

Treating coccidiosis

Good treatments for coccidiosis are available. Feeding baby chicks a starter feed medicated with *coccidiostats* (which kill coccidia) is advisable for the first month. You also can put certain medications into the chickens' drinking water. Amprolium and decoquinate are commonly available coccidiostats. If older birds seem to be infected, you can treat them with these medications as well.

External parasites

External parasites are the creepy crawlies found on the outside of the chicken. They can suck blood or eat feathers. The bloodsuckers can make chickens anemic if many of them take up residence. Feather-eaters irritate the birds, making them pick at their own feathers and skin, which damages their looks for showing. Birds with heavy external parasite loads often don't lay well or gain weight well. Young birds can even die from external parasites.

If you have a small flock that you handle frequently, external parasites like lice and mites may be unacceptable. You don't want external parasites on you, and you may want your chickens to be as comfortable and healthy as possible. Most external parasites that affect birds don't live on humans, but a few will take a bite out of you if they get on you. You also may want optimum production. These reasons are all good ones to choose to treat your birds for external parasites.

Wild birds are the source of many kinds of external parasites, which is good reason to keep them from nesting and roosting in chicken shelters. Signs of external parasites include seeing them crawling on the chickens or in the coop; being bitten by them yourself; noticing chickens with broken, chewed-looking feathers and reddened skin patches; and seeing chickens doing a lot of scratching and picking at themselves. Some signs are more subtle, such as a drop in egg production, anemia with pale combs and wattles, and birds that look fluffed up and sick.

In the wild, chickens help control external parasites by taking dust baths. Dust smothers and dislodges the parasites and cleans the body of oils, dust, and debris that some parasites feed on. Help your confined chickens keep parasites away by giving them a large, deep box of sand to wallow in. Free-range chickens make their own wallows (generally in the middle of a flower bed).

Lice

Several types of lice get on chickens. Unlike human lice, chicken lice don't feed on blood; they eat feathers or shedding skin cells. They spend their entire lives on chickens and die quickly when they're off the chicken. Some chicken lice specialize in certain areas of the body: Your birds may get head lice, body lice, or lice that live on feather shafts. Chicken lice may get on you, but they don't live or reproduce on humans.

Lice are long, narrow, tiny insects that move quickly when you part a chicken's feathers. The eggs are small dots glued to feathers. When your chicken has a heavy infestation, you can see the lice scurrying around on the bird.

Controlling lice

To control lice, you have to treat the birds directly — treating the environment doesn't work. Permethrin, ivermectin, natural pyrethrum, and carbaryl dust are effective insecticides for lice, but you must consult a vet for a prescription and directions.

Some medications come as sprays or dusts. To dust a chicken, place a small amount of the product in an old pillowcase or a garbage bag. Put the chicken in the bag with its head out and the bag closed tightly around the neck. Shake the bag around the chicken to dust it. When handling these dusts, you may want to wear a face mask.

Mites

Most mites are bloodsuckers. They burrow into the chicken's skin and feed on chicken blood. A few eat feathers. They sometimes feed on people, too, even though they prefer chickens. Some types feed at night on the birds and then hide in cracks of the environment during the day; others stay on the birds. They can cause anemia, decreased egg-laying, and damage to skin and feathers. Some types even invade the lungs and other organs. Heavy mite populations can cause death.

The following three types of mites commonly infest chickens:

- **Common chicken mite:** This mite feeds on chickens at night and hides in cracks during the day.
- **Northern fowl mite:** This mite remains on the chicken throughout its life and causes scabby skin, anemia, and much discomfort to chickens.
- **Scaly leg mite:** This mite gets on the legs of chickens between the skin scales. It causes the leg scales to stand out from the leg, making the skin look thick and scabby.

Mites are tiny rounded insects that you can see only through a microscope. The fact that they bite humans in some cases is a strong reason to get rid of them. You need to treat both the birds and the premises for mites.

Controlling mites

A good treatment for scaly leg mites is petroleum jelly, linseed oil, or mineral oil applied liberally to the legs; these products smother the mites. Ask a vet for a prescription for ivermectin if other treatments don't work.

For other mites, you can use the same insecticides that work on lice: ivermectin, pyrethrin, and permethrin. You will probably need a prescription for use on chickens. You can also add sulfur to dust baths to help control lice and mites; you can purchase it at garden stores without a prescription.

For mites, you must also treat the housing. Pyrethrin and permethrin sprays work, and neem oil is fairly effective. Neem oil is available without a prescription at garden stores. Be sure to clean out carriers, too. Remove all bedding and material in nest boxes and replace it with clean bedding.

Don't try to eliminate parasites by spraying your housing with old-time remedies like kerosene or fuel oil. These products are environmental pollutants that cause more harm than good, and using them this way is illegal. They also can have toxic effects on your birds because they can be absorbed into your bird's skin. Also don't use pesticides made for garden and household use because many of these are toxic to birds.

Fowl ticks

The fowl tick seldom feeds on humans, but in the South, where this type of tick is most common, it can cause serious illness and even death in chickens. In northern areas, cold winters generally don't allow large numbers of the tick to build up.

The ticks spend only enough time on chickens to get a good blood meal; then they drop off and hide in cracks, vegetation, and so forth. You may never see them because they usually feed at night. If you suspect ticks, go out and get a chicken several hours after dark and examine the skin closely in a good light. Ticks can be as small as a period on this page before a blood meal, but when they're filled with blood, they're large enough to see easily. Ticks cause anemia, weight loss, decreased egg production, and general weakness in chickens.

Controlling fowl ticks

Ticks are difficult to control. You don't treat the chicken; you treat its surroundings. Treatment means spraying housing and treating pasture areas, and trimming or removing weeds and debris around poultry housing. Ask a vet for a prescription for tick-control products for use inside the coop.

You can use one of the many tick-control products sold in garden stores for treating grass and pastures outside the coop. However, you need to keep the chickens off them for 30 days after treatment. Some of these products say they're safe for pets after a specified time, but they may not be safe for birds. Err on the side of caution and keep chickens off for 30 days.

Chiggers

Chiggers are nasty little bugs that don't mind feeding on humans as well as chickens. Adult chiggers eat plants, but immature chiggers need blood to eat. Chickens get chiggers when they roam grassy areas or come into contact with hay or straw that's infested with them. You can get chiggers from chickens or the same environment where they got them.

Chiggers attach to the breast, wings, and legs of chickens and eat until full, sometimes for a few days. Then they drop off. As they feed, they inject an irritating substance into the skin, which causes intense itching (as any human who has ever had them knows). The feeding leaves small red spots, which can be seen on chickens' skin when they're butchered.

Chiggers cause great distress to chickens. The birds may appear ill and have no interest in eating or drinking. Their feathers appear fluffed up, and they scratch their skin a lot. Young birds sometimes die from heavy infestations.

Controlling chiggers

The control of chiggers is the same as with ticks: You treat the environment. In addition, you may need to move or destroy any hay or straw stored close to chickens. Clean out and replace bedding in coops and nest boxes as well.

Pastures and grassy areas may need treatment for chiggers also. Use the same products for ticks and follow the same precautions.

Bedbugs and bat bugs

On rare occasions, chicken coops may be infested with the same bedbugs that infest human homes or a close relative, the bat bug or bird bug. These tiny bugs feed on chickens (or humans) at night until they're full; then they drop off and hide in cracks of the housing. Wild birds or bats may carry bat or bird bugs to the coop, so eliminate these animals from coops. Bedbugs can get there from items carried into the coop, like carriers that bedbugs crawled into or feed sacks.

Bedbugs and bat bugs may be hard to diagnose. They cause red marks on chicken skin and, eventually, symptoms of anemia. If you suspect one of these types of bugs, you may have to visit the coop at night and catch a chicken to see if that's the problem. Examine it for tiny dark insects on the skin. You may also notice tiny spots of blood on the walls of coops and on roosts.

Controlling bedbugs and bat bugs

Getting rid of these pests is extremely difficult, and you must be careful not to carry them into your home. Most pesticides won't work. You may have to call a professional exterminator or simply tear down the coop and burn it.

Recognizing and Dealing with Disease

You're unlikely to see most of these diseases, especially if you inoculate your birds against the common ones. They're good to be aware of, though, in case you do encounter them. Check out the following sections for tips on recognizing certain diseases and then dealing with each one.

Checking for signs of disease

Look over your flock daily as you feed or care for your birds and check for signs of disease. Catching a problem early can stop its spread and may give you a chance to treat chickens before they're too ill. Some chicken diseases, such as avian flu, can infect humans, so noticing illness in your flock can protect you and your family, too.

Some general signs of disease in chickens follow:

- Inactivity, drowsy or weak appearance, unusual tameness, inability to walk, reluctance to move around, tendency to sit on the floor at night when the bird normally roosts
- Feathers fluffed up and head tucked under the wing or hanging down
- Swollen eyes, discharge from eyes, cloudy eyes, cloudy spots on pupils of eyes
- Discharge from nose, scabby or crusted nostrils
- Labored breathing, breathing with beak open, fluid running out of mouth, combs and wattles that look blue
- Green, white, or bloody diarrhea; pasted, clumped feathers around vent (anal area); sore, swollen, or distended vent area
- Sores, blackened or red areas on skin
- Lack of interest in eating or drinking
- Sudden drop in egg production
- A lot of poorly shaped or colored eggs

This list contains symptoms: signs that's something is wrong. These symptoms may be common to many different diseases, but when you see them, you know your chickens are probably ill. When you suspect a chicken is ill, immediately remove it from the flock and quarantine it. In the next section, we talk about some common chicken diseases, along with their specific symptoms and what to do about them.

Understanding some common chicken diseases

In this section, we list some common chicken diseases, detail the symptoms, and describe the recommended treatment. The diseases are listed in alphabetical order. You may need a veterinarian to confirm that your chicken has a certain disease.

If you're caring for sick birds, care for them *after* you've cared for the rest of the flock — and wash your hands between locations. You may want to change your clothes and shoes after caring for them. Don't use the same tools or containers to carry food and water between locations without disinfecting them. Don't let anyone handle a sick bird without gloves. If many birds appear ill, you may want to remove the ones that don't seem ill to another location instead of moving the sick birds. Handle them with gloves also.

Avian pox/fowl pox

Avian pox is a viral disease that mosquitoes carry. It can also be transmitted from chicken to chicken and from contaminated surfaces. The symptoms are white spots on skin and combs that turn into scabby sores. It can also make ulcers in the throat and on the trachea. Hens usually quit laying.

Because avian pox is a viral disease, antibiotics don't help. Keep the affected chickens in clean, warm, and dry quarters. Give the birds soft food, such as moistened crumbles and tender greens, if they seem reluctant to eat. With good care, many affected birds will recover and then be immune to the disease. After recovery, the birds do not infect other birds and can return to the flock. A vaccine for avian pox is available.

Botulism

Botulism is caused by a toxin produced by bacteria. Chickens get botulism from eating dead animals, eating insects, drinking water polluted by dead animals, or eating home-canned foods tainted by botulism that have been discarded. The symptoms of botulism are tremors, followed quickly by complete paralysis. Feathers on the bird pull off easily. Birds usually die in a few hours.

An antitoxin is available from vets, if they have it on hand, but it's expensive and it must be administered soon after symptoms begin. No other medications can help. If you suspect botulism, immediately check your coop for any dead or sick chickens and then lock up all unaffected free-ranging birds. Next, find the source — a dead animal or discarded food — and remove it.

People can get botulism from handling dead chickens, dead animals, or contaminated food. A small amount of the toxin can be deadly. Wear gloves when handling dead chickens or other animals and be extremely careful.

E. Coli infections (Colibacillosis)

Escherichia coli bacteria consist of many strains, some harmless some harmful. They are found in every animal that poops. The harmful strains can cause many kinds of problems in both chickens and humans. E. coli is spread by anything contaminated by poop. Contamination can spread by human hands — always wash hands well after tending the flock — bird feet, insects, rodents, shoes, clothing, and equipment. A hen can pass E. coli bacteria in eggs.

If a bird (or human) has a healthy immune system, they may not have any symptoms from an encounter with bad E. coli. Birds that are stressed from poor conditions, other diseases, or that have weak immune systems can have serious symptoms or even die. E. coli infections have many symptoms.

Chickens may have diarrhea, and act very sick, hunched up, feathers fluffed and inactive. They won't move or fight back when bullied by other chickens.

E. coli can also cause *egg peritonitis*, an inflammation of the hen's reproductive tract that can cause death. It also causes omphalitis, a fancy name for a bacterial infection of a newly hatched chick's navel. The chicks will look drowsy and won't eat well. The naval area will be swollen, look bluish or bright red and smell bad. Most chicks will die but some may be saved with antibiotics in the drinking water.

Tetracycline, lincomycin, and sulfa drugs, are used to ease E. coli symptoms but some strains of E. coli are resistant to antibiotics. Probiotics and oregano are some of the other things you can try. Some commercial chicken farms are using feed treated with oregano oil to prevent E. coli infections. You can try using a few drops of oregano essential oil in the sick bird's drinking water. Probiotics are sold for chicken use or you can feed some live culture yogurt.

When an E. coli infection is suspected in the flock, wash and disinfect all drinking and feed dishes. Clean and disinfect brooders and incubators. People with compromised immune systems should stay away from the coop and chickens.

Fowl cholera

Fowl cholera usually affects chickens older than 4 months of age. It's a bacterial disease carried by raccoons, rats, wild birds, and other pests and predators. It's also transmitted from chicken to chicken and carried on shoes, clothing, equipment, contaminated feed, and water. Symptoms of fowl cholera are profuse greenish-yellow diarrhea, difficulty breathing, blue combs and wattles, and swollen joints that cause difficulty in moving.

Fowl cholera is extremely contagious and deadly. No treatment exists, and if any birds do recover, they remain carriers and can infect other birds for life. Destroy all chickens with fowl cholera and then report the disease to your state veterinarian or state department of agriculture. If a fowl cholera outbreak occurs in your area, the state veterinarian can make vaccines available for healthy birds. In case of an outbreak, make your flock off limits to anyone who has been around other chickens and don't buy or sell chickens.

Infectious bronchitis

Infectious bronchitis is often called a chicken "cold." It's a highly contagious viral disease that's more serious than a human cold virus because it often causes chicken death, especially in chicks less than 6 weeks old. Symptoms are sneezing, coughing, discharge from the nostrils, and loss of appetite. Hens either stop laying or lay eggs with wrinkled, thin shells. Infectious bronchitis spreads from chicken to chicken, through the air, and on contaminated surfaces. Other culprits are chicken-keepers' hands, shoes, and clothing, and shared equipment.

Antibiotics don't work on viral diseases. Make sure infected chickens stay warm and dry, and tempt them to eat with treats. Some older chickens will recover but can shed the virus for months, infecting new chickens. A vaccine is available for infectious bronchitis, but it must be given to hens before they are 15 weeks old.

Infectious coryza

This coldlike disease is caused by bacteria, and chickens generally have worse symptoms than with infectious bronchitis. Chickens with infectious coryza have a sticky discharge from the nostrils and eyes, and the eyes may become swollen shut. The head becomes swollen, along with the combs and wattles. A moist area may develop under the wings. Laying stops. This disease is spread from carrier birds that contaminate surfaces and water with the bacteria.

Infectious coryza symptoms usually appear after new birds are added to the flock or young hens that were in the brooder are added to existing flocks. Antibiotics can help save birds with the disease. Most birds will actually survive the infection but they remain carriers for life and continue to infect new birds or someone else's flock. We recommend destroying the birds, but, if they live, they should be quarantined for life. There is a vaccine, but home flock owners will probably not be able to obtain it.

Infectious laryngotracheitis

This is another respiratory cold-like virus that can cause grief in your flock. If you show chickens or frequently attend swap meets with chickens, you will probably have this in your flock at some point unless you vaccinate your birds. Infectious laryngotracheitis is spread by carrier birds that can appear healthy.

Symptoms include a bloody discharge from the nostrils and mouths and you may notice blood staining the feathers on the neck and breast. The chickens struggle to breathe and up to half the birds that get the disease may die. Hens will stop laying. There is no effective treatment. Keep birds warm and quiet during the illness. Birds that live through the infection remain carriers for life.

Marek's disease

Marek's disease is extremely common in chicken flocks because adult birds that have had the disease and recovered often look healthy. They remain carriers, however, and can infect chicks under 20 weeks, which are more likely to die or have serious effects from the disease. They can also infect other adult birds that haven't had the disease.

Marek's disease is caused by a virus, specifically a herpes virus. At least six strains of the virus exist, and some are more deadly than others. It's shed in feather dust and skin particles and then inhaled by other birds. That dust can remain infective on coop surfaces and equipment for months and can even travel by wind to other flocks. Wild birds like quail and turkeys can spread the disease as well. It's also spread by mealworms and the darkling beetles they turn into.

Chicks may show signs of Marek's disease starting at about a month of age, and Marek's disease can infect birds of any age older than that. Some chickens may not show signs of the disease, some strains of chickens have some genetic resistance, and some strains are more likely to get symptoms and die. Female chicks are more likely to get symptoms than males. The older chickens are when they're exposed to Marek's disease, the less likely they are to get symptoms or die. Most birds that develop any noticeable symptoms will die.

Symptoms of Marek's disease are varied. Besides looking droopy and not eating well, some chickens may get twisted legs and wings or be paralyzed. A chicken sprawled out with one leg pointing backward and one forward is a sign of Marek's disease. The birds can have tumors on the body and on internal organs. The tumors can cause breathing difficulty or swollen abdomens. Another common symptom of Marek's disease is a gray eye that doesn't react to light and has an odd shape. It can be in one or both eyes and causes blindness. Marek's disease may cause red areas around feather follicles or raised, crusty scabs on the skin under the feathers.

Birds that have tumors, twisted legs or wings, or a sickly appearance should probably be destroyed — there's no cure. Remember that healthy-looking birds can still be carriers, and all new birds that you've added to your flock may either get Marek's disease from your birds or give your birds the disease. The best way to prevent Marek's disease is to start with vaccinated chicks and then add only vaccinated chicks to established flocks. To be effective, vaccinations for Marek's disease must be given within the first few days of life (preferably on day 1).

Hatcheries usually offer Marek's disease vaccinations for a few extra cents for laying and fancy breeds of chickens, and chicken owners should always opt for this vaccination. Many hatcheries that supply broiler-type chicks automatically vaccinate them, but ask to be sure. You also can buy Marek's disease vaccinations to vaccinate chicks you hatch at home, and we advise you to do so. If you're opposed to vaccinations or you can't find vaccinations, keep all chicks well away from older chickens until they're at least 6 months old — and hope for the best.

Never eat any chickens that die mysteriously, and never eat any chickens with tumors that are inside or outside the body.

Moniliasis (thrush)

Thrush is a yeast infection that may cause chickens to eat more than usual but look droopy, thin, and inactive, with ruffled feathers. If you look in the throat, you may see a white, cheesy-looking substance. You may see a white, crusty build-up around the vent (rectum), or the vent may look red and inflamed. Hens may lay poorly or stop laying.

Thrush can come from moldy feed or dirty feed and water dishes. It often occurs after treatment of other disease with antibiotics, which also destroy helpful bacteria that keep yeasts in check. Stop using antibiotics if the chickens get thrush. A vet can prescribe nystatin or other antifungals to help, and you may want to give some probiotics or cultured yogurt. Clean and disinfect feed and water dishes, and get rid of any moldy feed.

Mycoplasmosis/CRD/air sac disease

Mycoplasmosis comes in two forms, a mild form that causes weakness and poor laying and an acute form that causes coughing, sneezing, breathing difficulties, swollen and infected joints, and often death. Chickens usually contract it from another bird, and a chick can be infected while still in the egg if the hen was infected. Wild birds can also carry the disease.

Mycoplasmosis can be cured with antibiotics prescribed by a veterinarian. A vaccination for this disease also is available.

Newcastle disease

Two forms of Newcastle disease exist, one much more deadly than the other (called exotic Newcastle disease). The exotic form isn't known to be present in the U.S. now, and hopefully it will never appear here. Newcastle disease can also cause a mild disease in humans, so be careful when handling birds you suspect of having it. Both types of Newcastle disease are viral diseases that are highly contagious; they can be carried by wild birds, on shoes, clothing, and equipment, as well as transmitted between birds.

The mild form of Newcastle disease is common and causes symptoms like sneezing, coughing, and droopy-looking birds. It causes hens to stop laying or lay poorly. Eggs that are laid may have rough, blotchy-looking shells with an odd, wrinkled appearance or thin shells. Newcastle disease has no cure, and birds rarely become healthy and productive again. However, you can vaccinate young hens before they begin laying so you never have the problem.

You will hopefully never see exotic Newcastle disease in your flock. Exotic Newcastle disease affects most systems in the bird's body rapidly and may cause death before you notice symptoms. Symptoms include twisted necks, wings, and legs; tendency to walk in circles or aimlessly; paralysis, swollen heads and necks; and watery green diarrhea.

If you have many birds die suddenly, or if you notice any of the symptoms and suspect exotic Newcastle disease, immediately contact your state veterinarian. The office will probably inspect your flock and then give you directions on what to do.

Omphalitis (mushy chick)

Omphalitis is a fancy name for a bacterial infection of a newly hatched chick's navel. It comes from staph or strep bacteria picked up from dirty surfaces in the incubator or brooder. The chicks looks drowsy and doesn't eat well. The naval area looks swollen, appears bluish or bright red, and smells bad. Most chicks die from it, but some may be saved with antibiotics in the drinking water.

Once some chicks have the infection, it multiplies on surfaces they contact and can spread to other chicks. Remove any healthy chicks to a clean brooder if you spot omphalitis. Be careful when handling chicks and cleaning the brooder; strep and staph can spread to humans.

Pullorum

Pullorum is a serious disease caused by two strains of salmonella bacteria. Chickens get pullorum from carrier birds or from contaminated shoes, clothing, and equipment. Wild birds can carry it as well. Infected hens will transmit the disease to chicks in the egg.

In chicks, pullorum causes high death rates. Chicks with pullorum are inactive and don't eat well. They may chirp loudly and appear to be distressed. They have pasted rear ends with white diarrhea and difficulty breathing. Some chicks may suddenly die without any symptoms.

Older birds may simply have coughing and sneezing, don't seem to do well, and lay poorly. No good treatments for the disease exist, and since birds who do recover are carriers, any bird that has it should be destroyed. Unlike some other diseases, a simple blood test can diagnose pullorum. If you think your birds aren't doing well, have them tested for pullorum.

The U.S. is trying to eliminate pullorum. Most poultry shows and sales require all birds to be tested for pullorum before they enter the premises. In most places, many people are certified to test for pullorum, and they often can test your flock for a small fee. To find someone certified to test for pullorum, check the website `www.aphis.usda.gov/animal-health/animal-dis-spec-/poultry/participants.shtml`. Always buy your chicks from a hatchery that has been certified as pullorum free.

Administering Medications

Parasites and some diseases are treated with medication. Some medications are given in the water or feed. Some need to be injected into the chicken, either under the skin or in a muscle. Chickens are rarely given medication to be taken by mouth unless it's added to feed or drinking water.

As a home chicken-keeper, you want to avoid treating chickens with any medications unless you absolutely have to. Don't use antibiotics as a preventive; use them only to treat diseases that come up. If you need to medicate your birds because of illness — whether by injection, medication added to the water or feed, or any other way — make sure you read and follow the label directions exactly.

Don't guess at amounts; measure. Discard medications when they expire, and store medications exactly as the label directs. Using expired or poorly stored medications can harm or kill your chickens. Keep medications in the original container with the directions for use. Always store medications, including medicated feeds, where children, pets, and other livestock can't get to them.

While many medications sold over the counter for large livestock or pets may be the same medications that could help your chickens, it's not legal to give an animal a medication unless that animal is listed on the label. Chickens are considered to be food animals, and special rules apply for medicating food animals. Birds' bodies also metabolize medicines differently than other animals, and some medications safe for other animals can kill chickens. A veterinarian can legally prescribe some of the bird-safe medications meant for other animals; it's called extra-label use. Your vet can give you the correct dosage to use and advise you on withdrawal time.

Withdrawal time is the time after you stop using the medication that you must wait before you can eat the eggs or meat from medicated birds. If you want to sell organic eggs or meat, you cannot use most medications on the chickens producing them.

Encountering Death

If you find yourself with a dead bird on your hands, immediately remove it from the flock. It's time to take a close look at that bird and the remaining birds for signs of disease. Move the dead bird to an area with good light so you can inspect it closely. Unless you like smelly things, you want to examine dead birds as soon as you find them.

When you're inspecting the bird, make the following observations:

- **Do you see any sign of a predator attack?**
- **Does the chicken have any wounds, large clumps of missing feathers, or an unusually twisted neck or legs?**
- **Do you see any signs of entry into or damage to your housing?**
- **Does the bird's neck appear broken?** Sometimes predators don't actually touch a bird to kill it. Their presence may have caused a bird to fall off a roost at night or fly into a wall or fence. In these instances, a broken neck is the cause of death.
- **If the dead bird has no signs of injury, is it a very old bird?**
- **Is it a hybrid broiler chicken?** Heart failure in broiler birds often leaves them on their backs with their legs extended.

If you see no signs of accidental death or predator attack and the chicken wasn't old or a fat broiler bird, it may have died of a disease. Read the disease section and see if you can spot any symptoms.

Some chicken diseases can be transmitted to humans. Anytime you handle a chicken or find dead birds you suspect were ill, you need to wear disposable gloves. If you don't have disposable gloves, use plastic grocery bags or something else that forms a barrier between you and the bird. Even if you've worn gloves to handle dead or ill birds, wash your hands thoroughly with soap and hot water. See Chapter 10 for more information about diseases people can get from chickens.

Avian or bird flu

Avian flu is endemic in wild bird populations — which means there are always birds with avian flu viruses around. There are at least 118 strains of Avian flu. Some of the strains infect only one species of bird, others infect many species. The mallard duck can get at least 89 different strains of avian flu. Some strains of Avian flu cause severe illness and quick death in birds, others cause only mild symptoms. Chickens are fairly resistant to avian flu but they can get some strains, including some deadly ones.

The problem with avian flu, as with any flu virus, is that it frequently mutates. When those mutations gain the ability to infect other species, like humans, there can be serious consequences. At least 2 avian flu mutated strains that chickens get are able to infect humans. These mutations are named H1N5 avian flu virus and H7N9 avian flu virus.

Chickens get avian flu from other chickens or wild birds. People get avian flu viruses from close contact with chickens. When people get H1N5 or H7N9 virus, there is a high rate of deaths and serious long-term illnesses. Only rarely do people catch the avian flu virus from another person, which helps limit the spread of the virus. However, the fear is that once avian flu viruses are in humans, they may again mutate and become easily transmissible between humans, causing an epidemic.

Close contact with chickens means living with them, working with them, handling them, or eating meat from infected birds that isn't properly cooked. So far, there have been no outbreaks of avian flu virus in humans in the U.S., but outbreaks have occurred in at least five other countries causing many deaths. Outbreaks of some strains of avian flu have occurred among chickens in the U.S. and many other countries though. This has caused thousands of chickens to die or be destroyed.

Some strains of avian flu kill chickens quickly; others like the new H7N9 virus give chickens only a mild illness but cause severe illness and death in humans. This makes the virus hard to detect if chickens aren't dying from it. Importantly, people should not keep chickens in the home, should always wash their hands after handling chickens, and should always cook chicken meat thoroughly. Wild birds should be kept away from chicken flocks as much as possible. See the section on biosecurity in Chapter 10 for more information.

If many of your flock die suddenly, within a 48-hour time period, contact your state veterinarian. Symptoms of avian flu in chickens are like many chicken illnesses: failure to eat; droopy, depressed behavior; coughing; nasal discharge; and decreased or absent egg production. Some really bad strains of avian flu cause sudden death without any symptoms the chicken keeper notices. Avian flu can kill your home flock and ruin commercial chicken production even if it doesn't infect humans, so it pays to be vigilant.

Symptoms of avian flu in humans are similar to any other type of flu: high fever, muscle aches and pains, sore throat, diarrhea, severe headache, cough, and breathing trouble. Avian flu can progress to serious complications and death in humans. If you are suffering from flu-like symptoms and keep chickens, let your doctor know. Vaccines are available for both humans and chickens to prevent avian flu, but they are not being routinely given in the U.S. The reasons for this are many, including that it could lead to new mutations of the virus and make it hard to detect outbreaks in chicken flocks, which could lead to human outbreaks.

Sometimes you need to send a dead bird to a vet or a poultry specialist for a necropsy, which is like a human autopsy. When many birds are dying with no signs of injury and you want a definite answer on what is happening, a necropsy may be necessary. Necropsy can be free or expensive, depending on what resources your state has for animal health. If the dead bird may affect public health, such as with suspected avian flu, the state health department generally does the necropsy or lab tests for free.

Private vets experienced with poultry may do necropsies; they also vary in price. Ask your county Extension office or a vet where you can send the bird, and then call that office and ask about the cost. You then can decide whether it's worth it to you to know why the chicken died.

After looking at the birds' internal organs and noting any signs of illness, the person performing the necropsy forms a diagnosis based on the observations. In some cases, lab tests along with the necropsy results are needed to pin down the exact cause of death. When you have a professional do a necropsy and you get either a definitive answer or a high probability of what's wrong, you can then get some advice on what to do with the rest of the flock and find out whether any treatments are available.

If you need to have a necropsy done, the bird needs to be as "fresh" as possible. In other words, it needs to be newly dead or stored properly. If it won't be looked at right away, place it in a plastic bag, seal it, and put it in a cool place away from the flock and other animals. Double or triple bagging is a good idea. Chill the specimen waiting to be examined to about 40 degrees Fahrenheit in an ice chest or refrigerator, but don't freeze it.

If a necropsy isn't necessary, you need to properly dispose of a dead chicken. Usually, you have three options: Burn it, bury it at least 3 feet deep, or secure it in a plastic bag or two and place it in trash that goes to a landfill. See Chapter 16 for more details on properly disposing of dead birds. If you're in an urban area and you have no good way to dispose of a dead bird, contact a veterinarian — vets usually have a way of disposing of dead pets and may let you bring in the bird for a small fee.

Reporting Diseases and Deaths

If your chicken had a disease that's legally required to be reported, whether because it has human health implications or is a disease that APHIS (Animal and Plant Health Inspection Service, part of the U.S. Department of Agriculture) requires be reported, the professional doing lab work or an necropsy will report it.

If you aren't using a professional to necropsy or treat your flock and you suspect you have serious disease problem that needs to be reported, you can contact your state veterinarian or your state department of agriculture. You can find a list of reportable diseases here. `www.biosecuritycenter.org/reportdisease.php`. To find a state veterinarian, check out this site `http://www.usaha.org/Portals/6/StateAnimalHealthOfficials.pdf`.

Part IV

Breeding: From Chicken to Egg and Back Again

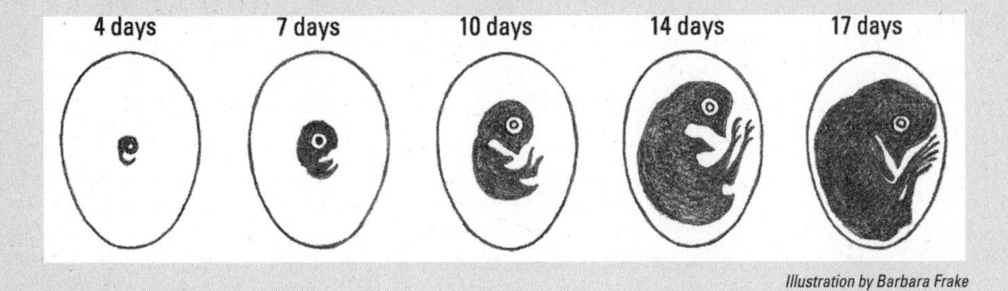

Illustration by Barbara Frake

Showcase your beautiful breeds with some tips on showing your chickens. Check out the article at `www.dummies.com/extras/raisingchickens`.

In this part . . .

- Love is in the air! We give you the basics on mating if you're interested in raising your chickens.
- Ready for a whole chicken family? Find out about incubating eggs and hatching chicks. We provide details on natural incubation by hens and with incubators, as well as how the hatching process works.
- Keep your chicks warm and protected. We show you how to raise chicks, utilize brooders, help hens help their chicks, and mingle children and chicks safely.

Chapter 12

Mating Your Chickens

In This Chapter

- Discovering how to sex chickens
- Checking out the reproductive system
- Looking at methods of mating and hybridizing chickens

When you experience the joy of chicken ownership, you may decide you want more chickens. What better way to get them than to produce your own? However, successfully mating chickens and producing chicks isn't as easy as some may assume.

Keep in mind that many modern breeds of chickens won't even sit on their own eggs — the desire to raise a family has been bred out of them. Therefore, you need either an incubator to produce new chicks or a hen from a breed that still has the brooding instinct. (We discuss incubation, both artificial and natural, in Chapter 13.)

In this chapter, we discuss mating chickens — how to choose mates, how to prepare chickens for breeding, and how to recognize mating and reproductive behavior.

Rooster or Hen?

You probably won't be surprised to hear that, to mate chickens, you need birds of both sexes. If you can't keep a rooster for some reason, you won't be able to raise any chicks from your hen's eggs, no matter how hard you try.

For some people, determining whether they have both sexes isn't so easy. We've seen many people waiting to get eggs from roosters, and an equal number waiting for their "rooster" to crow. The first order of business in mating chickens, then, is determining the sex of your birds — called *sexing.*

Sexing young chickens

Sexing newly hatched chicks is extremely difficult unless they're sex-/color-linked chicks. (We talk about sex/color linking later in the chapter.) Some breeds of chickens can be bred so that one sex is one color and the other sex is another color. This sex/color linking obviously makes chicks quite easy to sex as soon as they hatch. Unfortunately, however, many breeds cannot be sex-/color-linked and still remain purebreds.

Hatcheries often employ workers who have learned to sex baby chicks by examining their hidden reproductive parts. This work is delicate and exacting, but these hatcheries are usually pretty accurate when delivering sexed chicks. The problem is that if you order, say, 22 pullets (young hens) and 3 cockerels (young roosters), you probably won't be able to tell them apart when they arrive. Hatcheries don't mark chicks after sexing them.

By the law of averages, a batch of chickens sold as *straight run* or *as hatched*, or a batch that you hatch yourself, should be 50 percent male and 50 percent female. But in real life, it doesn't happen quite that way. If you toss a coin ten times, for example, you may get heads five times, or seven times, or two times. The same is true with a batch of chicks. Don't assume that every batch of chicks has equal numbers of male and female chicks.

Unless you have sex-/color-linked chicks, whereby each sex is a different color, sexing all chicks accurately is impossible before they get their feathers. As they grow and get feathers, it becomes easier. Here are some ways to do it:

- **Watch the hackle feathers.** Keep an eye on these feathers around the neck as they grow in. Males have pointed hackle feathers, and females have rounded tips on hackle feathers. These feathers tend to be shinier in the male as they near maturity.
- **Pay attention to patterns.** In many breeds of chickens, the females have a different color pattern than the males, and as chicks feather out, you can determine the sex by the pattern. You can look up the color patterns of the breed you're trying to sex to see what to look for.
- **Check out the combs and wattles.** Combs and wattles of males tend to grow faster and larger than those of females, although some females can have quite large combs. The combs of hens often flop over in large-combed breeds, but that can happen in roosters, too.
- **Consider size.** The legs and feet of males may grow faster. Overall, males also tend to be a little larger than females at any age.

Behavior can provide another clue. Males begin squaring off with each other from quite an early age. However, not all fighting chicks are males, and birds that are never observed fighting aren't necessarily females. All chicks go through a ranking or pecking-order process early in life. In a group with a lot of males, some males low on the ranking scale act more like females. Finally, many people believe males are generally more active as chicks than females.

Pick out birds that you think are a certain sex and mark them with reusable leg bands so you can observe the individuals more closely. You can get these bands from poultry supply catalogs. We also discuss identifying birds in Chapter 10. In time, the correct sex will be revealed and then you can judge how accurate your observations were. As you gain experience with chickens, particularly when you raise a certain breed for a long time, you'll get a feel for picking the sexes at a young age, although you may still get a surprise from time to time.

Sexing mature chickens

Although sexing adult chickens may be hard for the novice, most people who have been around chickens for a while can sex mature birds easily. First, start with the obvious:

- If it crows, it's a rooster.
- If it lays an egg, it's a hen.

Adult males rarely go long without crowing in daylight. Occasionally, we hear of a hen that supposedly crows, but usually it's a sort of choked, weak-sounding imitation of crowing. (On rare occasions, the loss of her ovary will make a hen take on masculine characteristics and she'll be able to crow. She/he will not be able to fertilize the eggs of another hen.)

Refer to Chapter 2 to see an illustration of the differences in roosters and hens. Here are some more clues:

- **Examine the feathers.** In roosters, the hackle, or neck, feathers and the feathers along the back, called *saddle* feathers, have pointed tips. In hens, these feathers have rounded tips. The hackle and saddle feathers may be a contrasting color from the body color in some breeds, and in roosters, the feathers have a shiny iridescence. In roosters, several long, thin, curved feathers, called *sickle* feathers, arch over the tail and are also iridescent.
- **Size up the combs and wattles.** The combs and wattles of roosters are generally larger and more erect than those of hens.

- **Check for spurs.** Roosters have a long, hard, toelike extension sticking out of the back of the leg, called a *spur.* Spurs are 2 to 3 inches above the toes and may curve slightly upward. These are apparent as the rooster becomes sexually active and grow quite large in older birds. Roosters use their spurs for protection and fighting. But beware: Old hens occasionally develop small spurs.

Reviewing the Reproductive System

In this section, we discuss the basic biology of the reproductive systems of both hens and roosters. The reproductive system of chickens differs from that of mammals and even some other types of birds. Knowing the biology of the reproductive system is helpful when you're trying to mate chickens and produce a new generation.

Remember that when we use the word *egg* in this chapter, we're referring to the familiar egg we eat, unless otherwise noted. The egg we eat does contain genetic material, like the eggs from other animals, but it also contains a food supply for a developing embryo because the embryo isn't attached to the mother as it develops. The egg we speak of here is a package, the genetic material from the hen and a food supply for the embryo in case the genetic material combines with sperm from a rooster and begins an embryo. In chickens, the true egg, or the female's genetic contribution to new life, is called the *blastodisk.* We discuss that more in a bit.

Roosters

A male chicken does not have testicles on the outside of the body. Instead, two large testes are located near the kidneys along the rooster's back. They're each attached by a tube or *sperm duct* to a small bump on the inside of the cloaca, called the intermittent organ. In the sperm duct just above the intermittent organ is an area called the *vas deferens*, which stores sperm produced by the testes until the rooster mates.

Unlike the sperm of most animals, chicken sperm are viable and active at the body temperature of the bird — about 106 degrees Fahrenheit. A young, mature rooster of good fertility produces more than 30,000 sperm each second. As he ages, or if his health declines, sperm production declines. As with all species, some roosters are more fertile than others, and some are infertile.

Roosters vary in age of sexual maturity. Fast-growing breeds may begin crowing, a sign of sexual maturity, at 20 to 25 weeks, but slower-growing breeds may take much longer. When a rooster is crowing loudly and frequently, he is mature enough to mate.

The pituitary gland of the rooster is stimulated by light, so fertility increases in the spring and early summer when daylight is long. Roosters can be manipulated by artificial light to keep them at top production. When you intend to produce fertile eggs outside that normal spring/early summer period, you want to give your roosters the same lighting treatments you give your hens. See Chapters 7 and 15 for information on lighting.

Roosters have plenty of sperm to mate numerous times a day with different hens. However, when a rooster has more than 15 to 20 hens to "service," he won't get to each of them every day. Roosters sometimes pick favorite hens or ignore some hens.

Hens

The female chicken begins early embryonic life with two ovaries, but the right one becomes dormant even before she hatches. The left ovary remains to produce eggs. The ovary is up along the hen's backbone, on the left side of the abdomen. Before a hen hatches, all the eggs she will ever have exist as tiny spots of genetic material in that ovary. She can have hundreds of them.

The ovary sits just above an organ called the *oviduct.* The upper end of the oviduct, called the *infundibulum,* is roughly funnel shaped, with protruding "fingers" along the edge. It catches the eggs released from the ovary. Fertilization of the egg occurs in this area. The rest of the oviduct is 10 to 14 inches long and runs along the hen's back, where it ends at an opening inside the cloaca. Pouches along the lower part of the oviduct collect and hold sperm for up to two weeks.

When the hen reaches sexual maturity — which varies with the breed and time of the year or lighting — the ovary starts forming egg yolk around various egg germ cells. Producing an egg yolk takes about 19 days. Yolk consists of fat, protein, carbohydrates, and water taken from the hen's body. Egg yolk production goes through several different stages in the ovary, as long as the hen is laying. When a hen is butchered, you can see these yellow yolks in clusters surrounded by the ovary membrane.

When the egg yolk is completed, the follicle holding it splits and releases it into the oviduct. The now ball-shape yolk is covered with a film called the *perivitelline membrane.* The genetic material to begin a new chick, called the blastodisk, is collected in a white spot just under that membrane on one side of the yolk. (Refer to Chapter 15 to see an illustration of the parts of an egg.)

Fortunately for omelet-lovers, even if a hen never mates and no eggs are ever fertilized, she will still produce eggs. A hen is sexually mature and ready to mate when she has produced an egg that is full size for her breed. If she comes from a line of production layers, she may lay regularly for about a year.

Hens don't lay an egg every day; it takes about 25 to 26 hours for each egg to make its way through the oviduct. The egg laying, therefore, happens later each day until it reaches about 2 or 3 p.m. At that point, the hen "resets" and skips a day or more of laying. She then begins laying early in the morning again. Each hen has her own individual cycles, even among egg-production breeds.

Hens of some breeds may lay in sporadic bursts, or only in the longest daylight periods of the year. If a hen gets the hormonal urge to sit on her eggs and raise a family, known as *broodiness,* she will lay around ten eggs, start sitting on them, and quit laying. Some breeds are more prone to broodiness; Cochins and Silkies are probably the top contenders for Mommy of the Year.

Hens stop laying when they *molt* — that is, when they shed old feathers and grow new ones. The molting period can last between a month and 10 weeks, depending on the breed. Poor nutrition, stress, and illness can also halt laying.

Hens continue to lay for several years, but egg production usually drops each year. Some very old hens (8 years is old) may still lay a few eggs in spring; others may stop laying for good when they reach 5 years or so.

How an egg forms

It takes about 25 hours for the egg yolk to proceed through the oviduct and be laid. All eggs, fertilized or not, follow the same procedure. As the egg goes along its route, it first comes to the *magnum*, a large area where about half of the "white," composed of water and protein, covers the yolk in a process that takes about 3 hours. The white part of the egg cushions and insulates the yolk.

The egg then travels to the *isthmus* and lingers there about an hour while it gets more "white" added and is molded into the familiar egg shape. It then goes to the uterus, called the *shell gland* in birds, and stays there for about 21 hours. The egg package gets some final "white" material, along with a membrane around that package and then a second tough membrane. Finally, the shell covers the whole package. Pigment glands may color the egg at this point. The egg then passes into the vaginal area for a short period before being laid. In this area, a clear, waxy coating covers the shell to help protect it from moisture loss and to lubricate its passage.

About 25 hours after it begins its journey, the egg passes from the hen's body through the cloaca. It leaves the hen's body at a temperature of about 106 degrees Fahrenheit, the temperature of her body.

Sperm Meets Egg: Fertilization

When a hen mates with a fertile rooster, sperm enter her oviduct and are stored in several glands near the end of the shell gland or uterus area of the oviduct. Some may also begin swimming up the oviduct looking for an egg. Sperm can be stored and remain capable of fertilizing an egg for more than 2 weeks, although fertilization isn't as likely to occur after 10 days; the best fertilization comes by day four. For you, that means a hen doesn't need to be mated every day to lay fertile eggs. Sperm from multiple roosters can be stored, and each day a different rooster's sperm may fertilize an egg.

When a formed egg on its way down the oviduct passes over a gland storing sperm, some sperm are "squeezed" into the oviduct and begin swimming toward the infundibulum, where the meeting of sperm and egg always takes place. By the time the passing egg hits the sperm pouches, the ovary may be ready to release a new yolk.

If a new yolk has dropped into the infundibulum when the sperm reach it, they have about 15 minutes to fertilize an egg. The blastodisk, or genetic material, is a tiny white spot just under the membrane surrounding the yolk. Only one sperm will unite with the blastodisk, but through a quirk of nature, many sperm need to penetrate the membrane — about 30 or so — for a good chance of fertilization to occur. Usually hundreds of sperm penetrate the membrane. Eggs are always fertilized before the whites and the shell are added to the yolk.

When the sperm unites with the blastodisk, it becomes a *blastoderm*, or embryo. The cells immediately begin dividing and making a baby chick, and they continue to divide until the egg is laid and cools below the hen's body temperature. An unfertilized egg has a yolk with a small, solid white spot, which you can see if you crack open an egg and examine the yolk soon after it is laid. A fertilized egg has a white ring with a clear spot in the center on the yolk.

Sometimes an ovary releases two egg yolks at the same time. They can both be fertilized in the infundibulum and then proceed down the oviduct, where both yolks are covered by one shell, forming a double-yolk egg. These eggs can be hatched, but the chances of both chicks surviving are slim.

Reproductive Behavior

In chickens, as with most animals, behavior is initiated and influenced by hormonal and chemical signals from the body, which get their clues from the environment.

Courtship and mating

Unlike many birds, the male chicken doesn't develop special plumage in mating season. Healthy, sexually active roosters may have a bit more iridescence to their feathers, and their combs and wattles may be firmer, darker, and glossier, but external plumage changes are slight.

Chickens don't have an elaborate courtship ritual, either. A male may circle a female in a strange, strutting manner, almost like walking on its tiptoes, occasionally dragging a wing. When he's really working it, he may pick up pieces of food and drop them in front of her. She then usually crouches down, spreads her wings out a little, and pushes her tail to one side. The rooster jumps on her back, holds on to the feathers at the back of her neck with his beak, and pumps his cloaca against hers for a few seconds.

Both birds' cloacas open, allowing the intermittent organ of the rooster to get close to the vaginal opening of the hen, and sperm is deposited. Deed over, the rooster jumps off, fluffs his feathers, and struts off. The hen also fluffs her feathers and then continues what she was doing before she was interrupted.

Hens are supposed to submit when the rooster tells them to, but occasionally they try to avoid the rooster or fight him. Depending on how many hens he has available and his mood, the rooster may accept this or may chase her down and forcibly mate her. Some hens actually seduce the rooster, crouching in front of him as an invitation. Some hens also crouch before human caretakers or other animals, especially if a rooster isn't present.

Although a rooster rarely avoids any hen altogether, he may pick favorites in a flock and mate with them more often than others. Hens also may pick favorites if they have a choice. Roosters vary widely in the quality of "husbands" they make to their flock. Some are solicitous and gentle with their hens, giving them the best foods, leading them to nest sites, and being fairly gentle during mating. Others are rough with hens and look after their own interests first.

Nesting and brooding behavior

All hens, even ones bred for maximum egg production, have a normal hormonal surge an hour or so before an egg is laid that causes them to seek out a nest and do a little homemaking. They sit in a nest, turning around to shape it, arranging bedding with their beak, and making crooning sounds. After they lay the egg, they may sit for a few more minutes and then get up and go on about their lives. This is normal behavior 95 percent of the time with laying breeds and a good deal of the time for dual-purpose and other breeds, whether the egg is fertile or not.

At some times of the year, however, usually when the days are at their longest, the hormonal surge doesn't drop much after the egg is laid in some hens — the hen is becoming broody. For the first 1 to 5 days, she may not spend all her time on the nest, but she'll spend a lot of time there or nearby, and she'll briskly defend it from any other bird or person — for example, you — who tries to remove eggs. After she has accumulated five to ten eggs in the nest, she sits tightly on it, leaving it only to eat and drink, and normally only when nothing she perceives as a threat to the nest is around.

Some hens are happy to share a nest with other hens. Silkie hens often choose one nest to lay in and crowd into it together to raise babies. Most egg-laying breeds don't get broody. Other breeds vary; some have a greater tendency than others to produce broody individuals. (We discuss natural incubation in more detail in Chapter 13.)

Mating Methods

You can breed chickens in two main ways:

- Group, or flock, breeding
- Pairs or trios (one or two hens and a rooster)

Unless you're just breeding chickens for fun — to let the kids see the miracle of life or for pets — the most important components of a mating or breeding program are the following:

- **A clear purpose or goal:** Do you want to produce meat birds, layers, or show birds?
- **Proper identification:** Any good breeding program needs to have individual birds permanently identified with leg or wing bands.
- **Good record-keeping:** A good breeding program needs written records; relying on memory just doesn't work. You need to keep track of *pedigrees* — the ancestors of your chickens — and the results you get when you cross the birds.
- **Knowledge:** Be sure you have a basic understanding of chicken reproduction, genetics, and biology.
- **Healthy chickens:** Start with well-nourished birds that are sexually mature.

You're then ready to produce new generations of chickens, whether for profit or personal satisfaction.

Flock mating

In flock mating, you have a group of birds that you allow to breed freely. For best results, limit your flock to about 15 birds, with 14 hens and 1 rooster. Putting more than 14 hens with a single rooster may reduce overall fertility because the rooster will breed each hen less often. Smaller groups are even more desirable when breeding chickens of the heavy, loose-feathered breeds because they tend to have lower fertility to begin with. Young roosters can handle more hens than older ones without a drop in fertility.

If you have more than one rooster in a group, the roosters will spend more time fighting and protecting territory and less time mating, so fertility may suffer. Even in a flock where the goal is just to produce as many meat birds as possible, having more than one rooster may complicate matters. You won't know if one rooster is doing much of the fertilization and the other is just eating feed. And if a genetic problem crops up, you won't know which rooster caused it.

A disadvantage of flock breeding is that it may keep you from identifying hens that aren't laying or hens that are producing less-than-ideal chicks for your goals. If a large percentage of infertile eggs are being produced, you may not be able to tell whether some hens are even getting mated or which hens are producing fertile eggs.

Flock breeding is best used when you want quantity rather than quality. It's also often used when space doesn't permit you to separate chickens into small breeding groups. You can use a purebred rooster with a flock of purebred hens of the same breed and produce fairly uniform purebred chicks. You can also produce hybrid chickens by using a different breed of rooster from the hens.

Pair and trio mating

Serious breeders of show birds and breeders trying to improve a breed put pairs or trios of birds — one or two hens and a rooster — into small pens or large cages. Keeping close records on this type of mating program helps you identify which hen produced which results with the rooster.

Remember that hens can store semen for 2 weeks or longer. Depending on her prior living arrangements, to make sure you're getting the offspring from a certain hen and rooster, you must isolate the hen for 3 weeks before introducing the rooster — or at least not save her eggs for hatching for that length of time.

If a hen hasn't been with a rooster for some time, the first one or two eggs she lays after being put with a rooster probably won't be fertile. Start collecting eggs for hatching with the third egg.

With pair or trio breeding, you can rotate a good rooster between cages every 3 to 4 days, and chances for fertilization will remain very high for the hens involved. If the rooster is with hens for at least 24 hours every 3 to 4 days, you should be able to use him with several cages of hens with good results.

When you keep hens in cages or small pens, make sure they have a good nest box with clean bedding. Producing a lot of fertile eggs does you no good if they get broken or become extremely dirty.

Artificial insemination

Artificial insemination isn't for the average flock owner. It can be done with chickens, but you have to keep the rooster close by anyway. Unlike sperm from other species, chicken semen doesn't store or ship well.

Neither collecting semen nor inserting it into hens is all that difficult, but artificial insemination is a specialized craft you need to observe and practice with someone who has experience. If you're considering the idea, ask at a university that does poultry research whether someone can demonstrate the procedure for you. Artificial insemination is sometimes done with large, heavy-feathered breeds that have difficulty breeding or with very special roosters who have lost a leg — chickens need both legs to mate successfully.

Selecting Birds for Breeding

Unless you just want a science experiment for the kids, take some time to choose the right combinations of hens and roosters. You also want to carefully prepare your birds for breeding so that optimum fertility results. In this section, we discuss the best ways to choose birds to breed and prepare them for the task.

Choosing the right combinations

No matter what kind of birds you intend to produce — show, meat, or layers — the parents of those chicks always need to be the healthiest and soundest chickens you have. Never breed birds with serious genetic faults, such as extra toes, deformed beaks, or drooping wings. Even birds with problems you aren't sure are genetic should probably be passed over for breeding purposes. For example, maybe that twisted leg was caused by slippery floors in the brooder and not by bad genes — but that bird still isn't a good candidate for breeding.

Both the hen and the rooster contribute genetic material equally to the offspring. A beautiful rooster with hens that aren't anywhere near his breed standards won't produce good-quality show birds unless you're very lucky. A beautiful bantam rooster won't produce good meat birds, even if he's bred to huge hens. Always breed the best rooster you have to the best hens you have for the best results.

Like begets like: If you want to produce good laying hens, try to get eggs from your best-laying birds. You want to put *those* hens with that good rooster — not the hens you don't mind separating from the flock because they don't lay well anyway. The rooster you choose to produce those new layers should come from an egg production breed as well.

Which are your best roosters? The most fertile, productive roosters are usually active birds that are a bit aggressive in protecting the hens. Remember that the rooster produces half of the genetic material that forms the new generation of chicks, so he should be healthy and of normal size for his breed.

Roosters need two normal legs to breed successfully.

Producing purebred chickens

If you're trying to produce new birds for showing, you naturally want to study the breed standards for that breed of chicken. Know which colors you're allowed to show, and breed only those colors. Often crossing two colors of a breed may get you odd-colored offspring that don't show well, so you need to study color inheritance in your chosen breed, too. Know what type of comb, what size bird, and what number of toes your breed should have, among other features, and breed good representatives of that chicken breed together.

The American Poultry Breeders *Standard of Perfection* is the bible for breed descriptions. We note the address and website of the American Poultry Breeders Association in Chapter 3. You may also want to attend poultry shows and look at good representatives of the breed you want to reproduce.

Sometimes you can compensate. If one bird has a fault, the other parent needs to be better than average in that area. For example, if one parent has feet that aren't feathered quite right, the other should have feet that are perfectly feathered for its sex. Don't forget about sex differences. Good tail carriage in a hen, even though her feathers lack the elaborate sickle feathers of the rooster, can mean better tail carriage in both the roosters and the hens in her offspring.

Even two great-looking chickens that match the breed standards may not produce good-quality chicks, though. Genetics can be funny. To see what birds produce what offspring, you need to keep good breeding records. When that ugly chick turns into a best-of-show winner, you need to know who its mom and dad were so you can produce more like it. You normally identify chickens using numbered leg bands. See Chapter 10 for info on identifying individual birds. Good breeders also keep pedigrees, a list of the ancestors of each breeding bird.

A breeding record can be an elaborate computer program or something as simple as a file card you write on. It should have the bird's identification number on it. You then list the rooster's identification number and the date when you put the hen with a rooster. (For roosters, you note the hen's identification number.) You may also want columns to record the number of eggs produced and set, when they were incubated, how many chicks hatched, and any other information that's important to your personal goals for breeding, such as how many chicks turned out to be "blue" colored. There's no one perfect record style; feel free to personalize records to fit your needs. Table 12-1 shows an example of a breeding record.

Table 12-1 Sample Breeding Record Hen B4402: Spunky/Silkie/Blue Splash/D.O.B. 3-15-13/Pen 2

Mated To	*Date Mated*	*Number of Eggs Produced*	*Number of Eggs Set*	*Date Set*	*Number of Chicks Hatched*	*Comments*
R304	3-30-14	12	11	4-25-14	10	All blue
R310	5-15-14	16	10	6-30-14	4	1 blue, 3 white; incubator problem

Inbreeding — the crossing of close family members — occurs often in the Animal Kingdom, but if you do it too often, the fertility and health of the off-spring start to suffer. Related — but not too closely related — individuals, such as "second cousins," are often the best choices for mating. Occasional *outcrosses* — breeding to totally unrelated birds that are carefully chosen to complement your own chickens — are also recommended.

When you're new to the sport of showing chickens, you may want to ask a long-time breeder of your breed to look at your flock and help you choose which birds to mate and which offspring show the most promise. Knowing how to select the best birds for breeding show birds is as much an art as a science. Taking the time to master it is part of the fun of the hobby.

When you produce show chickens, you need space to keep young birds until you can see what type of adults they develop into. You can cull some chicks at hatching or in the brooder, such as ones that have more toes or the wrong kind of comb, but usually you have to wait until the adult feathers are in and the body frame is filled out to truly judge how good the birds are. Some people kill culls; others sell them or give them away as pets. If they come from large breeds, you may want to eat them when they're large enough.

Producing hybrids

If you're interested in producing your own meat birds, or even some good laying hens, you may want to consider hybridizing. *Hybridizing* generally means mating two different purebred birds, although it sometimes involves mating a hybrid chicken to a purebred chicken or mating two different hybrids.

Crossing two purebred chickens often produces what's known as *hybrid vigor* in the offspring. The resulting birds are healthier, grow faster, and produce more eggs or meat. Of course, you have to cross two breeds that both have desirable traits for the kind of bird you're trying to produce. You can't cross a little Polish rooster with a Cornish hen and expect to get good meat birds just because the offspring are hybrids.

To get good meat birds, you need to cross two good meat breeds. For example, the most common cross is White Rocks and White Cornish. Alternatively, you may produce a good dark-colored meat bird if you cross the Dark Cornish with dark-colored varieties of Plymouth Rocks. Crossing heavy-bodied breeds that grow reasonably fast is the goal for meat production.

Be aware that your hybrid White Rock–White Cornish chicks may not grow as fast or with as much breast meat as the commercial White Rock–White Cornish ones that you purchase from a hatchery. That's because the parent birds of those commercial meat chicks have been carefully selected over many generations for high-quality meat production. Large companies own the particular strains of the parent birds, and they carefully cross the male and female lines and provide eggs for other hatcheries to hatch and ship to you as broiler chicks.

Not surprisingly, high-producing laying hens are often bred by crossing two high-producing egg-laying breeds. Such crossing occurs commercially most often with brown-egg-laying breeds, which are usually crosses of Rhode Island or New Hampshire Reds with other layers of brown eggs. White commercial layers are predominately White Leghorns. Of course, that doesn't mean you can't produce your own high-producing white layers by crossing Leghorns with Minorcas or other white layers.

Be aware that when you cross a bird that lays brown eggs with one that lays white eggs, you generally get offspring that lay brown eggs of some shade. If this result doesn't bother you, you can get many good crosses with Leghorns.

Of course, some people want it all. Getting good meat birds that also lay well is the goal of many home flock owners. This result means crossing a good laying breed with a decent meat breed. You'll probably never get a hybrid chicken that has fast-growing, deep-breasted, meaty males and hens that produce large quantities of large eggs — but you may get reasonably good meat and egg production.

Don't try to save commercial hybrids like Cornish-Rock broilers or Isa Browns and breed them to produce more like them. Breeding two hybrids produces unpredictable results: You get birds like both grandparents and a lot of others in between. The Cornish-Rock broilers are seldom good breeders if they're kept and allowed to mature — they're meant to be harvested when young.

Producing sex-/color-linked colors

The Isa Browns and many other high-production brown-egg-layers are examples of sex-/color-linked crosses. Breeders found that when they crossed certain colors of chickens, the baby chicks that resulted were one color if male and another if female. Therefore, these chicks can be sexed as soon as they are hatched. This consideration is most important in egg-laying breeds, in which the hens are allowed to grow to maturity and males are disposed of early.

Commercial hatcheries often kill male chicks of sex-/color-linked laying breeds shortly after hatching. Alternatively, they sell them as "frypan specials" — small, lightly fleshed meat birds even at mature weight. Incidentally, these chicks are often the ones sold as Easter chicks in brightly dyed colors.

Home flock owners can use the sex-/color link to their advantage. Although distinguishing the down color of chicks can be a little difficult in some crosses, most sex-/color-link crosses are easy to sex when hatched. After you sex the chicks, you can keep only the sex you want or separate the sexes early for different styles of rearing.

The following are two basic sex-/color-linking methods that work in most (but not all) breeds that have the colors mentioned:

- **Black sex links** are produced by crossing a red rooster with a barred-colored hen, such as a Barred Plymouth Rock. This cross produces chicks that are black — but the males have a white spot on the head. As they get feathers, the females stay all black, sometimes with a few red feathers on the neck. The males are barred black and white, with a few red feathers on the neck.
- **Red sex links** are produced by mating red roosters with hens that have what's called the *silver gene* or *Columbian* pattern. Some white birds carry the silver gene, but it's hard for the average chicken owner to determine this by sight. Instead, look for hens with white feathers tipped in black on the neck and tail (the Columbian pattern of, for example, the Delaware breed) or hens with the *silver lace pattern* (neck feathers that are black with white edges and body feathers that are white with black edges — the Silver Laced Wyandotte, for example).

  In the red sex link, male chicks are white (with very light yellow down) and females are buff or red (with dark yellow, reddish, or brown down). When mature, males are white with a few black tail feathers; females are red or golden red, some with a few white feathers or white down feathers under a red outer color, depending on the breeds that were crossed.

Don't try to keep sex-/color-linked chickens and breed them to each other, expecting the sex linking to continue. In the second generation, the chicks will be all different colors in both sexes.

Getting Birds Ready to Breed

After you choose the birds you want to mate, you need to optimize their health and condition to get top-quality eggs that hatch into healthy chicks. Examine them for parasites, frostbitten combs, and missing toes, all of which can affect breeding. Birds with frostbitten combs — combs that have black, shriveled areas — are often temporarily infertile, for example.

Breeding preparation is especially important when the birds are getting older. You need to do everything you can to optimize fertility. Nevertheless, some very nice-looking, lustily crowing roosters may prove to be infertile, and some hens that lay regularly may not produce eggs that can be fertilized. A home flock owner may discover those outcomes only after mating the birds under ideal conditions and getting no chicks. It's just nature's roulette.

Feeding future parents

If you're not feeding a properly balanced commercial ration, you may want to start doing so before you attempt to breed your chickens. It's important for hens to get a good layer ration for eggshell strength and quality and to keep the egg yolk rich enough in vital nutrients that it can sustain chicks until they hatch.

If your hens and roosters appear fat, they may also have problems breeding. Overly fat hens may develop liver problems and stop laying; roosters that are too heavy may have physical trouble mating. Feel the breast area of the bird in question. If it's very plump and the breastbone is difficult to feel, the bird is probably too fat. This often occurs when chickens in small pens get fed a lot of high-calorie snacks, such as corn or bread. The answer is to cut back on feed and increase opportunities for exercise.

If hens and roosters are thin, increase the rations and check for internal parasites. Thin birds feel light, and their breastbone is prominent, without much padding. You may need to worm the birds before breeding. Healthy birds that are thin may show an increase in fertility when rations are increased slightly. For more information on internal parasites and how to check for them, see Chapter 11.

You don't need to add vitamins or supplemental treats to encourage chickens to mate or ensure fertility, as long as their diet is well balanced.

Maintaining lighting and temperature

Lighting and temperature are crucial to maintaining fertility. Although chickens will mate year round, roosters may not be fertile when the days are short and the temperature is cold. Likewise, hens may not lay well when the days are short or the temperature reaches extremes.

If you want fertile eggs outside the normal spring/summer period, you need to set up artificial lighting in your shelters so that the chickens get 14 to 16 hours of light per day. If the temperature falls below 40 degrees Fahrenheit or rises above 85 degrees, you need to provide supplemental heating or cooling to get high-quality eggs for hatching.

The sexual maturity of chickens may be delayed if they reach 20 weeks of age during the period when the days are getting shorter. Those birds may not become fertile until spring — or until the flock owner adds artificial lighting. If chicks are hatched early, in late February or early March, they may become sexually mature with natural lighting in late summer. Egg-production breeds are more likely to become fertile in fall and winter than other breeds. If chicks hatch later than March, you may want to add artificial light to your housing to extend day length to 14 to 16 hours.

For more information about lighting, see Chapter 15. Managing hens that are producing eggs for hatching is much the same as managing egg-laying hens, except that you add a rooster.

Trimming feathers

In some cases, you need to trim feathers around the vent to allow the chickens to mate successfully. This is especially true of loose-feathered breeds such as Cochins. Examine both the rooster and hens of heavily feathered breeds, and either trim or pluck out feathers around the vent area so they don't inhibit contact between the two birds' vent areas.

Chapter 13

Incubating Eggs and Hatching Chicks

In This Chapter

- Understanding the incubation process
- Choosing your incubation method: Hens versus incubators
- Tracking the development of a chick embryo
- Knowing what to expect from the hatching process

Almost as soon as people get chickens, they want more. Some people start thinking about how much fun it sounds to let a mother hen raise some cute little chicks. Other people want to add interest to their chicken hobby by buying an incubator and hatching eggs for fun or profit.

In this chapter, we cover the two methods of incubation — natural incubation by hens and the use of an incubator. We also explain how chicks develop in the egg and fill you in on what to expect during the hatching process.

If bringing new chicks into the world is what you're after, this chapter is the one for you!

Plan for your baby chicks before you start incubating those eggs. You need either a space for a hen and her brood to be separated from the rest of the flock and protected, or a brooder to keep chicks warm and protected (see Chapter 14 for more on brooders). Before you get too far into the incubation process, read Chapter 14 for the full scoop on raising baby chicks so you know what to expect.

Making More Chicks: Incubation Basics

Incubation is the 21-day period from when a fertilized egg is laid to when a chick is big enough to survive outside the egg. Incubation is accomplished by the hen sitting on the eggs and *brooding* them (keeping them warm) or by some other source of heat that keeps eggs reliably warm, such as an incubator.

A long time ago, people figured out that some chickens were better moms than others, and they learned to use the best moms to hatch eggs from the not-so-good moms. This discovery allowed farmers to develop breeds of chicken that continually laid eggs — instead of stopping to sit on them after they laid ten or so, which was the norm.

When electric incubators were developed, chicken breeding got a real boost. They made it possible for farmers to multiply chickens rapidly by combining eggs from several hens in one place. Plus, they could raise eggs from any hen — even one that didn't want to sit on her own eggs. Over the years, incubators have become quite reliable and easy to use. Today even home flock owners can afford to buy incubators and learn to use them.

So why do you want to hatch chicks? The vast majority of home flock owners who are thinking of hatching chicks are doing so because they want to, not because they have to. Baby chicks are available from a number of sources, including many mail-order catalogs, so if all you want are baby chicks, you can skip the hatching process and buy them. If you're up for something a bit more challenging (and fun!), however, hatching your own eggs may be for you.

Some people hatch eggs because they can't buy chicks of the breeds they want — some rare chicken breeds may be available only this way. Other people want to make their own hybrids — for example, to make a good meat bird that also does well free-range, or to produce a hen that lays extra-large, deep brown eggs. Still other people want to replace old layers with hens just like them. Truth be told, most home flock owners are thinking, "We have this beautiful rooster and some truly charming hens — wouldn't they make some cute baby chicks?"

You may want to do your own hatching once, just to see how it's done. Or you may want to show your kids or grandkids a bit of how nature operates. Just don't be surprised if you get hooked on hatching!

Choosing Your Hatching Method

You can hatch eggs in two ways: You can let hens do it or you can use an incubator. In this section, we fill you in on the pros and cons of each method and help you choose the method that's right for you.

Don't believe those old wives' tales of putting eggs in your bra (or your wife's bra) to hatch them. Bras don't make good incubators, nor do fanny packs. Humans aren't meant to hatch eggs — human body temperature (98.6 degrees Fahrenheit) is much lower than the temperature of a hen (106 degrees). And whatever you do, don't use the microwave, even on low!

Looking at the two methods: Hens versus incubators

Before you can choose the hatching method that's right for you, you need to know a little bit about them. In this section, we explain each method.

Relying on mother hens

Nature has perfected hens to hatch eggs, and even the best incubator can't match the hatching ability of a good hen. But an incubator can be a great help to chicken owners, especially when you don't want your hens to take a vacation from laying so they can sit on eggs or when you have hens that don't want to sit on eggs.

A hen normally lays about ten eggs before she starts incubating them. This collection is called a *clutch* and represents the number of eggs the hen can cover when setting on them. The first few days after eggs are laid, the hen doesn't sit on the nest all the time, so the development of the first eggs doesn't get too far ahead of the last and the chicks don't hatch too far apart. After laying a certain number of eggs (it varies from one hen to the next), she then begins to sit tightly, leaving the nest for only a few minutes each day to eat and drink. She defends the nest fiercely.

During the hen's incubation period, you merely need to protect her from the elements and predators and give her food and water. She does the rest. If you want to give the hen every advantage, though, isolate her from the rest of the flock in her own safe little area so that the other birds don't disturb her or break eggs.

If you want to incubate eggs the natural way, pick hens from breeds that are more inclined to brood than others (see the section "Understanding why some hens brood and others don't," later in this chapter, for more on which breeds brood best). Keep a few of these hens around just to hatch eggs, if you want.

For more information on the hen method, see the section "Letting Mother Nature Do It: The Hen Method of Incubation," later in this chapter.

The incubator method

Even the most advanced incubators require more attention than a hen that's hatching eggs, but incubators still offer some advantages over hens:

- You can use an incubator at any time of year. Some hens lay eggs all year but won't sit on them in any season but late spring or summer.
- You can use an incubator when you don't have hens that are interested in sitting on eggs.
- Most incubators hold many more eggs than one hen can sit on.
- Incubators hatch chicks in a clean, protected environment.
- You can observe the hatching process, especially if you use an incubator with windows.

When you use incubators to hatch eggs, you store fertile eggs until you have the number you want to hatch (see the section "Taking care of fertile eggs before incubation," later in this chapter). Then you place them in the incubator. You need to closely monitor the heat and humidity in the incubator, and if you don't have an automatic egg turner in your model, you need to turn the eggs at least twice a day.

We go into more detail about the incubator process in the section "Going Artificial: The Incubator Method," later in the chapter.

Determining which method is best for you

Deciding whether to let a hen incubate your eggs or use an incubator depends on several factors:

- **Whether you have room for an incubator:** Incubators need to be placed in a heated room for best results. If you don't have room for an incubator in a heated place, going the hen route is your best bet.

- **The breed of your laying hen:** If you have hens of a high-production layer breed, they're not going to sit on their own eggs (check out Chapter 3 for breed information). The urge to do so has been bred out of them. If you have room for more hens, you can get a hen or two of a breed that's known for sitting and put fertile eggs from the other hens under them (see the section "Understanding why some hens brood and others don't," later in this chapter, for more info). If you can't have — or don't want — more hens, you need an incubator.

  If you have breeds of chickens that will sit on their own eggs, you can wait until they feel broody and let them raise a family. We discuss broody behavior in more detail later in this chapter.

  Hens stop laying while they're sitting on their eggs and while they're caring for their chicks. So if you want to keep egg production high, or if you want to raise chicks at a time when the hens don't feel broody, you need to use an incubator.

- **Whether you want to see the hatching process:** If you or the kids want to observe eggs hatching, use an incubator. You won't be able to see much as a mother hen hatches chicks because they hatch beneath her, and too much disturbance will cause her great stress and may harm the chicks. You can, however, observe a mother hen caring for chicks and see their antics in a natural environment after the hatching is complete and she takes her babies into the world.

- **What time of year you want the eggs to hatch:** Left to nature, most chicks hatch in late spring through summer, and hens may not begin laying again until the following spring. If early laying is important, use an incubator to hatch eggs early in the spring.

  If you want chickens for meat, the time the hen picks to hatch the eggs may not be the best growing time for your climate. Of course, you can still eat the excess chickens from any type of hatching and rearing, but if you want consistent meat production at convenient times for you, use an incubator to hatch meat chicks.

  If you want show birds to be at a certain age for a show, you may need an incubator to plan hatching at the best times.

  If you want to sell chicks, it's best to plan for them to hatch in spring and early summer, which can mean using an incubator to catch the highest demand period for chicks.

Sitting hens are great for the home flock owner who isn't good at keeping track of things like turning eggs and who isn't in any rush to have chicks at a certain time. Sitting hens are the green way to increase your flock, especially if you need only a few chicks at a time, because they don't require any electricity to do their job.

Letting Mother Nature Do It: The Hen Method of Incubation

Some hens can lay and incubate eggs without your help. In fact, some may sneak off and surprise you with a cute little brood. But with a little help from you, a hen has a better chance at success.

After you get things going, though, step back and try not to interfere. Too much fussing by the human caretaker may cause a hen to abandon her nest.

Understanding why some hens brood and others don't

Hormones affected by the lengthening daylight hours trigger the nesting instinct in hens. Hens have to be laying actively to get broody, so hens must also be in good health and must be receiving good nutrition. For about an hour before an egg is laid, and for 30 minutes afterward, a hen of any breed gets a bit broody, going to a nest and moving nesting material, sitting in it, and crooning. In some breeds at certain times of the year, the hormones that influence motherly behavior don't go down after the egg is laid; instead, they intensify over a week or so until the hen is sitting firmly on a nest.

In certain breeds of chickens, man's breeding practices have prompted them to produce fewer of the hormones that influence broody behavior. This selection process was done to keep them producing eggs because when a hen starts sitting on a *clutch* (group) of eggs, she stops laying. Table 13-1 shows some breeds of chickens, listed according to how often they get broody.

In breeds that sit on their eggs, laying a certain number of eggs and having the right environmental conditions may trigger broody behavior. Still, even in these breeds, each hen is an individual, and some hens are better moms than others. Some people think hens that were naturally hatched and raised by their own mothers take better care of their nests. This trend has been proven true in other species, so it's probably partly true in hens, too.

Hens of naturally sitting breeds seem to be influenced to some extent by the hens around them, and several hens may start sitting at one time. Several hens may actually sit on the same nest, although you want to discourage this behavior because it often results in broken eggs. Instead, have plenty of nests and enclose each hen as she settles down to sit firmly.

Table 13-1 Chicken Breeds and How Often They Sit

Breeds Unlikely to Sit (<10% Chance)	***Breeds That Might Sit (50% Chance)***	***Breeds That Often Sit (>75% Chance)***
Any color Leghorn	Australorp	Ameraucana
Hamburg	Japanese	Araucana
Isa Brown and any sex-/color-linked production layer	Maran	Brahma
Minorca	New Hampshire	Cochin
Polish	Old English Game	Jersey Giant
Rhode Island Red or White	Plymouth Rocks	Orpingtons
White Rocks	Wyandottes	Silkies

Note: This table doesn't cover all breeds — just some of the common ones. Breed descriptions generally note whether the breed sits. See Chapter 3 for more breed descriptions.

Encouraging your hens to brood

If you have a hen from a breed of chicken that's known to sit on eggs, how do you encourage her to get started? Natural environmental conditions, such as long days and warm weather, seem to help. If you want a hen to sit in conditions other than these types, it helps to manipulate the light so that the length of the hen's day gradually increases.

Don't encourage sitting in the coldest months of the year unless you can heat the coop to about 40 degrees Fahrenheit. Hens don't sit tightly for the first several eggs, so those eggs may get chilled enough to die. Plus, extreme cold may keep even the later eggs from hatching.

Hens also like dark, comfortable nests with plenty of nesting material. Just getting in one of these nests makes some hens ready for a family. You may want to set up some of these nests away from regular egg collection nests so you can enclose the hen to protect her as she sits.

Some people keep hens for sitting in their own special housing, especially when the hens are a different breed from the hens producing most of the eggs. Active breeds sometimes pick on sitting hens, and hens and chicks need special protection from predators that a regular coop may not be able to offer.

Hens don't need roosters to get the broody feeling. They'll sit on infertile eggs — or even on junk or rocks. But some home flock owners notice that when a rooster is around, he may encourage a hen to sit on a nest, even occasionally standing guard for her. On the other hand, some roosters may harass sitting hens, especially if few other hens are around.

Adding eggs to the nest

You may want to give a hen eggs that she didn't lay because the hen that laid them isn't a broody hen. If so, you'll be glad to know that most hens readily accept eggs from other hens. Either remove the eggs that the hen laid and replace them with the ones you want to hatch, or just give her a few new eggs to sit on and remove a few of hers.

The eggs you add need to be very close to the same stage of development as the ones the hen laid. Don't give a large hen more than 10 to 12 eggs, and give a bantam breed no more than 6 to 8 large eggs or 10 to 12 of her size if you want optimal hatching (see Chapter 3 for what constitutes a bantam breed).

If you want to save eggs to hatch from another hen, start collecting the desired eggs about the time you notice the broody hen starting to lay. Keep the collected eggs in a place between 45 and 65 degrees Fahrenheit and place them with the small ends down. Don't shake them or subject them to rough handling. Turning them by tilting them one way one day and then the other way the next day is a good idea. Don't store eggs more than ten days; after that, the eggs may not hatch as well. (See the section "Taking care of fertile eggs before incubation," later in this chapter, for more on how to properly store fertile eggs.)

If you're buying eggs to set under a hen, don't order them until you have a sitting hen. Even if it takes a few extra days for her to hatch the new eggs, you're better off using fresh fertile eggs than you are storing fertile eggs for a long time.

You can delay a hen's sitting by leaving just one or two eggs in the nest. Remove those eggs before adding the new ones. Hens will remain sitting a few days beyond the original hatch date, but if one or two chicks hatch more than two days before the rest, the hen will probably abandon the late eggs so she can lead the chicks to food and water.

You can either replace each egg the sitting hen lays with a fake egg or leave the eggs in the nest until you're ready to remove all of them. You can buy plastic, wood, and ceramic eggs from poultry supply catalogs (we list some hatcheries that sell supplies in Chapter 4); you can often find fake eggs in craft supply stores as well. Try to use natural-colored fake eggs. Don't use Styrofoam eggs. If you just remove the eggs without replacing them with fakes, the hen may not settle down to brood the eggs right away and may even abandon the nest. You want a hen to be sitting firmly on the nest before you give her eggs you bought or saved to hatch. She will sit firmly when she thinks the clutch is big enough, and an empty nest gives her the wrong signals.

When you're ready to set the eggs, reach carefully under the hen and remove the eggs she has. She may peck at you, grumble, and slap you with her wings. Then, protecting the new eggs in your closed hand, slide them under her carefully. That's all it should take.

Occasionally, a hen throws out a new egg. Sometimes she does so when the egg is a different color or size from hers. If it isn't cracked, put it back in the nest.

Giving a sitting hen what she needs

Isolate a sitting hen from the rest of the flock, or at least from nonsitting flock members. Chickens that aren't protected may be picked on or disturbed by other hens, disturbed by the rooster, or picked off by predators because they won't move off the nest. Other hens may crowd into the nest with the sitting hen, adding more eggs than she can sit on or breaking eggs in a shoving match.

If you find a free-ranging hen sitting somewhere, try to protect her from the environment and surround her with some kind of small mesh fencing to protect her from predators. Put a board or tarp on top, and she'll be fairly well protected. If you have a hen sitting inside, also try to protect her from other flock members and predators.

A hen doesn't need a very big enclosure while sitting. She needs enough room to stand and stretch and flap her wings when she gets off the nest, and she needs room for small feed and water dishes. About 3 square feet for full-size hens and 2 square feet for bantams does the trick. You can make the enclosure with a circle of fencing with small openings that keeps chicks in, or you can use a cage with small wire openings.

Some hens don't object to a careful move of the whole nest to a protected place, but it's best to get the hen sitting in a place where she can be protected from the start of incubation.

If you need to move a hen, do it at dusk or at night. Try to slide something (such as a big cardboard box) under her and the eggs, and move the whole thing to the new spot.

Be careful: Wilder hens may fly off the nest as you move it and be hard to catch. The hen won't know where you moved the eggs and won't find them. Move slowly and quietly as you pick her up and confine her with the eggs.

By morning, a confined hen should've decided to stay put, but make sure she can't get to the old area. Some hens may abandon the eggs and go sit where the nest once was. If you must move a hen quickly during the day, get a good nest area ready, move the eggs and the hen into it, try to darken the spot, and then leave — don't fuss about her. Many hens will settle right down, but if yours doesn't, you can't do much about it.

Hens don't carry eggs to a nest, although they may roll eggs a short distance back to a nest by reaching out from the nest with their beaks. If you see eggs out of the nest, mark them with a marker or pencil and put them back in the nest. If the marked egg turns up back outside the nest, the hen may be trying to discard it because she knows it's no good, especially if no other eggs are out of the nest. Sometimes the hen is just clumsy. Other times, other chickens or predators may be disturbing the nest. If an egg comes out of the nest a third time, discard it.

If eggs crack or get dirty from manure or mud, don't wash them. You can try to clean the egg with a soft, dry cloth, but washing it can cause bacteria to invade the shell. Don't handle the hen's eggs if you can avoid it. They will get chilled every time you remove them.

Make sure the hen always has food and water available. She won't eat or drink very much — that's normal; she isn't active, and eating too much food creates the need to get off the nest more often. Place the food and water a little ways away from her so she has to get up to eat — getting up once in a while is good for her. Clean the pen area if it gets too dirty. Food, water, and a clean pen are all the care she needs.

Caring for a hen and chicks

Twenty-one days have passed, and the big day is at hand. If a hen has remained sitting firmly on the nest until now, you've got a very good chance that at least some of the eggs are going to hatch. Hens hear and respond to chicks peeping while the chicks are still in the shell. Now isn't the time to disturb the hen frequently — you want her to stay on the nest until all the healthy eggs have hatched. So listen carefully and watch the hen for tiny heads popping out from under her, but don't shoo her off the nest to look.

Hatching day

Chicks hatch under a hen. You'll know it's happening by some clues:

- Pieces of eggshell outside the nest
- Peeping sounds
- A tiny head sticking out from under the hen

The chicks hatch out over the course of 36 hours or so. The hen stays put until all the chicks are hatched and dry or until about 2 days after the first chick hatches.

The hen knows she must move then so that the hatched chicks can get food and water. If she's in a small area, she may go back to the nest with the chicks, and some eggs may still continue to hatch if it's warm and they didn't get too chilled in the hen's absence. It's better, however, to have the eggs hatch as closely to each other as possible by setting the eggs all at the same time.

If you intend to collect the chicks and care for them in a brooder, collect them on the second or third day so that you don't disturb the hen and any eggs that are continuing to hatch (see Chapter 14 for more on brooders).

For the first day or two, chicks spend a lot of time hiding under their mom. Keep the kids or other visitors quiet when they visit, and don't allow them to pick up the hen or try to scare her away from the babies so they can see them. The chicks will be just as cute in a few days, but the mom will also be more relaxed and the chicks will be more active.

The early weeks

Baby chicks need food and water they can reach the moment the hen moves from the nest with them. Place chick starter feed in shallow pans and have a water container with a narrow opening nearby. You can leave a dish of the hen's food with her, but don't have any large open water dishes the chicks can drown in. The hen will show them how to eat and drink.

The chicks will hide under the hen if they feel scared or cold. The hen will fluff up her feathers to help cover them all. They don't normally need supplemental heat other than the hen, but if the temperature is below freezing, a heat lamp suspended over one corner of the pen can be a real help. Also make sure there's some bedding between the cold ground and the chicks.

Keeping the chicks in a dry area is the best you can do for them. If they get wet and chilled, even the hen may not be able to warm them enough, and they won't do well. If the pen is outside on the ground, heavy dew in the morning can be a problem; mow short any vegetation that's inside the pen.

We don't believe in just letting hens roam the yard with their babies. Far too many dangers may lay in wait for the little ones — cats, kids, snakes, hawks, and even crows and blue jays pick off chicks. Roosters rarely hurt chicks, but other hens often do. The mother hen will try to protect them, but she can't always do so.

If you've ever tried to catch baby chicks that are even a few days old, you know that catching them in a large area is difficult, so you're better off keeping them in a smaller, more manageable area. A hen with chicks needs at least 6 square feet of space until the chicks are feathered; after the chicks are feathered, they may need more space. Several hens with their own chicks may share a large area, but some hens try to steal other hens' chicks.

Make sure the chicks can't get out of the fencing or shelter on their own. The mother hen can't protect them if she can't get to them. Don't think you can keep chicks confined with a low barrier — they quickly learn to hop up and over. They're also good at finding any small opening they can squeeze through.

Plans for transitioning teens

Chicks can be separated at any time from their mother, but your best bet is to wait until they've fully feathered out at about 4 to 6 weeks so they don't need supplemental heat. If it's really cold — say, below 30 degrees Fahrenheit — and you really want to separate them from the hen, they may need some heat until they're 3 months old. You can put young chickens in their own pen, or you can take the mother hen out of the rearing pen and put her back with the flock.

When the chicks have all their feathers, you can let them out with their mom to free-range, if you keep your chickens this way. They're still a little vulnerable at this stage, though, so be prepared to lose some chicks. If you have penned chickens and you want to introduce the mother hen and her fully feathered family to the flock, put them in the adult housing with the hen around the time the other birds are going to roost.

Keep an eye out the next day to see whether any young birds are getting picked on. If they're wounded and bleeding, immediately separate them from the flock. As long as they're not wounded, let them learn their place in the flock. Sometimes roosters don't tolerate young *cockerels* (male chicks), so you may need to pen them separately.

For more information on raising chicks, see Chapter 14.

Going Artificial: The Incubator Method

You may opt for artificial incubation when a hen won't sit on her own eggs, or you may just find an incubator easier. Incubators have come a long way since they were invented in 1843, and chicken producers have learned a lot about incubating eggs in the last 50 years or so.

With artificial incubation, you collect the eggs and put them into a container (the incubator) that keeps them warm and maintains the proper humidity until they hatch. If the incubator has windows, you can watch the hatching process.

Choosing an incubator

You can find incubators for almost all budgets, and you can even build your own incubator with some parts you can purchase. Poultry supply catalogs, farm stores, and pet stores carry incubators — deluxe models that are almost "set and forget," as well as some cheaper models that require more careful attention from you. Many secondhand incubators are for sale in newspapers, in magazines, and online.

Before you buy an incubator, you need to consider the following factors:

- How often you'll likely be using an incubator
- How many eggs you want to hatch at a time
- What type of poultry, if any, besides chickens you want to raise
- What your budget can afford
- How much you want to be involved in the project

If you or your kids want to see the eggs hatching, you want to get an incubator with an observation window.

If you think the egg incubation project will be a one-time experience or something you do maybe once a year, buy a cheaper incubator or a good secondhand one. You can always upgrade later. If you're already deep in the chicken hobby and you think you'll be hatching eggs frequently, buy a better incubator. If you want to hatch eggs of turkeys, ducks, or geese in addition to chicken eggs, make sure the model you buy can accommodate the bigger eggs.

Most incubators require electricity (although you can buy gas-powered ones) and are meant to be used in a heated room.

The following list describes the three main categories of incubators (Figure 13-1 gives you a couple examples):

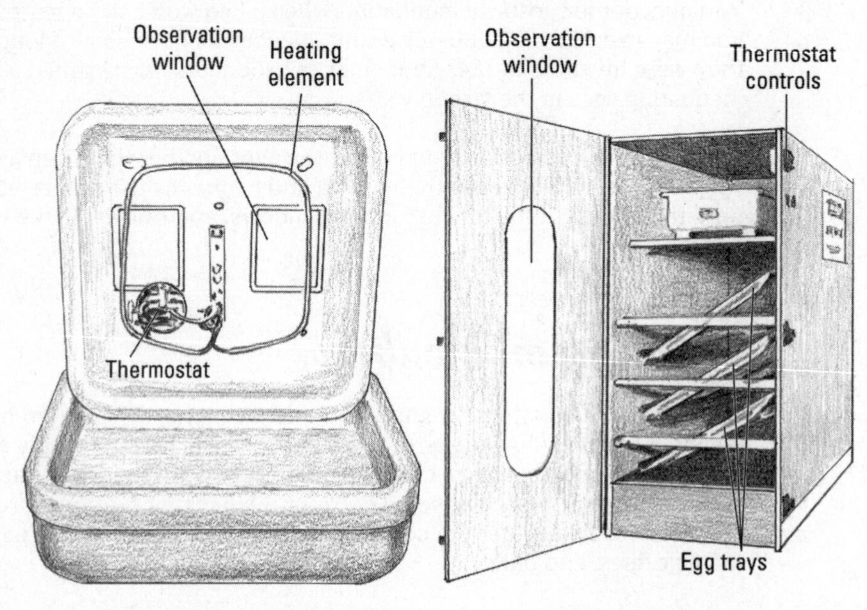

Figure 13-1: Types of incubators.

- **Still-air incubators:** Still-air incubators are the cheapest models of incubators. They're usually made of Styrofoam or plastic. They range from models that can hatch just 4 eggs to models that can hatch 36 or more eggs, depending on egg size. Many of these have clear tops or observation windows.

  Air in these incubators moves through the incubator by cooler air being drawn in from the bottom and rising as it's heated through some ventilation holes on the top of the incubator. Some models include fans to move air. Most models have a reservoir in the bottom that you add water to in order to provide humidity. Some models have automatic egg turners; otherwise, you need to turn the eggs by hand.

  These incubators require careful attention daily, and you need to follow the instructions exactly. They can provide a good hatch if you pay attention to details.

- **Forced-air incubators:** Forced-air incubators move heated air through the incubator with some kind of fan system. They're more expensive than still-air incubators, but they generally provide for a better hatch because they eliminate the hot and cold spots that still-air incubators sometimes have.

Many forced-air models also come with automatic humidifiers and egg turners, and some deluxe models have warning buzzers and other bells and whistles. Some attention is still needed, but it's less than the still-air models require. Some come with observation windows. They range in capacity from 12 to 50 or so eggs.

These models are probably the best type for most home chicken flock owners.

- **Cabinet-style incubators:** These incubators are large units with forced air, automatic humidifiers, and egg turners, and they have several shelves to accommodate eggs at different stages of hatching or of different sizes. They generally don't have observation windows, and they cost you a considerable amount unless you find a good deal on a used model. These units are for the serious chicken hobbyist.

Automatic egg turners are standard in some incubator models but can be added to others. They work by tilting the eggs in a different direction every few hours — they don't actually roll the eggs over. Automatic egg turners increase the number of eggs that hatch because the eggs are turned consistently and gently. They have to be turned off 3 days before the chicks hatch.

Accessorizing your incubator

You need to buy a few incubator accessories if the incubator itself doesn't supply them. Some necessary accessories include the following:

- **Thermometer:** The thermometer needs to be at a level near the top of the eggs. You can place your thermometer on a stand or hang it to achieve this level. Make sure the thermometer has measurements large enough to read from the observation window, if you have one. Position it in the center of the incubation space so that it doesn't touch the eggs or the walls of the incubator.

  Typically, when an incubator has a window, you read the thermometer remotely from outside the incubator because the minute you open the incubator, the temperature drops. Many remotely read thermometers are on the market — you place a probe or sensor in the incubator, and it transmits a reading to a device located outside the incubator. You can find such thermometers made specifically for incubators or for general household use.

  The correct temperature is critical to success with incubation, and the thermometer must show each degree. Even a difference of 1 degree can affect hatching. Buy a good thermometer and handle and store it carefully.

- **Hygrometer:** You can buy a hygrometer, which measures humidity, specifically for incubators or for household use. Many of the remote-style thermometers also measure humidity.

  Humidity is crucial to a good hatch, so make sure you buy a good instrument. Many of the better incubators have built-in hygrometers, some of which actually automatically add water from a reserve tank to adjust humidity.

You may also want an *egg candler*, a light that helps you look inside an egg to see whether it's developing. You can buy or make these instruments. (See the section "Looking Inside the Egg," later in this chapter, for more on candling.)

Setting up and caring for your incubator

After you buy an incubator, carefully read the directions that come with it, and then save those directions so that if you decide to use the incubator again in a year or so, you have the directions to refresh your memory.

Your next step is to find a place to set your incubator. It needs to be located near an electrical outlet, and it'll be there for a month (so you probably don't want to put it in the middle of a high-traffic area). Choose a room that's heated to at least 60 degrees Fahrenheit and away from windows, heating vents, and doors that may cause wide fluctuations in room temperature. Also select a place where it won't get knocked off or jostled too often.

If you have children or pets, you may want to put the incubator where it's out of sight and out of mind. Children are prone to opening the incubator too often, handling eggs, or fooling with controls. Pets can hear the chicks peeping in the shells in the last days of incubation and may damage the incubator trying to get to the eggs. They may also jostle controls while inspecting the incubator or, in the case of cats, sleep on top of it, blocking the vent holes.

Have the incubator set up and running for at least 24 hours before you add eggs, to make sure it's working correctly.

Cleaning your incubator

Every time you use the incubator — even if it's brand new — you need to clean it. Remove any debris and then wash the incubator with hot, soapy water and rinse. Make sure to wash and rinse the screens the eggs lay on and the egg-turning racks, too.

Consult your manual to see whether it's safe to submerge any parts in water before doing so.

If it's a nice, sunny day, letting the incubator spend a few hours open in the sun is a good way to kill germs and dry the incubator at the same time. Otherwise, dry the inside of the incubator with clean paper towels.

You can use a solution of one cup unscented household bleach to four cups of water to wash the incubator with. Then rinse it and allow it to dry before use. Don't use other strong disinfectants or cleaning solutions on incubators. If any chemical residue remains, it can harm the embryonic chicks by passing through the porous eggshell.

Adjusting the temperature

Having the right temperature — the entire time of incubation — is one of the crucial steps of good incubation. For still-air incubators, the temperature needs to be 102 degrees Fahrenheit. In most other incubators, the temperature needs to be 100 degrees, but read your incubator's instructions for information on the correct temperature.

Make sure the temperature is at the correct setting before you add eggs to your incubator.

The outside temperatures, how often the incubator is opened, and the stage of development the chicks are in can affect the temperature inside the incubator. If the incubator has a thermostat, as most do, it will likely maintain the correct inside temperature automatically, but still check the thermometer often and adjust the thermostat right away if the temperature is incorrect.

Unless it happens too often, a short period of cooling, as when turning eggs manually, doesn't harm the chicks. Even a short period of time at temperatures above 104 degrees, however, can kill most or all of the chicks.

When chicks get near hatching, their bodies actually produce some heat, and this change may cause the temperature to rise in the incubator. Pay close attention and adjust the temperature, if necessary, close to hatching time. At hatching time, people often open the incubator frequently to check on the chicks. Doing so can chill them and may even cause them to die during hatching. If you have eggs in different stages of development in the incubator (more than 2 days apart), it's helpful to have a second incubator so that you can move the eggs that are scheduled to hatch together.

Paying attention to humidity

Chicks require the right humidity to develop correctly in the egg and to be able to hatch. The humidity in the incubator needs to be about 55 percent until the last 2 days, when you need to bump it up to 65 to 70 percent.

You can usually increase humidity by adding water to some sort of reservoir. In still-air incubators, the reservoir is generally a pan under the screen the eggs sit on. Other types of incubators have different reservoir systems. When you add water to these reservoirs, make sure you heat the water to the incubator temperature first.

If water is dripping off the inside of the incubator or the eggs seem wet, the humidity is probably too high, which can kill your chicks. You need to increase the ventilation by opening additional vent holes or even cracking the lid for a short time — just keep track of the temperature to make sure it doesn't drop while you're doing so.

Making sure the incubator has adequate ventilation

Chicks require oxygen to breathe. They pull oxygen in through the pores of the egg before they hatch. All incubators must have some way for fresh air to enter the incubator and stale air to leave. Still-air incubators usually get their fresh air through small openings you can open or close; other incubators pull in air with a fan through a vent. As chicks grow, they require more oxygen, so you may need to increase ventilation. As chicks hatch and breathe air, ventilation is critical.

Determining whether the incubator is ventilated correctly is difficult. Read your incubator instructions carefully. They tell you how to adjust the ventilation, if necessary, based on experiences with that incubator. Unless the instructions direct you to do so, don't cover any ventilation holes in the incubator. Most incubators require that you open additional vents near the hatching date. If instructions are missing, look for capped vents and open them on the 18th day of incubation. Adjust the temperature and humidity if opening vents causes them to fall too low. If your incubator has no capped vents, the incubator probably supplies enough ventilation.

Finding and storing fertile eggs

When you have the incubator ready, your next step is to get fertile eggs. Finding fertile eggs may take a little time if you don't have your own hens and a rooster, so start looking well before you want to hatch eggs. Even if you have your own hens and a rooster, you may not be getting fertile eggs. (In Chapter 12, we talk about how eggs are fertilized and how to make sure your own chickens are laying fertile eggs.)

Fertile eggs are usually available in the spring and summer, but they may be hard to find at other times if you're not producing your own.

Finding fertile eggs

Try to find a local source of fertile eggs, if you can. Sending eggs through the mail and getting a high percentage of them to hatch is difficult — the average rate of hatch under ideal conditions for fertile eggs sent by mail is only 50 to 60 percent. Normally, you don't get any guarantees when fertile eggs are mailed to you because the seller can't control the shipping temperatures and the way the eggs are handled in transit.

Here are some good places to get fertile eggs:

- **Mail-order catalogs:** Most catalogs that sell baby chicks also list eggs for sale. See Chapter 4 for a list of hatchery catalogs.
- **Backyardchickens.com and other online chicken forums:** These sites are good sources for breeders looking to sell fertile eggs.
- **People who sell show stock and rare breeds:** These people may also sell fertile eggs.
- **Friends:** If you have a friend who has both hens and roosters, he may be able to save fertile eggs for you.
- **Local farmers:** Some farmers sell free-range eggs at farm markets. If you have a market near you, ask the owner whether she can save you some eggs for hatching. Eggs that have been handled correctly — for the intended purpose of hatching — are more likely to hatch than eggs that have been washed and cooled for consumers to eat.

If you can't find eggs nearby, call your county Extension office. You may find a 4-H poultry club that can help you, or the Extension educators may know where you can get eggs. A vocational high school that teaches agriculture may be able to help you as well.

If you're saving your own eggs or you have the chance to pick and choose which eggs you want, choose the cleanest ones. You can brush off dirt with a dry cloth; some people even use fine sandpaper or fine steel wool. If eggs are heavily soiled, it's best not to use them. Discard any cracked eggs and any eggs that have very thin shells or are oddly shaped; oddly shaped eggs seldom produce good chicks.

Eggs you buy from the supermarket won't hatch because those hens aren't kept with roosters. Even the eggs labeled as organic are probably from hens without roosters. We've heard of eggs that were bought at farm markets from free-range hens hatching, but don't count on hatching eggs that were meant for eating.

Taking care of fertile eggs before incubation

The most important point to remember about fertile eggs meant for hatching is that they're living things — they must be handled gently and kept at the right temperature, or they'll die. Don't shake them or toss them around — if you do, you may kill the embryo.

Don't wash fertile eggs intended for hatching! Washed eggs are far less likely to hatch than unwashed eggs, and the unhatched, washed eggs may also affect the hatching of eggs around them.

When you wash eggs, you remove the protective coating that the egg gets as it leaves the hen, and the egg's pores often draw in bacteria. The warmth of incubation turns the contaminated egg into a factory for bacteria reproduction, often killing the embryo and spreading to nearby eggs.

The fertilized egg starts dividing to become an embryo as it travels down the oviduct on its journey out of the hen. When it's laid and cools below her body temperature, it goes into a state of suspended growth until conditions are right again. Eggs being stored for hatching need to be stored at temperatures between 45 and 65 degrees Fahrenheit. If the egg temperature drops below 45 degrees for very long, the embryo will probably die; if it goes above 65 degrees, it may start to grow.

Growth at less than optimum temperatures makes weak chicks that seldom survive the whole incubation period and can cause deformities in the chicks that do survive. In cold or very warm weather, collect the eggs from your hens for hatching as soon after they're laid as possible and then move them to the right conditions. If you're collecting your own eggs, you may need to put them in a cool basement or even in the bottom vegetable drawer of your refrigerator to suspend growth.

Store eggs for hatching with the small end down. An egg carton is good for this purpose. You can store eggs for about a week without much drop in vitality, but after a week of storage, the percentage of eggs that will hatch drops sharply. After 2 weeks of storage, few eggs will hatch.

Eggs that you're storing for incubation need to be rotated from side to side twice a day. Rotating keeps the early embryo from sticking to the shell in a bad position.

If you're getting eggs from anywhere but your own hens, have your incubator set up and ready to put them in as soon as you receive them. If you're collecting eggs from your hens, store eggs until you have about a week's worth; then set up the incubator and put the eggs into the incubator all at once. Using a pencil or nontoxic marker, mark the eggs with the date you set them.

If you're going to be hand-turning the eggs, put an *x* on one side so you know which eggs you've turned.

It's best not to have several different hatch dates in the same incubator unless you have a cabinet incubator with multiple shelves or drawers. Eggs at different stages of incubation require slightly different care.

Caring for eggs in the incubator

A mother hen seems to know instinctively what her eggs need. If it's very hot, she gets off the eggs to let them cool a little; if it's cold, she sits tightly. Her body provides the perfect humidity, and she fills it with water herself. When you take over the job of incubation, you can never be as good as a hen, but with careful attention to details, you can have a successful hatch from an incubator.

Turning eggs

Hens don't actually turn their eggs with their beaks on a regular basis, as many people think. (They do occasionally rearrange them with their beaks, but it's usually for their own comfort.) Instead, their comings and goings from the nest and their shifting of positions to get comfortable alter the position of the egg several times a day.

Some debate swirls about turning eggs, but most experts believe the position of eggs needs to be changed two or three times a day for the first 18 days of incubation. Automatic egg turners can do this for you, or you can do it yourself by rolling the eggs to a new position. The turning keeps the embryo from becoming attached to the outer membranes and the eggshell. If you're turning the eggs yourself, do it quickly so you don't chill the eggs too much.

If you have egg racks for turning eggs in your incubator, place the eggs in the racks with the small end down. If you're using an incubator without racks, lay the eggs on their sides. Cluster them in the center of the incubator, if they have a lot of room.

Wash your hands before handling eggs. Oil or bacteria from your hands can cause hatching problems. Warm hands are much friendlier to eggs than cold ones. (How do *you* feel when someone touches you suddenly with cold hands?) And be sure to wash your hands again *after* touching the eggs. Eggs can have harmful bacteria on them, too.

On the 18th day, stop turning the eggs. If you're using an automatic egg turner, be sure to turn it off. The chicks are getting in position to hatch, and they don't have much room to move around anyway. If you change the position of the eggs at this point, the chicks have to reposition themselves for hatching, and doing so wastes valuable energy and may even make it impossible for them to hatch.

Hatching eggs

On the 18th day of incubation, you also need to increase the humidity in the incubator to 65 to 70 percent. You may want to increase the ventilation — refer to your incubator's directions to see whether doing so is advised. Get your brooder set up and warmed on the 20th day so you can transfer the chicks to it (see Chapter 14 for more on brooders).

Eggs that you put in the incubator at the same time likely will hatch within 18 hours of each other (see the section "Knowing what to look for: Stages of embryonic growth," later in this chapter, for more on embryo growth). Chicks struggle to get out of the egg, and it may take some time for a chick to fully hatch. If a chick requires help hatching, it usually isn't a strong, healthy chick.

When chicks start hatching, people get excited, and they want to open the incubator and handle the chicks. Stop right there! Leave the chicks alone until they're dry and fluffy. They're fine in the incubator for a few hours while the others hatch. Remove the dry, fluffy ones every 6 hours and put them in the brooder. Every time you open the incubator, you lower the temperature and humidity and make it harder for the ones still hatching. If any eggs are left after 18 hours from the time the first chick hatched, you can leave them for another 24 hours, but after that, examine them for signs of hatching and throw out any that aren't beginning to hatch.

Looking Inside the Egg

When you're incubating eggs, you may want to know what's going on inside them. If you're new to incubation or you have children, you may want to open an egg every couple days and look inside. Doing so kills the embryo, of course, but it gives you a fascinating look at the miracle of a chick forming in just 21 short days — from a glob of cells to a baby chick that can run around and feed itself. If you decide to open eggs and look inside, we suggest opening them on the 3rd, 7th, 12th, and 16th days of incubation.

You may even want to set extra eggs so you have eggs to sacrifice for this little biology lesson. You may get some eggs that didn't develop embryos, so set even a few more. For example, if you intend to open four eggs, you may want to start eight more eggs than you want to end up with.

If you don't want to open eggs and sacrifice the chicks inside just for a biology lesson, you'll be relieved to know that you can get a glimpse inside without opening an egg. In this section, we show you how.

Egg ultrasound: Candling an egg

Candling is a way to look inside an egg without opening it and killing the embryo. Candling involves shining a bright light on the egg in a darkened room and seeing a shadow inside. You can see the size of the air cell, veins in an egg with an embryo, and the dark mass that indicates the yolk and the embryo.

Candling works best with light-colored eggs, but you can candle brown eggs, too. If you do so quickly in a warm room, you won't harm the developing chick. You can buy candlers from poultry supply places, but they're also relatively easy to make (see Chapter 4 for a list of reliable poultry supply places).

To make a candler, follow these steps and refer to Figure 13-2:

1. **Line a cardboard box with aluminum foil.**

   The aluminum foil makes the light stronger.

2. **Place any good, strong source of light inside the box.**

   An LED flashlight or reading lamp works, as does a 60- to 75-watt incandescent light bulb.

3. **On one side of the box, make a ½-inch hole.**

To candle an egg, follow these steps:

1. **Wash your hands.**
2. **Turn off the lights in the room.**
3. **Quickly remove an egg from the incubator and hold it outside the box against the hole in your candler box.**

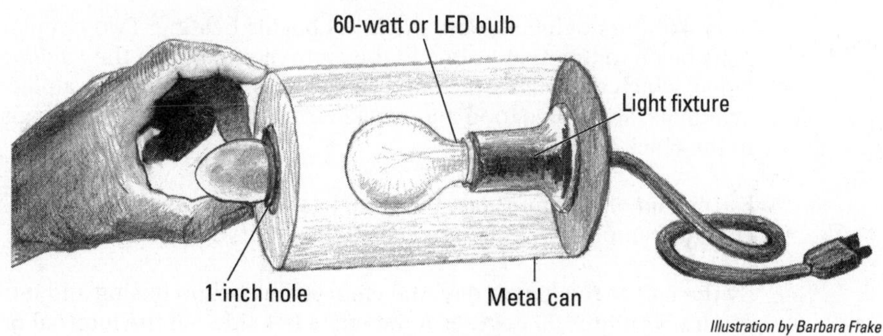

Figure 13-2: Making an egg candler.

Illustration by Barbara Frake

The beam of light will shine into the egg, allowing you to see shadows inside the egg. You're looking for the size of the air cell, dark embryo masses, and blood veins inside the egg.

Don't keep the egg out of the incubator too long — no more than 5 minutes. Mark the egg before you put it back in the incubator so you know which eggs you've looked at. If you intend to candle a few eggs more than once, you may want to number or otherwise identify each egg. You can candle an egg as many times as you want, but the more you handle the egg, the less likely it is to hatch.

Knowing what to look for: Stages of embryonic growth

A tiny chick develops very quickly. Almost as soon as the sperm penetrates the female reproductive cell, called the *blastodisk* or *true egg*, which is a small white spot on the yolk, the cells begin dividing to produce a chick. (For more information on fertilization, see Chapter 12.) By the time the egg is laid 26 hours or so later, a ring of cells has already begun to form. If the egg is incubated by the hen or put in an incubator, cell growth rapidly continues. But most eggs cool and go into a state of suspended growth for at least a few days.

As a chick develops, the air space in the egg enlarges, and the egg actually loses weight. Looking at the egg's air cell changes during candling is one way to see whether a chick is developing (see Figure 13-3).

After 24 hours of incubation at 100 to 102 degrees Fahrenheit, the chick's head, eyes, digestive system, and backbone have begun to form. They're still quite small — you need a magnifying glass to see these structures well, but rest assured that they're there (see Figure 13-4).

After 44 hours of incubation, the heart begins beating. Two circulation systems begin to function — one in the embryo and one in the *vitelline membrane*, which surrounds the embryo and acts something like a placenta in mammals, absorbing food and oxygen from the egg white and transferring it to the chick.

By the end of the third day, the beak has begun to form, limb buds are present, and temporary "gills" appear on the embryo.

By the end of the fourth day, the embryo has begun flexing and moving and generally rotates 90 degrees to lie on its left side. All the internal organs have formed by the end of that day.

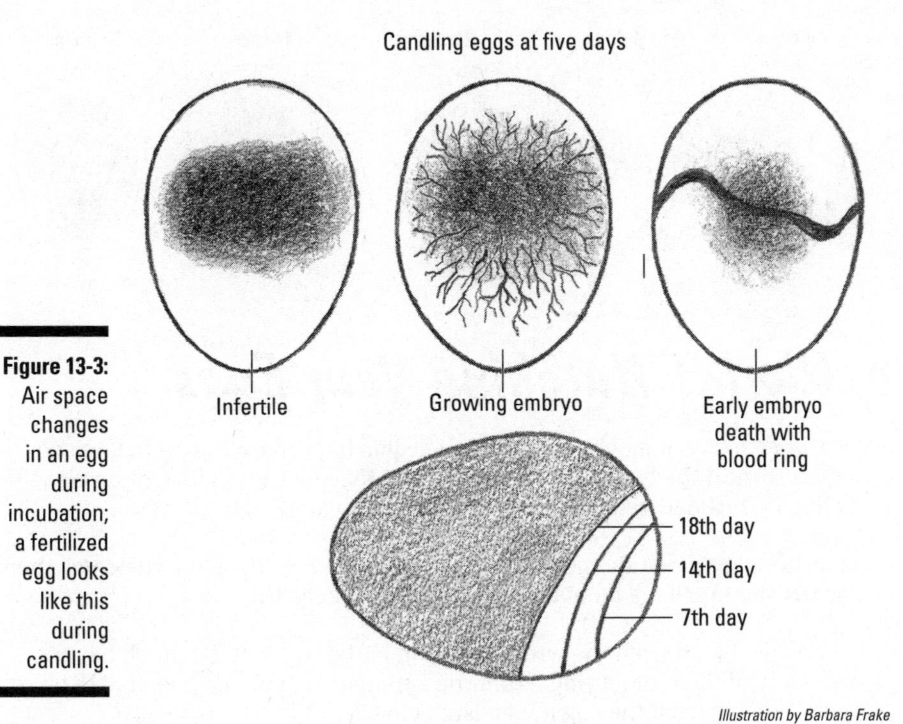

Figure 13-3: Air space changes in an egg during incubation; a fertilized egg looks like this during candling.

Illustration by Barbara Frake

By the end of the fifth day, the reproductive system has formed and begins to differentiate into male or female, although you can't tell the sex by looking at the embryo.

By the end of the seventh day, digits appear on the wings and feet that will make toes and wing sections.

By the end of the tenth day, the tracts where feathers will grow have formed, and the beak hardens.

By the end of the 14th day, the toenails have formed.

By the end of the 16th day, the embryo occupies much of the space inside the egg and has begun to use the egg yolk as a food supply because the egg white has been used up.

By the end of the 19th day, the yolk has been drawn up into the chick's body in preparation to hatch.

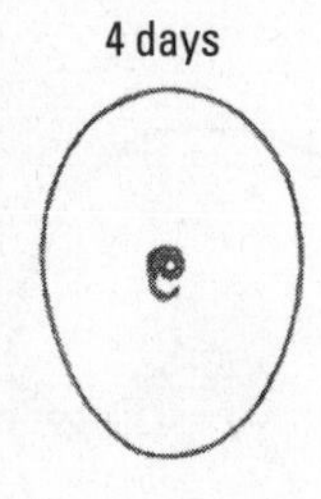

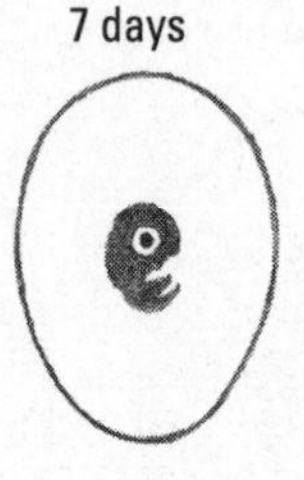

Figure 13-4: The stages of embryonic growth.

Illustration by Barbara Frake

Hello, World! Hatching Your Eggs

On the 18th day of incubation, a chick begins to prepare to hatch. Its head moves toward the large end of the egg, and the chick curls its feet up close to its head. The head gets tucked under the right wing.

You shouldn't move the position of the egg because, if you do, the chick then has to orientate itself all over again, which weakens the chick.

Chicks can hatch from the wrong position in the egg, but it's much harder, and many of them die trying. From the 18th day on, you may be able to hear peeps coming from the egg if you listen closely.

On the 20th day, the chick pierces the membrane around it, sticks its beak into the air space in the big end of the shell, and begins to breathe. As it uses the air up, on the 21st day, the chick uses its *egg tooth* (a hard area on the end of the beak) to break a hole in the eggshell, letting in outside air.

The chick chips away at the shell in a circular pattern, frequently stopping to rest. This process can take from 30 minutes to several hours. When the shell is completely cut around in a circle, that top circle pops off and the chick uses its legs to push off the back part of the shell. At this point, the chick is wet and tired, and it may lie there for an hour or more to rest and dry off before getting up and moving around.

Playing doctor: Helping a chick hatch

Sometimes people feel sorry for a chick that appears to be having trouble getting out of an egg. It may have *pipped* (made a tiny hole) and have its beak out, but it may seem unable to proceed any farther.

The temptation is great to help these chicks, but doing so can cause more harm than good. You can't just pull off the shell. Hatching is a slow process even in ideal conditions — you have to be patient.

Generally, when eggs pip but then fail to hatch, the temperature, oxygen level, or humidity is too low. The temperature may drop because someone is constantly opening the incubator to check on the chicks' progress. Poor ventilation may cause the chicks to become weak because of a lack of oxygen. And humidity can get too high and actually drown hatching chicks because they can't breathe. These problems may be correctable and the chicks then will finish hatching. But even if conditions seem to be fine, many experienced chicken owners believe that chicks that can't hatch on their own are doomed to either die anyway or live a weak, unhealthy life and so don't help them. New chicken owners usually want to help the chicks and it doesn't hurt to try. Just don't be too upset if they don't make it.

Sometimes the membrane has dried out too much around the chick, or the chick is in a bad position for hatching. If the pipped hole isn't in the large end of the egg, the chick is in the wrong position. If you see no hole but you hear peeping inside, the chick also may be in the wrong position. If you see a hole in the right area but the chick can't seem to finish hatching, the membranes may be dried out.

In these cases, helping — slowly and carefully — can save a healthy chick. Make sure the chick is still alive — it will move or peep if it is. If you've decided to try to help the chick, start by making a warm operating area with a padded, clean surface. Sterilize a small pair of nail scissors and a pair of tweezers with rubbing alcohol, or boil them for a few minutes. Have some clean, warm water nearby.

Under the shell is a thick membrane loaded with blood veins. The chick has already pierced the membrane in one spot if it has started to hatch. If you tear the membrane too early, it will bleed profusely and either weaken the chick or kill it.

With the scissors and tweezers, carefully pick little pieces of shell off the membrane, around the hole in the egg. Ideally, you work like the chick would, circling the large end of the egg and removing half the shell. Around the air cell, there may be no membrane, allowing you to remove shell pieces easily. If the chick has started a hole in the side or small end of the egg, the chick is in the wrong position, which is why it's having difficulty hatching. You need to be even more careful in this case.

Be careful not to cut the chick. If you cut a vein in the membrane and it bleeds, stop at once and put the chick in the incubator. You can't do anything to stop the bleeding. The bleeding will probably stop, but it weakens the chick. You can start again in an hour or so if the chick is still alive and not free.

If half the shell is removed and the head and neck are exposed, the chick should become active and wiggle out of the rest of the membrane and the egg. Moisten the membrane that's left with a little warm water and place the chick, with some shell still attached, back in the incubator. It will probably just lay there for a while. If the chick isn't up and walking within an hour or so, it's probably too weak to survive.

Handling the bad hatch: When things go wrong

If eggs hatch early, either the temperature was too high in your incubator or you counted the days wrong. If the eggs hatch late, the incubator was a little cold or you counted the days wrong. If the eggs are hatching over a wide time span (more than 18 hours), your incubator may have had hot or cold spots, or you may have improperly stored the eggs before you put them in the incubator.

If many eggs don't hatch by the 22nd day, pull some out and open them. If no embryo is inside, they probably weren't fertile or they were stored too long or improperly. If embryos started forming and then died, the temperature or humidity of the incubator wasn't right, the incubator didn't have adequate ventilation, or the chicks died from bacterial contamination or a disease. Not turning the eggs enough or handling them roughly can also cause embryos to die.

If you open eggs and find some living embryos, put the unopened eggs back and give them some more time. If many eggs don't seem to be hatching, always check a few eggs before you discard all of them.

If chicks hatch but they appear "sticky," with egg goo on them, the humidity may have been too high, the temperature may have been slightly low, the ventilation at hatching may have been poor, or they may have an inherited condition that causes this symptom.

Malformed or crippled chicks can be caused by slightly high temperatures during incubation, disease, chemical contamination of the incubator, or failure to turn eggs enough during incubation. Bad legs can also be caused by hatching chicks on slippery surfaces.

If hatched chicks have large, bloody, or mushy naval areas when hatched, the incubator temperature may have been slightly low, the humidity may have been too high, or the chicks may have a disease called omphalitis, which is caused by bacteria invading the naval area. For this reason, chicks should hatch on a dry, clean surface.

Chapter 14

Raising Chicks

In This Chapter

- Getting to know brooders
- Helping the hen help the chicks
- Feeding and watering chicks
- Knowing what's normal when raising chicks
- Combining chicks and children safely

All babies need food, water, warmth, and protection, but some need more help than others. Many baby birds are hatched naked and helpless, but in the pheasant family — to which chickens belong — when the chicks hatch, they're already wearing warm suits of down (also called *fluff*). After merely an hour or so, the chicks are on their feet and moving around, and they're fast on their little feet after just a few hours. They're also able to feed themselves. Their major needs from someone else are warmth and protection.

Because being warm and protected are primarily what chicks need outside help with, we begin this chapter with those subjects. Later we discuss the proper feed for chicks and the growth stages of chicks. We also take a look at any adjustments in care they may need as they grow.

You can give warmth and protection to chicks in two ways — by putting them in a *brooder* or by helping a hen protect them and letting her warm them. Because most people use a brooder, we start with that topic. If you have a hen raising your chicks, you can skip this brooder section and go right to the section "Helping a Hen Provide Warmth and Protection."

The Basics of Brooders

Baby chicks need to be kept quite warm their first few weeks of life. Without the proper warmth, they won't eat or drink and will soon die. In natural conditions, they snuggle under their mama's breast and belly, where the feathers have thinned out and they can get their backs up against her warm body. They leave her warmth to find food and water and do a little playing, but as soon as they get cold, back under her they go.

Chicks raised without their mama need a source of warmth, too, and that's where you come in. Unless you have the money and means to heat a whole room to 95 degrees Fahrenheit, you need to provide a brooder for your chicks. A *brooder* is an enclosed area that allows chicks to find that perfect warm spot but also gives them a little space where the air is cooler. A brooder also protects the chicks in the absence of a mother hen.

Many types of brooders are on the market (see the section "Buying a commercial brooder" for more details). Most use a heat lamp, but some use gas or propane to heat the unit. Alternatively, you can make your own brooder. Fashioning one from materials found around the home or from local stores is relatively easy to do (see the "Making Your Own Brooder" section for specifics).

Knowing when chicks need a brooder

Chicks need to go right from an incubator to a brooder. (We discuss incubators in Chapter 13.) If you're getting your chicks from another source, put them into the brooder as soon as you get them. Set up the brooder a few days in advance of putting chicks in it so you can see how it operates and adjust the temperature as necessary to keep it around 95 degrees Fahrenheit.

Your chicks need a brooder for at least a month, depending on the weather and where they're housed. If you're raising chicks at a very cold time of the year, they may need to be kept warm even longer, especially at night.

If the weather is very hot (above 90 degrees Fahrenheit), take special care to see that the brooder doesn't get too hot. Baby chicks can overheat. Temperatures above 100 degrees are too warm. Older chicks may be uncomfortable at temperatures above 90 degrees, especially if the humidity is high. In this case, you need to provide some ventilation or other cooling to lower the temperature. The chicks may still need the brooder to confine them and keep them safe, however.

Choosing the brooder size and shape

Most home flock owners raise fewer than 100 chicks at a time, which means they need some form of container in which to brood chicks instead of using a building like large-scale chick raisers do. You can partition off a corner of a room or building to brood larger groups of chicks, however.

A round ring of cardboard for the brooder wall can keep chicks from piling up in a corner and smothering each other. A 4-foot-diameter ring provides adequate space for 50 chicks for 3 to 4 weeks.

Average-size chicks need about 36 square inches per chick for the first month. This space looks like a lot of room at first, but chicks grow rapidly, and it won't look like too much room when they're a month old. Some brooders are designed to be expandable so they can grow as the chicks grow.

Brooders need to be at least 18 inches deep, especially if the source of heat is overhead, as is the case with heat lamps. If your brooder uses heat lamps, be sure to adjust them upward as the chicks grow so their heads can't touch them. Propane- or gas-heated brooders don't need to be as deep as heat lamp brooders because the heat source generally isn't overhead. However, a deeper brooder can help prevent older chicks from jumping out, and the brooder should have some sort of cover to prevent the chicks from escaping.

Chicks need to be able to move from an area of optimal warmth to a cooler area within the brooder, if possible, so they can find just the right temperature that makes them comfortable. Rectangular brooders allow the heat source to be at one end, with progressively cooler temperatures toward the opposite end of the brooder. With square or round brooders, the heat source needs to be in the center.

Getting the temperature just right

For the first week, baby chicks need a temperature of 95 degrees Fahrenheit. Measure the temperature at the height of a chick's back, about 2 inches or so from the floor. Every week, lower the temperature by 5 degrees until you reach about 70 degrees. If temperatures fall below 50 degrees at night, you may want to turn on the heat at night for a few more weeks.

Gas or propane brooders generally have a thermostat. If a heat lamp is your brooder's source of heat, you need a good thermometer in the brooder to keep track of temperatures. Your thermometer needs to be a nonmercury one (in case it breaks) that has markings that are large enough for you to read from outside the brooder. Brooder thermometers are sold in poultry supply catalogs, but any thermometer that you can read with ease is workable. Check the thermometer often in the first days of using a brooder so you know how the temperature inside fluctuates with the temperature outside. Usually you adjust temperature by adding more ventilation, raising or lowering the heat lamp, or changing the size or wattage of the heat lamp bulb.

Infrared heat lamps may not register the heat correctly on a thermometer because those lamps are designed to heat the chicks, not the air. With an infrared lamp, you need to watch the chicks to see whether they're warm enough. We discuss infrared heat lamps in more detail later in this chapter.

When using heat lamps, you may want to hang two different-size light bulbs in the brooder so you can quickly switch them on and off if the birds may undergo big changes in environmental day and night temperatures.

Most brooders have solid sides to help hold in heat. If the temperature keeps getting too hot, a few holes drilled in one or more sides of the brooder, a few inches from the bottom, can help airflow. You access the brooder from the top and ventilate it there, too. Hot air rises, and cool air is pulled in as hot air rises. Adjusting the open area at the top helps regulate the temperature.

Gauging temperature by your chicks' behavior

After you gain a little experience, you won't need a thermometer to tell you whether the temperature in a brooder is right. The actions and sounds of the chicks will let you know if it's too hot or cold. Watching the chicks for a few minutes will tell you whether everything is all right (see Figure 14-1). Keep a few tips in mind:

- If the chicks are casually and quietly walking around, eating and drinking, and some of them are peacefully snoozing near the heat source, the temperature is probably fine.
- If the chicks are huddled in a pile near the heat source or in a corner of a propane or gas unit, cheeping shrilly, they're cold. Cold chicks also eat less and are less active.
- If the chicks are spread along the edges of the brooder far from the heat source and have their beaks open, panting, they're too warm. Hot chicks drink more, eat less, and aren't very active.

Eventually, if the temperature isn't corrected, you'll begin losing chicks.

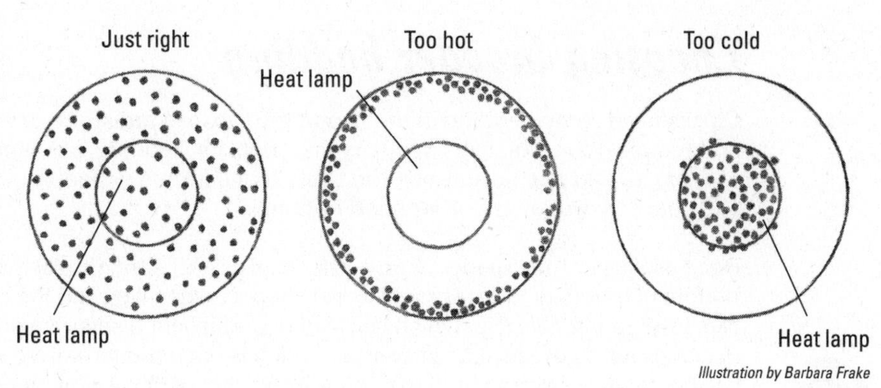

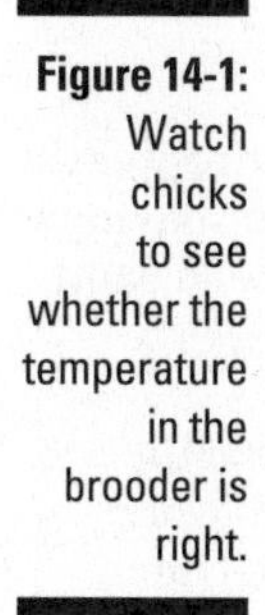

Figure 14-1: Watch chicks to see whether the temperature in the brooder is right.

Lighting the brooder

If you use regular light bulbs or heat lamp bulbs for heat in a brooder, additional lighting probably isn't necessary. If your heating comes from infrared or red heat bulbs, propane, or gas, you may want to add some overhead lighting. Make sure additional lights aren't too close to the brooder, or they'll affect the temperature, unless they're fluorescent bulbs.

For chicks designed to be pets or layers, keeping the overhead lights on for about 12 hours a day is sufficient; however, for modern broiler chicks, the lights need to be on at least 18 hours a day. These chicks need to eat more to sustain their rapid growth, and chicks don't eat in the dark.

If you're not in the habit of turning off the lights while the chicks are in the brooder, they may be thrown into a panic if the lights go off and it's suddenly dark. You can train them a bit by occasionally turning off the lights for a short time and then turning them back on. A dim light may keep them from panicking. Just don't forget to turn the lights back on, or the brooder may get too cold.

As the chicks get older and are ready to transition to a pen, get them used to less light at night. You can dim room lighting each night by using a dimmer switch, using heat bulbs of lesser wattage in the brooder if heat is still needed, or moving the heat bulbs farther away. If the housing you're transferring the chicks to gets dark gradually at night, the birds will probably transition quite well without any elaborate preconditioning. If, however, the housing gets very dark with the flip of a switch, you need to either install a night light or do some darkness preconditioning.

Choosing brooder bedding

Chicks need some bedding in the brooder to absorb moisture, dry droppings, and reduce odor. You can use many materials for bedding, but pine shavings are easy to find and seem to be the best choice for many people. Clean sand and coarse sawdust are other good options.

Don't use paper (newspaper or any other kind), cardboard, or plastic on the bottom of brooders. These materials get slippery when wet, and the chicks can develop *spraddle legs*, which can lead to permanent disfigurement. Other materials not to use include kitty litter (which is often treated with dyes and scents and may swell if ingested), soil, vermiculite, perlite, cedar shavings (which sometimes irritate eyes and lungs), cotton batting, hay, leaves, and straw.

Hay, leaves, and straw, in particular, may contain harmful substances, get slippery and moldy when wet, and aren't absorbent. Also avoid shavings made from unknown wood or from hardwoods. When moist and warm, hardwood shavings can develop a harmful mold. Anytime you notice mold on bedding, remove the bedding at once.

How often you need to change the bedding in a brooder varies with the type of litter, the age and number of chicks, the type of brooder, and many other factors. In general, if the bedding is damp and smelly, you need to change it completely. If it appears dry and fluffy but dirty, just add a fresh layer on top. After a while, the litter gets either wet or too deep for the brooder and needs to be changed.

You usually need to remove the chicks to change the litter. Place them in a large box where they'll be safe as you work. Unless it's really cold, they'll be warm enough for a short time without a heat source. If you wash the brooder with disinfectants or strong soaps, make sure you rinse the brooder well and dry it before adding bedding and chicks.

Always scrub a brooder between batches of chicks. Use a good disinfectant, such as common bleach (1 part bleach to 4 parts water), rinse it well, and let it air dry.

Buying a commercial brooder

You can buy brooders, or parts to make them, from poultry supply catalogs, but you also can put together very functional brooders at home with common items you can buy at a hardware or farm store. If you're brooding a few chicks once or twice a year, a homemade brooder is probably the most economical and practical way to go (we discuss ways to build a homemade brooder in the next section).

If, however, you want to raise many batches or large batches of chicks throughout the year, you may want to look into a commercial brooder. Shipping for a brooder can cost almost as much as a brooder does, so you may want to check newspaper ads and ask around locally for a new or good used model.

One type of commercial brooder is the *battery* brooder, which was once very popular. Battery brooders are small cages in a stacked arrangement, with the back of the cages enclosed and heated. The heat source generally runs up the back of the unit. Battery brooders are still available but are expensive. Each little cage section holds a limited number of chicks. The chicks don't have much head room and need to be moved to larger quarters in a short time.

Other commercial brooders generally consist of a heat source — either electric, gas, or propane — enclosed in a metal ring or *hover,* which the chicks get under to access the warmth. You place this unit inside a larger confined area that needs to be lighted because the heat source doesn't emit light. If you use anything but electricity to heat brooders, put a carbon monoxide detector in the room to protect you and the chicks. Be sure to place all brooders in a weatherproof building.

Making Your Own Brooder

Whatever kind of brooder you build needs to be easy to move and clean, large enough for the batches of chicks you intend to brood, strong enough to keep out predators, and not easily ignited or melted by a heat source.

Homemade brooders, of course, need to be inside a dry building with electricity. The heat source for homemade brooders is generally a heat lamp. If the building is heated, running your brooder in cold weather costs less.

You probably don't want a brooder in your house. Baby chicks smell and raise dust, and a brooder creates an added fire hazard. Chicks can also be a source of salmonella bacteria. A brooder may work in the basement or garage, but a barn or shed is better if you have one.

Some people like to set the brooder right on the floor, but we like them elevated so we can reach the bottom without stooping. If you do set yours on the floor, place something like cardboard, layers of newspaper, or insulation under the brooder to insulate it from the cold floor or ground.

The tops you use on brooders need to be strong enough to keep out any predators, as well as strong enough to support the weight of other farm animals, such as barn cats, which are often attracted to the heat of the brooder and may sleep on top of it. You need to run the electric cord for your heat source either out a corner of the body or through the top. Tops need to be adjustable to provide ventilation.

Building the body of the brooder

The most common way to build a brooder body is simply to make a box out of wood. Use fairly thick wood for good insulation. You can either put a bottom on the box or set it on a tray or the floor. Use a piece of small wire mesh in a removable frame for the top. You also need pieces of wood, glass, or Plexiglas to arrange on top to cover as much of the opening as necessary to keep heat in but still allow for ventilation. Brooders with solid sides work best so the chicks aren't exposed to drafts. However, if you're brooding chicks in very hot weather, a wire side or two may help with ventilation in the daytime.

Here are some common brooder body ideas:

- **Kiddie pools:** You can also make a quick brooder out of a kiddie swimming pool. Choose a hard plastic- or metal-sided one — not a blow-up pool or one made from flimsy plastic. And look for one at least 18 inches deep. Again, leaky pools are fine. Keep your eyes peeled for discarded pools. You don't need the liner for metal pools, just the siding.
- **Plastic storage tubs and containers:** With a little care, you can use these containers for small batches of chicks. But they may melt if a heat bulb is too close to the sides. Don't use a top without holes for ventilation!
- **Stock tanks:** You can make an excellent brooder from a *stock tank* — a large tank found in farm stores to water livestock. These tanks come in various sizes and are made of metal or heavy-duty resin material. Pick a fairly deep tank in the size that allows about 6 square inches per chick. A tank about 4 feet long by 3 feet wide is big enough for 25 chicks. You need small mesh wire for the top and, once again, lumber, glass, or other solid material to cover part of the top. Stock tanks are usually oval or round in shape. Oval tanks are easier to fit covers on. Stock tanks can be expensive but last many, many years. Look around for used tanks for sale at yard sales or farm auctions. You may even find one that's been discarded. A leaky metal tank works just fine for a brooder and is often free.

Don't use cardboard boxes or rings of cardboard. Boxes are hard to clean, and they get wet, deteriorate, and can catch fire. Rings of cardboard tend to get damp on the bottom and start to fall apart. Also, never use containers that once held toxic materials, such as pesticides — some containers absorb some of the toxic contents and then emit them as a gas, even if washed. It doesn't take much to kill baby chicks.

If you're really into recycling, by all means, think outside the box, so to speak. For example, check out these ideas:

- We've seen an old chest freezer used as a brooder. It's well insulated, so it holds heat well, and it's easy to clean. You can remove the motor to make it easier to move (though it's still a bit heavy). You can lower or raise the lid to regulate heat. You can also put in a wire top under the lid. Attach two blocks of wood to the edge of the top opening so the lid won't ever shut all the way by accident and shut off the air supply.
- We've also seen dryer drums, old bathtubs, and stacks of tractor tires used as brooders.
- Large aquariums also work for small groups of chicks. Make sure you cover the glass bottom with bedding.

Use your imagination, and you may find the perfect brooder.

Heating the brooder

Most home flock brooders use a light bulb for warmth. Using a light bulb is a simple solution, and chicks seem to actually prefer light with their warmth. They're attracted to light, which helps them find the warm spot.

Chicks also seem to like the heat to come from overhead. This preference may be because their backs are against a hen's warm body when they're under a hen, so overhead heat feels natural. And because the chicks' lungs are close to the ribs and backbone and since blood circulation runs through the lungs, using overhead heat is an efficient way to warm the body.

Floor heat isn't a good way to heat chicks. Don't set a box on a heating pad.

Looking into reflector lights

Almost every hardware or farm store carries a *reflector light*, a bulb socket with a metal hoop around it to reflect light and heat. For use in a brooder, choose one with a ceramic socket, a UL-listed (Underwriters Laboratory) tag, and a ring on the back of the socket to suspend the light. Those reflector lights made for brooding or as a heat source also have guard wires in the front of the reflector to keep the bulb from touching bedding if it falls.

If you need to use a high-wattage bulb — more than 150 watts — to keep the brooder at the correct temperature, don't use a reflector light with a plastic socket because it may melt. Always check the listed maximum wattage for the socket, and don't use bulbs over that maximum. The socket either needs to have a tag should or be embossed with the maximum wattage.

Choosing bulbs

You may need to buy special heat lamp bulbs or reflector bulbs. Regular incandescent light bulbs work if the brooder is in a fairly warm area and doesn't need much heating, but you may have a hard time finding bulbs in high-enough wattages to warm a brooder in cold temperatures. Fluorescent bulbs of any kind don't work because they don't give off heat.

Heat lamp bulbs come in infrared and regular light wavelengths. Infrared bulbs give off less light with the heat, and new chick raisers may find it difficult to regulate heat with them because although the chicks feel the heat, it doesn't register well on a thermometer. Using infrared bulbs means you have to rely on watching the chicks to be sure the temperature in the brooder is right (see the section "Gauging temperature by your chicks' behavior" for more details).

Don't worry about the light regular bulbs give off. Yes, it's bright right under the light, but chicks don't seem to mind. Reflectors point the light down, and the other areas of the brooder aren't quite as bright.

The number and wattage of bulbs you need depends on the temperature of the room the brooder sits in, the size of the brooder, what the brooder is made of, and how many chicks you're going to brood. A 250-watt heat light bulb is usually the highest wattage a reflector lamp is rated to hold. This size bulb in a brooder of about 15 square feet sitting in an area where the temperatures don't fall below 45 degrees provides enough heat for 25 chicks. It may be enough for 50 chicks in a 25-square-foot brooder if the outside temperature is warmer.

Don't use reflector bulbs or other bulbs that are treated with a nonstick coating; they are usually labelled "shatter resistant." These bulbs emit dangerous fumes when heated that can harm or kill your chicks.

You may need to experiment to find out what size bulbs and how many you need in a homemade brooder. That's why it's wise to set up the brooder before you have the chicks. Keep in mind that you don't have to heat the whole brooder to the optimum temperature — only an area sufficient to let all the chicks rest under it without piling on top of each other.

A small batch of chicks probably needs just one lamp. If you need more than one lamp, you can use 150-watt bulbs in two lamps, one 150 and one 250, or some other combination. Hang the lamps close together instead of at opposite sides of the brooder. You want to concentrate the heat in one area and let another area be cooler.

When you're buying bulbs, pick up an extra bulb or two. Bulbs can fail, and it often seems to happen at a time when you can't easily find another. Usually, though, one bulb lasts the entire brooding month, even when it's on 24 hours a day. Buy another bulb one size smaller, or dimmer, than you think you need, in case the chicks appear too warm — it's also good for the end of the brooding cycle, when chicks need less heat.

Installing heat lamps

Suspend the lights over the chicks so that they point down. Don't attach them so that they reflect light sideways. Chicks get the greatest warmth when the light reflects directly downward. Raise and lower the height of the lights to adjust temperature. Hang the lights in the center of a round or square brooder, or at one end of an oval or rectangular brooder.

Don't suspend the light from its cord — it may pull out of the receptacle and fall, and the light will be difficult to position if the plug receptacle is near the floor. Always use a chain or wire to suspend the light from the ring on the back of the reflector (a small, cheap dog tie-out chain works well). Usually you have to run the chain or wire up through a hole in the brooder lid.

If you don't have something strong above the brooder to suspend the light from, you may need to lay a pole or rod across it. Make sure it can't roll or fall into the brooder. If you have a sturdy lid, you may be able to suspend the light from it, but if you raise the lid every time you need to put food and water into the brooder, this arrangement will be awkward.

Start with suspending the lights about a foot from the brooder floor. If the temperature is right for the chicks, leave them there. Never put the lamp so close to the chicks that they touch it. If you have to raise it to above the top of the brooder to make them comfortable, use a smaller-wattage bulb.

As the chicks grow and need less heat, raise the bulbs, reduce bulb wattages, or adjust the heat by opening or closing more of the top. Having a wire top in a frame facilitates this by allowing you to place something solid over it to conserve warmth or remove part of the solid material to let heat out. The wire top also keeps chicks from getting out and predators from getting in.

Putting safety first

Unfortunately, heat lamps and other methods of warming animals have caused many home and farm fires. Always use extreme care when using such sources of heat. We've mentioned some safety tips already, but here's a rundown:

- **Always use UL-approved equipment.** UL stands for Underwriters Laboratory, which rates the safety of electrical items.
- **Never exceed the wattage listed for a bulb socket.**

- **Check cords for frayed and broken areas before use.** Don't use lamps or heating equipment with damaged cords.
- **Use proper extension cords, if needed, but don't run extension cords outside unless they're for outdoor use.** Keep cords out of wet areas and anywhere passersby can trip over them.
- **Don't suspend lights by their cords.** Make sure you suspend them from a sturdy wire or chain and they can never fall into the bedding. Leave the guards on the reflector.
- **Keep heat bulbs away from anything that can melt or ignite.** Avoid paper, cardboard, and the sides of thin plastic containers.
- **Keep water dishes as far from the bulb as possible.** If cool water splashes on a hot bulb, it will break.
- **Keep cords out of brooders and away from where animals can chew on them.** Chicks rarely peck at cords, but they sometimes try to roost on them. Damaged cords can electrocute you, other animals, or chicks and can start fires.
- **When using gas or propane heat, make sure the ventilation is good, and use a carbon monoxide detector in the area.**
- **Keep anything combustible from touching a heat source.**

Helping a Hen Provide Warmth and Protection

If mother hen is present when the chicks hatch, she'll take care of their warmth. She needs just a little help from you to protect the chicks. If they're penned up, you need to separate the mom and her chicks from other chickens. In a coop, make sure the babies aren't able to get out and be harmed by other chickens. Roosters seldom bother chicks, but not all hens are motherly types, and they may hurt chicks.

In a free-range situation, if a hen hides and hatches eggs somewhere, the other chickens probably won't bother her — but her babies will be at big risk from predators. Mama hen will try to protect them, but often she's no match for sneaky or overpowering predators. Mother hens may also steal babies from other hens if the babies are about the same age. A hen does have a limit to how many she can care for, and if chicks end up with surrogate moms, you may never know who's related to whom.

To save as many chicks as possible, enclose mama hen and chicks in a cage or pen that the chicks can't squeeze out of, at least until the chicks are about half grown and well feathered — about 3 months old. If the chicks do manage to get out and the mom can't, she has no way to protect them.

In pens or cages where baby chicks are present, make sure the chicks can reach water, but also ensure that the dish is too shallow for them to drown in. Chicks need finely ground starter feed instead of adult feed. The water and feed issues are another reason why separating chicks from older birds is a good idea. If they're free-ranging, they may find enough to eat in nature, but it doesn't hurt to offer some chick feed in a shallow dish.

Feeding and Watering Chicks

Baby chicks don't actually *need* to eat or drink right away. They have the remains of the egg yolk in their abdomen that will sustain them for a couple days — that's how they can be shipped in the mail to you. However, if you hatch them at home or get them soon after hatching from a local source, it's a good idea to offer feed and water as soon as the fluff has dried and the chicks are moving around.

Nature built in a little time before the chicks have to be fed, to give other eggs more time to hatch. In nature, a few chicks often hatch before the rest. The hen stays on the nest waiting for more eggs to hatch, and the hatched chicks usually stay under her. After a day or so, the hen leaves the eggs that didn't hatch and leads her young ones out to forage.

Starter feed choices

Baby chicks need chick starter feed, which is available at farm and pet stores. It's formulated for them and sized right for tiny beaks. Adult chicken feed, feed for other animals, bread crumbs, and so on may keep chicks alive, but they won't grow well with these types of food.

Chicks grow fast. The broiler-type chicks especially need a high-protein starter feed, or they'll quickly start having problems with their legs. The feed is labeled as "broiler feed" or "meat bird feed." You can also use game-bird starter feed. Whatever you use, make sure the protein level is a minimum of 22 percent. If chicks don't get enough protein to sustain their rapid growth, their legs can become weak and twisted, and their wings may also grow distorted.

If you have only pet and layer-type chicks, you can use regular chick starter feed. When you have a mixture of meat and layer birds, use meat bird starter. It doesn't hurt the others, although you want to switch the layer chicks to a lower-protein feed when you separate them and move them from the brooder around 5 weeks. Regular starter feed needs a protein level of 20 percent.

Most baby chick starter rations don't require added grit. *Grit* is a fine gravel that helps birds digest food. If you don't use commercial feed, you need chick grit — unless the chicks are free-ranging with the mother hen and finding their own. You can purchase baby chick grit in feed stores, or you can use canary or parakeet grit from a pet store.

Medicated feed

Another choice to consider is whether to buy medicated or unmedicated feed. To get baby chicks off to a good start, medicated feeds include antibiotics and medications to control parasites and bacterial diseases. When chicks have heavy infestations of parasites like coccidia they are more susceptible to viral diseases like Marek's.

We strongly recommend medicated feed for the first month of a chick's life, especially if you don't vaccinate chicks. But we also know some chicken keepers worry about raising animals on antibiotics and other medications. If you choose to use unmedicated feed you may lose more baby chicks and your chicks may not grow as well as those on medicated feeds, but it's a personal choice.

Meat birds may need to be off medicated feed for a few days before you butcher them, to avoid meat with medication residue. Some modern starter medications can be used right up to slaughter. Check the feed label for how long meat birds need to be off medicated feed, and always follow the recommendations. In general, if you have enough chicks — say, 25 — you'll use up a 50-pound bag in a few weeks. You can then switch to unmedicated starter feed, if you like. If you're having problems with disease or your chicks don't seem to be doing well, leave them on medicated feed longer. Stop using medicated feeds in chicks that will be laying hens when they're 16 weeks old.

Medicated feed must be labeled as such, but feed bags often look similar, so always check the label. It may be printed on the bag or attached to a paper tag sewn into the top or bottom seam.

Alternatively, you can buy medications for coccidiosis or antibiotics to add to the chicks' water. However, don't use medications in the water of baby chicks that are eating medicated feed without a veterinarian's recommendation. Also don't use medicated starter feeds if your chicks were vaccinated for coccidiosis at the hatchery because it will make the vaccination fail.

The feeding process

Place the feed in fairly shallow containers or dishes at first. Use dishes that are long and narrow or that have a slotted cover to prevent chicks from walking in the feed or scratching it out. As chicks grow, so must their feed containers. Have several feed containers or one large enough that all the chicks can eat at one time. You may need to add feeders in addition to switching to bigger ones as chicks grow. Many chick feeders and water containers are colored red, which is thought to attract chicks, but they'll learn to eat and drink from containers of any color. Figure 14-2 shows some common feed and water containers.

Meat-type chicks need to have feed before them for at least 16 hours a day. But a few hours without feed, usually achieved by darkening the brooder so they can't eat, makes the broiler chicks grow a bit more slowly, which makes them healthier. For other chicks, a dish that's empty for a few hours is okay, as is feed available all the time. But if your chicks act ravenous and swarm the feeder when you add feed, you're not giving them enough.

Keep feed dishes clean. If you're using a self-feeding, hopper-type feeder, make sure you check it frequently to see that it has feed and is working correctly.

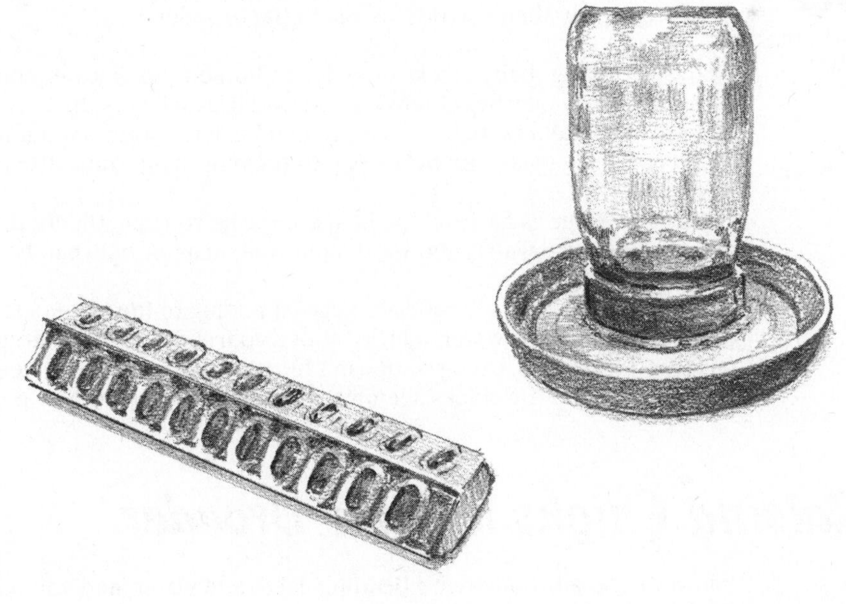

Figure 14-2: Examples of typical chick feed and water containers.

Illustration by Barbara Frake

You can lead a chick to water ...

Always have water available for your chicks. A special waterer for baby chicks with a shallow, narrow opening is best. You can buy inexpensive rings that screw onto a wide-mouth jar or purchase larger, more elaborate water holders.

Baby chicks will drown in deep water containers, so don't use anything more than a couple inches deep.

If you use open containers, such as saucers, lids, and so on, add marbles or small stones so that there's little open surface. You want to prevent chicks from walking through water or perching on top of water containers. Keep the watering station clean. The area around the dish generally becomes wet and messy, and water dishes often get shavings or other bedding kicked into them. Scrub out any algae or scum that develops in containers. Then disinfect with a bleach solution of nine parts water and one part bleach, and then dry. If you want to inhibit mold growth without harming the chicks, use four to six drops of unscented household bleach per gallon of drinking water.

Placing the water container on a large, shallow tray or lid to contain spills helps keep the litter dry and keeps bedding out of the water container. Some people elevate the water container on a small block of wood. But even when you do so, you may still need to change the bedding under the water area more frequently than you do the rest of the brooder.

As they get older, baby chicks often try to hop on top of water containers. This high perch may help them escape, and it usually results in fouled water and water containers that you hate to handle. If the water container has a flat top, try gluing a plastic funnel on top to prevent them from sitting there.

Place your water as far from the heat source as you can. Chicks don't like to drink really warm water, and water splashed on a hot bulb can break it.

For the first day, some hatcheries suggest adding to the water a teaspoon of sugar per quart of water, a little bit of a sports drink, or some special mix they've devised. Doing so doesn't hurt but generally isn't necessary for healthy chicks. If the chicks seem very weak, adding sugar may perk them up.

Raising Chicks in Your Brooder

You've set up a brooder with bedding, feed, and water, and you're checking the temperature frequently to see how you need to adjust it for your conditions. Now you're ready to check out some other considerations for getting chicks off to a good start.

What to do the first hour

The brooder is nice and warm, and you have a box of chirping chicks in your hands. What do you do now? If your feed and water dishes aren't already full and waiting, fill them. Use room-temperature water if you're filling the water dishes just before placing chicks in the brooder so you don't lower the brooder's temperature.

Remove the chicks from their shipping container one at a time. Hold each one next to the water container and gently dip its beak in the water. Just dip — don't hold its beak in the water. Chicks are often dehydrated from shipping and need water more than they need feed. This taste of water encourages them to drink. Spending a bit of extra time dipping beaks can get your chicks started right.

If you have a chick in the container that looks very weak and doesn't stand on its own, try to get just a drop or two of warm water into its beak and then place it under the heat source. That's the best you can do. Promptly dispose of any dead chicks by burying them.

The first few days

The first day, place some large, shallow containers (large enough that the chicks can walk on them) on the floor of the brooder with some feed scattered on them. Keep the regular feed dishes in the brooder, too. Some people use the lid of the shipping container for this large container; others use the tops of egg cartons, deli trays, or paper plates. The chicks quickly find out what feed is because pecking at the ground for food is natural for them.

Getting the chicks settled in

The baby chicks will be a little confused and noisy at first, but if the conditions are right, they'll soon settle down. For the first couple hours, check on them frequently, but don't disturb them, if at all possible. The first day or two isn't the time to let the kids play with the chicks. They're under stress and need time to adjust to their new home.

Sprinkling a bit of feed in front of the chicks attracts them by movement, as does tapping a pencil or small stick in the area of the feed. In nature, the chicks would be watching their mama as she pecks for food and then imitating her. Chicks copy each other, so after a few are eating and drinking, you can relax. All the other healthy chicks will soon follow their lead.

Spotting health troubles

A dead chick or two in the first few days is pretty normal and can be attributed to shipping stress or genetic weaknesses. If many chicks die, or if chicks are still dying after a week, something is wrong and you need to check all conditions in the brooder.

Observe the chicks to see whether they're eating and drinking. Is the water too warm? It shouldn't feel warm to the touch. Can they reach the food? Try putting in some flat, shallow dishes again, with feed scattered on them. Do they appear cold? Or too hot? Check with a thermometer on the floor. If you're using propane or gas heat, do you have enough fresh air ventilating the brooder to prevent carbon monoxide buildup?

If you think the conditions are good, you may want to contact the seller of the chicks and say that you're having a problem. Don't be accusatory; just tell what's happening. The seller may be aware of the problem and replace the chicks, or may offer you advice on what to do.

If you get replacement chicks, don't mix them with the previous chicks, if any are left. If they were diseased, they can infect the new batch. If all the chicks from the first batch are gone, clean and disinfect the brooder before you place the new chicks in it. If some of the old chicks are still using the old brooder, you'll need another brooder for the new arrivals.

Inoculating for diseases

Pullorum and fowl typhoid were once serious poultry diseases, but the National Poultry Improvement Plan (NPIP), started in the 1930s by the U.S. government, largely eliminated them — although occasional outbreaks still occur. NPIP is a voluntary plan to test breeding birds and then eliminate any birds that are ill or that are carrying the disease. When you buy chicks, always buy them from a hatchery that has an NPIP-certified pullorum/typhoid-free status. No vaccinations exist for these diseases.

Make sure your chicks are vaccinated for Marek's disease, if the hatchery offers it. This inoculation is almost always done at the hatchery at 1 day of age. Sometimes you're charged extra for the vaccination, but it's worth it. If you're unsure whether the hatchery vaccinates chicks for Marek's disease, ask someone who works there. Marek's disease is a common and highly contagious disease caused by a herpes virus, and it can kill unvaccinated chicks. Chicks from all over the country need to be vaccinated.

Some hatcheries now offer vaccination for coccidiosis, a common parasite that can cause poor growth in chicks or even kill them. If it's offered, get the vaccination. But don't use medicated starter feed on chicks vaccinated for coccidiosis because it will cause the vaccination to fail.

If Newcastle disease occurs in your area, you may want to vaccinate day-old chicks for this viral disease, too. Newcastle is a respiratory and neurological disease carried by wild birds that causes large losses in young birds. The United States is working to eradicate Newcastle, and vaccinating chicks is recommended. Hatcheries sometimes start the vaccinations, but vaccination requires three or more doses, depending on the vaccine used. Vaccines are easy to administer and can be put in the eyes or nose.

Home breeders who hatch their own chicks can vaccinate for Marek's or Newcastle disease, but it's difficult to do on your own and expensive to pay a vet to do it — if you can even find one who's willing to do so. For very rare or expensive chicks, the expense may be worthwhile. Vaccinations make chicks cost a little more, but the peace of mind is worth it.

To vaccinate the chicks that you hatch at home, you need to order the vaccines from a poultry supply company well before you hatch the chicks: They need to be vaccinated in the first few days after hatching. (See Chapters 10 and 11 for more on poultry diseases and vaccinating at home.)

Trimming beaks

Some chick hatcheries offer beak-trimming services. Chicks' beaks are trimmed to keep them from pecking at each other. After blood has been drawn on a chick, the other chicks may peck it to death. This type of behavior is more prevalent in certain breeds, such as game fowl and other aggressive breeds. Crowding and other stressful conditions may also prompt pecking.

If stress and crowding are kept to a minimum, home flock owners don't often encounter deadly pecking. Chicks do experience stress from the beak-trimming procedure. We believe beak-trimming should be avoided unless you've had problems in the past.

Home breeders can trim beaks if pecking becomes a problem in chicks. You can purchase an electric de-beaker if you have a lot of chicks to do — or simply cut off the tip of the upper beak using dog nail clippers. But we suggest you focus on providing comfortable, stress-free conditions for chicks and isolate any that become injured before you take these measures.

Preventing disease

It may be cliché, but the adage is true: An ounce of prevention is worth a pound of cure. Vaccinations are one form of prevention, but even if your chicks are vaccinated, immunity to certain diseases can take up to 10 days.

You can take the following precautions to keep your chicks healthy in the meantime and beyond:

- **Keep baby chicks away from adult birds and older chicks.** Of course, if you have a hen that has hatched chicks, she'll be near them.
- **Don't mix batches of chicks bought from different places.** If one batch has a problem, it'll spread to all the chicks.
- **Wash your hands between caring for different batches of chicks and between caring for older chickens and chicks.** Hand washing helps prevent the spread of disease among chickens, just like it does among humans.
- **Don't interchange feed and water containers.** Separation is especially important if you've had disease problems with your other birds.
- **Keep your brooder and chick-growing areas clean.** Promptly remove wet or moldy bedding. Keep your feed and water containers clean, too.
- **Observe your birds daily for signs of illness.** Remove and isolate any that appear ill.

People who show birds or who buy, sell, and trade birds are more likely to carry diseases to chicks on their hands, clothing, shoes, and equipment because they're in contact with all kinds of birds.

Watching the Stages of Growth

Baby chicks with the correct nutrition and warmth grow quite quickly. Different breeds may grow at different rates, with the modern broiler-type birds growing the fastest.

Baby chicks are covered in down at hatching, and they develop their first *pinfeathers* in the wing tips by the end of the first week. Feathers develop in the wing and tail areas first and then on the back and neck. You'll see some down still peeking out between feathers for at least a month.

As your chicks grow taller, they start to look like their adult selves. Don't count on the way they look now, though, to tell you what they'll look like as adults. Some feather colors and patterns will keep changing as they grow. And yes, they'll grow out of the ugly stage.

Don't put the young chicks with adult birds. They'll be bullied and may even be killed. They need their own secure pen until they're mature — about 6 months old.

One month: Tween-agers

At about a month, when the birds are fully feathered, they're ready to move out of the brooder to larger and cooler quarters. However, some bantam and slow-growing breeds may need a little more than a month in the brooder if they don't seem well feathered and the temperatures are cold.

If temperatures are below 50 degrees Fahrenheit, the chicks still need a brooder lamp in their larger quarters. If the outside temperature climbs above 50 degrees in the day, you can turn off the lamp, but turn it on again when the temperature falls.

At a month, chicks need a minimum of 2 square feet of floor space each. They can go outside if the weather is dry and warm, but they must be well protected from predators and shouldn't be turned loose to free-range yet.

Remember to change the size or number of feeders and water containers as your chickens grow. In warm weather, 25 chicks that are a month old or older may drink more than 1 gallon of water a day.

Six weeks to maturity: Teenagers

At 6 weeks, change the feed of broiler-type chicks to *grower* or *finisher* feed, with 20 percent protein, until they're butchered. They may be almost ready to butcher at 6 weeks, depending on how large you like your meat birds to be. Don't let broiler-type hybrid chicks grow too long. The bigger they get, the more prone they are to sudden death and to blisters on the breast and other health problems. They'll likely all be ready to butcher by 10 weeks.

Give other types of chickens — pets and layer pullets, for example — grower feed of 16 to 18 percent protein at 6 weeks. Switch pullets that are going to be layers to layer feed, with 16 percent protein and additional calcium and other nutrients, at 22 weeks or when laying begins. If the chickens are from large, slow-maturing breeds like Cochins or aren't meant for heavy egg production, you can hold off switching to layer feed until you know they're laying eggs.

Most chicks start to *roost* (perch off the floor at night) as soon as they get feathered, so when you move them out of their first brooder to other quarters, provide them with roosts that are off the floor but low enough for them to hop to. For most chicks, that's be a foot or two off the ground. Meat birds don't require roosts. Check out Chapter 7 for more about roosts.

I'm a big chicken now: Young adulthood

When chickens have reached full size and have come into sexual maturity (which varies from 18 weeks to 25 or more weeks, depending on the breed), they're ready to move to their adult home — if they aren't already there. If they're crowing or laying eggs, they're mature. However, don't suddenly throw young birds into housing with older birds. They'll be bullied, pecked, and maybe even killed — especially when you put young roosters with older males.

Instead, gradually introduce the old and young birds to each other. If you allow your birds some free-range time, that's a good time to introduce new flock members. (See Chapter 10 for more on introducing new birds.)

After a cockerel is crowing well (and maybe even before), he's capable of mating with hens and fertilizing eggs. If you don't want this to happen, separate cockerels from hens by 20 weeks. For more about mating chickens, see Chapter 12.

Young pullets start "playing house" as they near the 18- to 22-week mark, especially if it's spring or summer. Winter weather and long nights may delay maturity, and some breeds also take longer to mature. Pullets explore nest boxes, sit in them and croon, and arrange the nesting material before they begin laying.

It's wise to put in nest boxes for young hens — if they don't have any — by 18 weeks. The practice of playing house tends to encourage good egg-laying habits later. The first eggs the hens lay will be very tiny and may be oddly shaped, but in a week or two, they'll be normal sized. For information on training young layers, see Chapter 15.

Chicks and Children

Children love fuzzy baby chicks, and young children often think of them as toys. But of course, they're not toys, and the decision to buy chicks for children should never be made on a whim. If your children are begging for the baby chicks they see in the farm store, don't buy them that day (unless that was your secret intention).

If the children are old enough to help you plan, get them involved with looking up information on caring for chicks and making a brooder together. You should have already made some decisions — and several of the earlier chapters in this book can help you — about why you want to keep chickens and how you're going to do it. Only then should you buy the chicks.

Older children who want the chicks for school, 4-H, or FFA projects should be very involved in planning and setting up a brooder and choosing what breeds of chickens to raise. But don't fool yourself: We can tell you in advance that, no matter how motivated and responsible kids seem to be, you have to supervise the care of or actually care for the chicks.

If children want to show the birds in 4-H, FFA, or poultry shows, the farm store probably isn't the best place to get the chicks, unless you can preorder the breeds you want. When you order from a hatchery and say that the birds are for showing, knowledgeable people there may be able to select premium chicks — ones that appear to be robust and have no obvious show faults.

If you've succumbed to temptation, your children surprised you, or a relative thought chicks would make a good birthday present, go back and reread this chapter. And good luck.

Human health issues

Think of all those cute photos out there of children kissing chicks or rubbing them on their faces. Those pictures always make us cringe. Several diseases can spread from chicks to children with such close contact.

Bird flu, or *avian flu*, is often the first disease people think of when they think about diseases they can get from chicks. However, unless you live in a country where a certain strain of the virus (H5NI) has been confirmed in domestic poultry, the chances of you or your children getting avian flu are extremely slim. (We discuss avian flu in more detail in Chapter 11.)

Your children are much more likely to get salmonellosis than avian flu. The chicks don't look or act ill, yet chicks can transmit salmonella bacteria in their feces, which then is transferred to other parts of their bodies. Salmonellosis causes gastrointestinal distress and can become quite serious or deadly in some people, particularly the young, aged, or immune compromised.

Everyone needs to thoroughly wash their hands in hot, soapy water after handling chicks or chickens, their food or water dishes, soiled bedding, roosts — anything to do with chickens. Children are more likely to eat or touch their mouths or eyes without washing their hands and, therefore, are more likely to become ill from handling chicks. When toddlers are around chicken coops, prevent them from putting anything in their mouth that may have been contaminated. After handling chicks or adult chickens, children especially need to thoroughly wash their hands before doing anything else.

Lessons on proper handling

Tiny chicks are fragile, and even the loving squeeze of a small child can prove deadly. If the chest and rib cage of chicks are held so tightly that the chest can't move, the chick will suffocate even if the mouth and nostrils are clear. Children don't mean to squeeze, but they're often afraid of having the chick get out of their hands. So teach them: No squeezing.

Other problems result when chicks fall from the hands of children and are injured, or when they're picked up by their legs or necks. Adults should pick up chicks by scooping them up from underneath and carefully transferring them to a small child who, preferably, is sitting down. The child should be taught to cup both hands to hold the chick and not to hold it with one hand.

Even children who are old enough to properly pick up and handle chicks shouldn't be allowed to play with them too often. Taking the chick out of the brooder is stressful to it, and if chicks are too frequently handled, they may not grow as well and may become more susceptible to disease.

Children should also be taught not to bang on the sides of the brooder or do other things to disturb the chicks or make them run around. The more stress the chicks experience, the less likely you are to successfully raise them.

Part V
Special Management Considerations

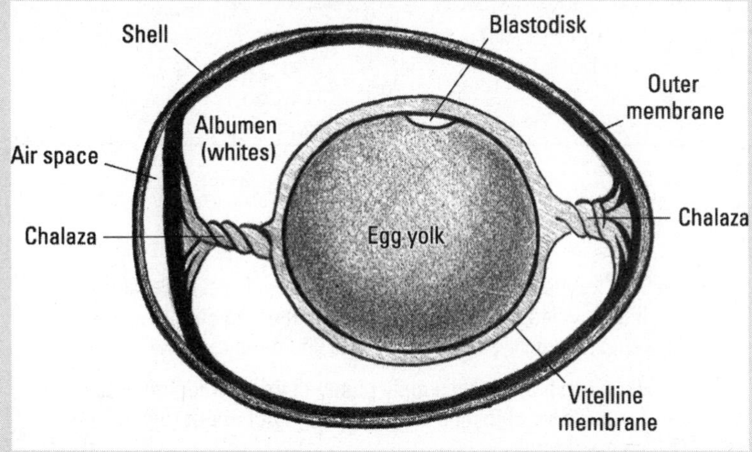

Illustration by Barbara Frake

When raising chickens for meat, you need to be able to package the product safely. Go to `www.dummies.com/extras/raisingchickens` for an informative article on packaging home-butchered poultry.

In this part . . .

- Get the lowdown on figuring out your hens. From understanding their life cycles, to handling egg production, to checking out the eggs, we give you all the details you need to know.
- Chicken for dinner! We supply plenty of information on butchering your own chickens so you can find out about raising and killing meat birds.

Chapter 15

Managing Layers and Collecting Eggs

In This Chapter

- Understanding your hens and their life cycles
- Managing hens for egg production
- Collecting and inspecting eggs
- Knowing how to handle egg production problems

Getting hens to lay well and keeping them laying takes a little more care than many people think, but something about producing some of their own food makes many people want to try. In this chapter, we discuss how to manage laying hens — whether you have two hens or many. We also address problems you may encounter with your layers.

Our focus is on producing eggs to eat rather than incubating them to produce more chicks. However, if you're trying to get a certain breed of chicken to reproduce, you may also want to read this chapter. Getting hens to produce eggs for hatching and getting them to produce eggs for eating have many similarities. Here we discuss hen management and problems you may encounter with egg laying — and the only thing breeders of eggs for hatching need to do differently is add a rooster!

Knowing What to Expect from Your Hens

How many hens you need to keep to satisfy your egg needs depends on your appetite for eggs, your breed of chicken, and how long you intend to keep the hens.

If you have production egg layer breeds, such as Isa Browns, Pearl Leghorns, Cherry Eggers, and so on, and they're well managed and healthy, they'll probably produce five to six eggs a week in their first year. Other breeds, such as Orpingtons, Plymouth Rocks, and so on, will probably produce four to five eggs per week. Some fancy breeds, such as Polish, Modern Games, Houdans, and so on, produce fewer eggs in a more seasonable pattern — higher in spring and summer, and lower or absent in fall and winter. If you have mixed-breed birds, your egg production will vary widely.

Because it takes more than 24 hours for an egg to work its way from being released from the ovary to being laid, a hen can't lay an egg 365 days a year. Some high-producing hens have been known to lay 300 eggs the first year, but your home-raised hens will probably lay between 200 and 250 eggs their first year of laying if they're a breed developed for egg production.

Young pullets lay smaller-than-normal eggs very irregularly for the first few weeks. As they settle into a routine of laying, their own pattern of individual egg size, color, and production rate develop. Some hens are better layers than others, even in egg-laying strains.

In the long days of late spring and early summer, egg production may be slightly higher in all breeds. And in the deep winter, egg production may cease, or at least decrease, for a while. As hens age, they produce fewer eggs. You need to decide whether you want to keep them or replace them with better layers. Most home chicken owners are kind-hearted and become attached to their ladies, so even after they quit laying well, their owners keep them around. If you expect your egg collection numbers to remain the same, you need to increase the size of your flock as your hens age.

So a family that's keeping four hens from a good egg-laying breed can expect to collect about 18 eggs a week, at least in the first year. You may get lucky and collect up to two dozen eggs. Remember, some eggs you collect may be cracked or otherwise unusable.

The layer's life cycle

When a hen hatches, she has all the eggs she's ever going to produce in her ovary as immature cells. As she nears sexual maturity, hormones tell those eggs to begin developing. In laying breeds, we expect laying to begin about 18 to 24 weeks after hatching. Other types of chickens may take slightly longer to begin laying. Remember that hens will lay these eggs whether or not a rooster is present.

After hens begin laying, both laying and dual-purpose breeds generally continue fairly regularly for at least a year — usually the first 2 years — unless they're affected by severe stress from very bad weather, illness, poor nutrition, or other factors. Some ornamental breeds, however, may lay eggs for only a short time, even in the first year. They may take a break from laying during the first molt. Then they'll likely begin laying again. After the third molt, egg production drops to low levels or, in some birds, ceases. Many older hens still lay an occasional egg in their golden years.

Each hen is an individual, and her egg-laying ability varies as time goes on. A young hen is influenced by many internal and external factors to begin laying or even to stop laying after she's already started. Next we talk about what influences laying.

Internal factors that influence laying

Many factors influence the laying ability of a hen. We start with some issues that influence the hen that are beyond your ability to control (with the exception of the molt period, which you can manage):

- **Age:** Hens begin laying anywhere from 18 weeks to one year of age, depending on breed and what time of year it is when they mature. In breeds bred for egg production, laying usually begins around the 22nd week. Old hens may stop laying altogether or lay few eggs. All types of hens lay more eggs between 20 weeks and 2 years old than they will later.
- **Genetics:** Some breeds have been selected over the years to lay more eggs than others. Laying breeds of chickens also have been selected to molt quickly and then to resume laying.

  If you're serious about producing eggs, you want to choose your hens from breeds that lay well. Even if you don't need a top-producing strain, you do want hens that lay reasonably well all year. Dual-purpose breeds may be right for you. Although all breeds of chickens lay eggs that taste the same, some breeds, including all bantam (miniature) breeds, lay very small eggs, and some may only lay in the spring and summer. For more about breeds, check out Chapter 3.

- **The molting period:** Although an external influence triggers molting, the molting process is biological, so we include it here. Molting is triggered by decreasing day length and generally happens once a year, in the fall. The hen loses her feathers gradually and then replaces them over a period of a few weeks. Growing new feathers is energy intensive, so hens stop laying while molting. Some breeds don't resume laying after a molt until the days start getting longer again. You can manage molting, to some extent, by using supplemental lighting, which we discuss a bit later. We discuss molting in more detail in Chapter 10.
- **Sex:** If you want eggs, you need hens. You don't need any roosters to get eggs. Sexing some breeds of chickens can be difficult. For more information on sexing, see Chapters 2 and 12.

External factors that influence laying

We can control, at least to some extent, some of the environmental factors that influence laying. Seasonal daylight variations, stress, and nutrition all affect laying. Domestic hens have been selected so that they're less sensitive to these factors than their wild relatives, but some seasonal variation affects laying unless you effectively manage all these factors.

Here are the most influential external factors:

- **Exposure to light:** Day length is a major environmental factor affecting laying. Chicken hormones that affect sexual activity, fertility, and egg laying are strongest when the days are getting longer — through spring and early summer. Shortening days, such as in late summer, tend to trigger broody behavior in some hens: They want to sit on eggs and raise a family before it's too late. When a hen starts to sit on eggs, she no longer lays eggs. (Some hens have this desire to raise a family more strongly than others and, thus, try to sit at any time of the year.)

  The seasonal effect is most pronounced in young pullets that are just beginning to lay and in older birds, but all breeds and ages are affected. Basically, increasing day length stimulates laying, and decreasing day length depresses laying. We discuss managing lighting a bit later in this chapter and also in Chapter 6.

- **Stressful environment:** A stressful environment causes pullets to delay the start of egg-laying and may stop laying hens from laying. Stress from predator attacks, frequent harassment or handling, fighting in the flock, crowded conditions, a move to a new shelter, and so on can all slow or stop laying. Just like people, some breeds and individual chickens handle stress better than others. We discuss managing stress a bit in the next section and in Chapter 10.
- **Temperature:** Extreme hot or cold weather can delay the start of laying or cause laying to stop. Managing the temperature can optimize laying. We discuss heating and cooling the coop in Chapter 6.

Managing Your Hens' Laying Years

In this section, we discuss actions a flock owner can take to optimize egg production, whether the eggs are for table use or for hatching. All breeds of egg-producing chickens benefit from these management strategies. Some strategies to manage layers are similar to what you do to manage chickens for meat or show, but they have their differences, too.

Getting young hens ready to lay

Good egg production means starting with good birds. To get healthy young pullets (young hens) to the verge of "henhood," you need to pay attention to their nutrition and overall health and well-being. Healthy, happy birds begin laying sooner than stressed and poorly maintained birds.

Remember, you don't want to start laying rations until you start getting eggs or at 22 weeks, whichever comes sooner. Growing pullets need a diet that's about 16 percent protein, and they don't need the calcium and minerals at the levels contained in layer feed. See Chapter 8 for more on nutrition.

After the pullets that are designed to be layers leave the brooder, you need to separate them from other chickens, such as meat or show birds, so you can manage them for egg laying.

Helping your pullets avoid stress

Chickens establish a pecking order, and you may see a lot of fighting and stress while that order works itself out. After an order is established, though, it helps reduce stress because every hen knows her place. If your pullets have been together since they were hatched, a pecking order is in place well before they begin laying.

If you're assembling a group of pullets that weren't all raised together, try to get them together by the 15th week so they have time to establish a pecking order before laying begins. Young pullets that are thrown into an established group of older hens will be bullied and will feel a lot of stress for a while, which will delay laying.

If possible, pullets should go directly from the brooder to the housing they will occupy as adults. Doing so gives them a long time to adjust to their surroundings and reduces stress.

We don't recommend allowing pullets free range until they've established a good laying pattern. We talk about why establishing good laying patterns is important later in this chapter. It's okay if pullets have some access to the outdoors in an enclosed run.

Providing encouragement

Young pullets like to play house as their hormones begin to prepare them for laying. If they have the proper nest boxes available, they'll try them out by sitting in them, arranging nesting material, and practicing crooning lullabies. Dark, comfortable, secluded nest boxes attract them. Try to have nest boxes in place by the 18th week.

Although a rooster isn't necessary for keeping laying hens, if you have one (particularly a mature and experienced one), he can be a great coach for young pullets. Roosters often find a good nest box, sit in it, and call the girls over. Then they scratch and turn around in it, showing the hens just how much fun it is. When a hen enters, the rooster stands outside, talking to her. Playing house with a boy is much more fun than playing with just girls.

Using lighting to encourage laying to start

If your pullets are "coming of age" when the days are still long, exposure to natural daylight will probably trigger laying between the 18th and 24th weeks. So your pullets need to be hatched early in spring — early March is ideal and early April is the latest — so they reach that age before the days get short.

However, if you have the ability to artificially light the coop, you can push the pullets' hatching date to later in the spring. Likewise, you can spark the start of laying later in the year instead of waiting to start the next spring, as hens would do without artificial light if they hatched in late spring through fall.

You want to light the coop for pullets just like you do for adult hens — 14 to 16 hours of light and 8 to 10 hours of darkness. You can read more about how to manage adult hens later in this chapter. Begin supplementing natural daylight with artificial light when you take the pullets from the brooder — or as soon as you get them. Doing so allows laying to start as early as the pullets are physically ready.

Encouraging Egg Production After It Has Begun

After hens begin laying, you want them to continue laying while remaining healthy and happy. Hens that don't lay well eat as much as ones that do, so you want to encourage every hen to produce to her full potential. We discuss management techniques you can use to achieve that goal in this section.

Providing supplemental lighting to keep hens laying

You can keep adult hens laying more reliably when you use artificial lighting to supplement natural light. The length of the day and the intensity of light stimulate the hormones of hens and prompt egg-laying. In the winter, hens that get only natural light may stop laying, but if you add artificial light, you can keep your hens laying. Fourteen to 16 hours of bright light followed by 8 to 10 hours of dim light or darkness is the ideal lighting ratio to keep hens laying. The lighting needs to be pretty reliable — not just when you remember to turn on the lights in the morning. Many people buy a timer for the coop lighting so that they don't need to get up early or go out late to turn lights on and off. A small coop may need only one light, but having two is better, in case one burns out.

We prefer to have a dim light in the shelter all night so that if hens fall from the roost, they can find their way back. Plus, they can better defend themselves from predators. We explain the difference between dim and bright light in Chapter 6.

You can adjust supplemental lighting to fit your schedule, to some extent. In the shortest days of midwinter, if you can tend the coop before natural darkness falls or shortly afterward, you should have lights come on at 4 a.m. and go off at 6 to 7 p.m. Studies have shown that adding light in the morning, before natural daybreak, works better than extending it in the evening. If chickens can see the outside, they tend to go to bed when it starts getting dark, whether it's light inside or not. But if you need to tend the coop after natural darkness has fallen, leave the lights on until then, as long as it isn't past 8 to 9 p.m. That way, the hens won't be quite as disturbed, although they may already be roosting when you come in. You may want to start the hens' day early anyway, somewhere between 5 and 6 a.m. Remember, in late spring and early summer, you don't need to leave the lights on as much, unless your birds aren't exposed to natural light.

Keeping up a routine and minimizing stress

Just as most animals are at their best in calm, comfortable surroundings, your hens lay better in those kinds of situations. Make sure your hens have everything they need to be comfortable — good feed; clean water; dry, clean surroundings; nest boxes; roosts; and maybe a sandbox to bathe in.

They need enough room to move around comfortably, scratching and pecking, flapping their wings, and conversing with friends, but also be able to avoid their enemies. Yes, in large egg factories, the hens are packed in small cages, wing to wing, and they do lay. However, as a home flock owner, you don't need or want to subject your birds to those kinds of conditions.

Laying hens can be pets, but you need to teach children not to chase them and provide guidance on how to handle them gently. Kids should stay out of the shelter, especially in the morning when hens are laying. When they do enter the shelter to collect eggs or help you feed, they need to stay quiet and calm. Don't allow other pets to chase hens or to annoy them from outside their pen, either.

Don't disturb hens too much in the morning before laying. It's not the time to bring in visitors, catch birds, or clean the coop. You can do your normal feeding and watering, but you may want to avoid providing extra treats that cause everyone to come running.

Try not to add and subtract hens from your flock too often. Not only do you risk bringing in disease, but the birds need to reestablish a pecking order each time, and this process can be stressful. A group of older pullets or hens that are suddenly put together must do some squabbling and fussing to establish a pecking order and may not lay well for a few weeks. We talk about behavior and the pecking order in Chapter 10.

Retiring old birds when the laying days are done

As your flock ages, you have a decision to make. In earlier days, old hens were sent to the soup pot, but today many people wouldn't be hungry enough to eat those tough old birds, even if they were strong enough to kill them. Instead, many people keep these old hens and feed them, adding young hens to give them eggs. Doing so may be a problem for you if you're limited in the number of birds you can keep. As an alternative, you may be able to sell old birds or give them away for pets.

In many small flocks, losses to predators and accidents make the decision easier, and you can add new flock members. If you're not sentimental or you need to keep egg production high, getting rid of all the old birds and replacing them with new birds sometime after the second molt is a good plan. Commercial egg farms often get rid of all hens after the first molt, but for home flocks, the second year of laying is usually good enough to keep them.

Mixing young hens with older ones can cause some disturbance in the henhouse. We discuss some ways to handle introducing new flock members in Chapter 10.

Collecting, Cleaning, and Storing Eggs

Harvesting your crop of hen fruit is one of the joys of chicken-keeping. Fresh eggs do taste better — ask a friend to compare supermarket eggs with your fresh eggs. But to keep that fresh taste advantage, you need to know how to correctly collect and store eggs.

And while we're at it, let's define *egg* here. The word egg can refer to the female reproductive cell, a tiny bit of genetic material barely visible to the naked eye. In this chapter, egg refers to the large stored food supply around a bit of female genetic material. Because eggs are detached from the mother while an embryo develops, they're not able to obtain food from her body through veins in the uterus. Their food supply must be enclosed with them as they leave the mother's body.

The egg that we enjoy with breakfast was meant to be food for a developing chick. Luckily for us, a hen will continue to deposit eggs regardless of whether they have been fertilized to begin an embryo.

Children (and even some adult friends) may have a lot of questions about how the eggs are laid, such as "Where do they come out?" They may get confused if they see baby chicks hatching from some eggs in an incubator and then see you cracking eggs in a pan for breakfast. Take time to explain that not all eggs will turn into chicks and educate them on how egg production works.

Getting your eggs in one basket

In the previous section, we talk about training hens to lay in nest boxes, which makes collecting the eggs easy. If you have free-range hens, you may find eggs popping up in all kinds of unexpected places. The problem with this approach is that you don't know when the egg was laid in many cases. Trying to produce farm-fresh eggs makes no sense unless you know they're really fresh eggs.

If you ever crack a spoiled egg, you'll never want to open another. If you're ever unsure about how long an egg has been in a location, it's best to discard it — carefully. Rotten eggs full of sulfur gas can explode like a mini stink bomb, leaving you gagging your way to a faraway location.

Get yourself one nice basket, pan, or bowl to handle your largest egg collection, and faithfully collect your eggs each morning. If you have free-range hens and you can't confine them until after they lay in the morning, get used to checking several locations each day where the hens are known to lay.

If you find eggs and you're not sure how old they are, you can try one trick. Fill a large bowl with water and gently put the eggs in it. Eggs that sink on their side are probably fresh. Eggs that stand upright or that float are probably old and need to be discarded. As an egg ages, it loses moisture and the air space gets bigger, which causes the egg to float.

Hens lay most eggs within a few hours after sunrise, which, of course, varies with the season. If your coop is lighted for egg production, your hens will lay most eggs before 10 a.m. Try to collect eggs soon after your hens have finished laying. This keeps them from being broken, which keeps hens from learning that they're good to eat. It also gives other animals less time to raid nests. If you can't pick up the eggs soon after they're laid, at least pick them up once a day. If you have a lot of hens, you may want to schedule morning and evening collection times.

Picking up eggs soon after they're laid avoids another problem, too. In very cold weather, eggs left in the nest too long can freeze and crack. These cracks can be hard to see when you bring the egg to room temperature. Carefully check all eggs collected in very cold weather and discard any that are cracked; the cracks allow disease organisms into the egg. Even if frozen eggs don't crack, the quality goes down. Yolks may thicken and become tough, and the whites will be watery.

Eggs can sit in warm temperatures without spoiling for many hours. Even in very hot weather, eggs collected later in the day should be fine. However, any free-range eggs that sat in the sun for several days can be another matter: They're probably spoiled.

In mild weather, with temperatures above freezing but not above 60 degrees Fahrenheit, eggs will remain fresh for many days, even in a nest. However, it's still a good idea to discard eggs if you have no idea how long they've been sitting somewhere.

Cleaning your cache

Some people recommend storing eggs without washing them, but even if you work hard to keep fresh, clean nesting material in your nests, some eggs are going to get dirty. Usually they're dirtied by chicken poop, but sometimes a broken egg, muddy hen feet, or other contaminants will soil them. We think that keeping dirty eggs in the refrigerator is a good way to spread harmful bacteria. Even if they look clean, we recommend washing all eggs before storing them.

Chicken egg shells are porous, but the shells have a natural "oily" coating that helps prevent bacterial contamination. Water cooler than 20 degrees below the internal temperature of the egg causes the pores to shrink and may draw surface contaminants into the egg.

Hot water and soap remove the egg's natural coating and also open shell pores. Rinse eggs in running, mildly warm water instead. Don't soak eggs in water, either. You can buy special egg-washing soaps in poultry catalogs, but you can also use an unscented dish-washing soap. Less is better; use soap sparingly. Don't use abrasive soaps. Strong scents may give the eggs an "off" flavor.

Rinse all sides of an egg. Don't scrub too hard — sometimes eggs have rough spots or "pimples," and if you scrub these off, you damage the shell and allow the inside to be contaminated. Use a paper towel to scrub eggs; never use abrasive pads. Discard towels after each egg, for best food safety. Sometimes an egg has flecks of blood or pigment on the surface, and washing usually removes them.

After washing the eggs, dry them with paper towels before placing them in storage. Wet eggs can stick to cartons or other containers, and then when they're picked up, a piece of the shell may come off.

If eggs are really heavily soiled, it may be better to just discard them. Lightly cracked eggs or eggs with soft shells, very rough shells, or other oddities may be cooked and fed to pets or back to the chickens. We talk more about storing eggs later in the chapter.

Assessing Egg Quality

When you buy supermarket eggs, the eggs in the carton are similarly sized and the eggs are all pretty "egg shaped." In most areas of the country, supermarket eggs are chalk white. When you collect eggs from your own hens, you're going to notice many differences among the eggs in size, color, shape, and even surface texture. These differences occur in commercial eggs, too, but imperfect eggs are cracked and sold in huge batches of liquid eggs.

Just because an egg is different looking doesn't mean it isn't good to eat. Most egg oddities mean nothing — they're superficial differences. If you're going to sell some of your excess eggs, you'll probably want to save those odd eggs for home consumption.

Eggs can also have internal quality issues. Most of these are also harmless, but novice fresh egg users may find them shocking. And some internal and external egg-quality differences mean you have a management problem, disease, or other problem to correct. In the next section, we discuss egg quality, both internally and externally.

Identifying parts of an egg

Take a look at the diagram of an egg see Figure 15-1. It helps to know what each part is as we talk about it. If you have an egg to spare at home, it also helps to look at one with a scientific eye — dissect it and find the parts. A store-bought egg also works just fine. When you understand the normal parts of an egg, you'll be better able to determine what's abnormal.

Looking at the outside

Begin by looking at the outside of an egg. It's quite a miracle of nature, actually. Inside that shell is a food supply meant to sustain an embryo through 3 weeks of development. The firm shell protects the developing embryo yet allows gasses to pass through minute pores.

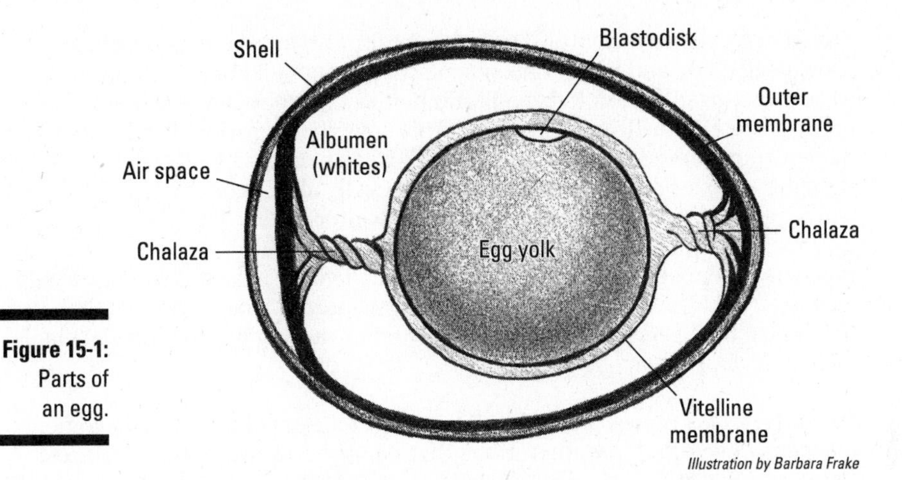

Figure 15-1: Parts of an egg.

The color of eggshells ranges from chalk white to deep chocolate brown. Some chicken eggs are green, blue, olive, or khaki colored. Light brown eggs may have a rosy tint, but you won't find any red, orange, or pink eggs unless the Easter Bunny leaves them. Even within the same breed, each individual can have a different shade of egg. As hens age, eggshell color tends to lighten in birds that lay brown or colored eggs.

Chicken eggs normally aren't spotted or flecked. Sometimes eggs have small dots of blood or pigment on them; this variation is generally not a permanent color for that hen. Often you can wash these spots off the shell.

The important point to remember about eggshell color is that it has nothing to do with taste or the nutrient content of an egg. It also has nothing to do with the internal yolk color. We think brown eggs have gotten a reputation for tasting better because they're often the eggs produced at home or locally and are therefore fresher.

If you like the look of white eggs or your kids refuse to eat any eggs that aren't white, nothing is wrong with producing white eggs at home. Your home-produced white eggs will taste much better than store eggs and will taste every bit as good as brown eggs.

Egg color

Researchers believe that all bird eggs originally had a white shell, including eggs from chicken ancestors. A mutation may have produced eggs that were spotted or brown, and then because predators were less likely to spot these eggs, they gained a natural advantage. In places where the mutation occurred, the odd-colored eggs soon became more common than white ones.

Brown egg color is a very old mutation in domesticated chickens. Since brown egg color and larger chicken size are genetically linked, people who were selecting chickens to breed for a heftier bird may have been inadvertently selecting for brown egg color. This selection is probably why so many dual-purpose breeds lay brown eggs. Chickens who lay brown eggs are also thought to be less dependent on the light cycle, also a genetic trait inherited with egg color, and lay better through the winter.

Pigment is deposited on egg shells in the shell gland near the end of the egg-forming cycle. The brown pigments are protoporphyrins (pigments that also help make blood red) and can be deposited in varying degrees on egg shells during development.

Recently, it was discovered that blue and green egg colors found in some chickens developed from mutations that occurred in two different places, China and South America, between 300 and 500 years ago — in domesticated chicken history, that timing is relatively recent. The blue and green colors actually result from a retrovirus that modified DNA so that, in the shell gland, a natural pigment in animal bodies, called biliverdin, is deposited on eggs as varying shades of blue and green. The virus isn't harmful.

Several chicken breeds developed from the birds that laid green or blue eggs. As we mentioned before, the shell color has nothing to do with egg nutrition or taste. If you breed chickens that lay brown eggs to chickens that lay blue or green eggs, the eggs from the offspring usually have both pigments, resulting in eggs colored olive or khaki. When birds that lay white eggs are crossed with birds that lay blue or green eggs, the offspring usually lay pale, pastel-colored eggs.

Egg size

Egg size varies tremendously among chicken breeds, from the tiny eggs of bantam breeds to the jumbo eggs of some production-laying strains. All eggs are equally good to eat, regardless of size. Even in a flock of the same breed, hens lay slightly different sizes of eggs.

When pullets first begin laying, their eggs are small, sometimes the size of a large marble. These eggs are practice eggs, and as the hen's body matures, the eggs become sized normally for the breed and for that hen. As hens age, the eggs sometimes become fewer but larger. When a hen hasn't been laying for a while, the first few eggs she produces when she resumes laying may also be smaller than normal.

Rough shells

Occasionally, chicken eggs have rough patches, "pimples," ridges, or other strange textures in their shells. These flaws are usually temporary problems, just quirks in the egg-formation process. The eggs may look ugly but are fine to eat. If these eggs occur throughout the flock or frequently, you may have some problem with management, nutrition, or disease.

Diseases such as infectious bronchitis and Newcastle disease can cause eggshells to have rough, odd textures or odd shapes. The unusual eggs occur frequently and throughout the flock. If you have other symptoms of illness or you suspect disease, have the flock tested by a veterinarian and don't eat the eggs.

Soft or thin shells

Sometimes hens lay eggs with a rubbery, soft shell or a paper-thin shell that cracks easily. These flaws are common when pullets first begin to lay and nothing to worry about. These also occur more frequently when hens near the end of their laying days. If they occur at other times, it may mean a nutritional problem, generally a mineral imbalance. Sometimes it can also signify a disease problem, especially if other symptoms are present.

Soft-shell eggs also occur when hens are exposed to long periods of hot, humid weather, which increase the hen's metabolism rate, making calcium in the blood less available for shell formation.

Cracked shells

Small, fine cracks may occur as a hen lays an egg. Cracks can also occur from the jostling of hens fighting over a nest. They happen when an egg freezes. And they also occur when the egg collector drops them or slips them in a pocket and forgets about them.

However cracks occur, you don't want to use cracked, unwashed eggs for human consumption due to the possibility of bacterial contamination. If the eggs are washed and refrigerated and you crack them just before you use them, then, of course, they're okay!

As you wash your eggs, check them carefully for cracks. You may need to bring frozen eggs to room temperature to see the cracks, or they may be very apparent, with bulging, frozen egg white. If you pick up a stored egg and notice cracks you missed when washing it, you'll probably want to discard it. You can cook these cracked eggs and feed them to pets or feed them back to the hens without a problem.

Oddly colored or sized eggs

If you notice an odd-colored egg in an established laying flock from time to time, it's probably just an odd quirk in the egg-laying process. Remember that some breeds — Araucana, Ameraucana, and Easter Eggers — lay greenish or bluish eggs. Some mixed-breed hens with genes from those breeds may lay oddly colored eggs as well.

The color of an egg doesn't affect its taste or nutrition, with one exception. If you collect eggs from a newly discovered nest and you don't know how old they are, white eggs that look darkened or grayish are probably spoiled. Brown eggs don't show a color change, but they may look dull and dry.

Eggs can sometimes be laid with odd shapes: long and narrow like a torpedo, very round, or lopsided. You don't want to use these eggs for hatching because they aren't the right shape for a developing embryo, but they're fine to eat if they're fresh. These odd shapes are just another unknown quirk, and most are a one-time occurrence, but a hen can have a defect in the oviduct that causes her to lay oddly shaped eggs.

Looking at the inside

The eggs of hens that eat only a commercial diet have pale yellow yolks. Free-range hens or hens that eat a lot of greens or foods like carrots and yellow squash, which have lots of carotene pigments, produce eggs with yolks that are golden yellow to almost orange. Some commercial egg producers feed their chickens marigold petals or other sources of red and yellow pigments to darken the yolk and then sell these eggs for premium prices. If your hens have any time to collect some natural food, their eggs will have that nice golden color. The color of a yolk doesn't tell you whether an egg was fertilized.

Around the yolk is a membrane, and you may also notice the white spot on the yolk, which is the blastodisk or blastoderm. On either side of the egg yolk, you may notice a white, cordlike strand. These strands are called the chalazae, and they keep the egg yolk suspended in the center of the egg.

The gelatinous "whites" of an egg surround the egg yolk and are almost clear in fresh eggs, but as an egg ages, they look whiter due to evaporation of moisture from the shell pores. There's a thin and a thicker layer of *albumin* (whites). The outer layer of the egg has two membranes, which are more apparent when the egg is cooked. One is close to the white; the other lies just under the shell and is thick and tough.

Commercially, an egg is graded AA, or top quality, if the yolk is large, compact, and distinct; the whites contain more thick than thin albumin; and the egg holds its shape well when cracked and slid unto a smooth surface.

Characteristics of a fertile egg

Commercial egg factories look inside eggshells before they pack eggs so that consumers don't get surprises when they crack eggs. They look inside an egg by shining a bright light on it in a darkened room. You can look inside an egg, too, although it isn't necessary if you're using the eggs at home. We talk more about candling an egg in Chapter 13.

Many people want to look inside an egg to see whether an embryo has started growing. Unless an embryo has been growing for a few days, you can't tell a fertile egg by candling it. If the egg was refrigerated promptly, any early embryo is killed and isn't noticeable, nor can it harm you.

Remember two points here. Fertilized eggs don't taste differently from unfertilized eggs and have no extra health benefits. And just because a rooster was with a hen doesn't mean every egg she produced was fertilized.

We've heard stories of people cracking an egg and seeing what they thought was an embryo. If you're collecting your eggs promptly and storing them correctly, you won't see anything resembling an embryo developing. If you actually see a chick embryo, either you didn't collect the egg soon enough after it was laid or you have a very warm refrigerator. What people think is an embryo is usually a foreign object in the egg, the chalazae or yolk chords, or a blood spot.

Blood spots

Blood spots don't mean the egg was fertilized. They can happen in any egg, fertile or not. They simply result from a blood vessel breaking somewhere along the hen's reproductive tract before the shell was applied to the egg. This occurrence doesn't seem to affect the hen's health in any way. Blood spots may look nasty, but they won't hurt you. You can eat these eggs or use them in baking if they don't offend you too much.

Other objects in eggs

Occasionally, other objects turn up in eggs. Little pieces of debris, usually from the lining of the reproductive tract, are sometimes seen. These eggs are safe to eat. Items a chicken eats cannot pass from the digestive tract into an egg.

Very, very rarely, you may find a worm in an egg. It means your hens have worms and may need to be treated. A worm or worms penetrated the reproductive tract by moving into the cloaca and then up the oviduct. You shouldn't eat the egg, of course, although other eggs from the flock are safe to eat if they're free of worms until treatment begins. You'll need to treat the entire flock and discard eggs until the directions on the worm medication say it's safe to stop.

Different yolks for different folks

Sometimes two egg yolks are passed from the ovary at the same time and are enclosed in a shell. This double-yolked egg is generally larger than normal, as you may expect. Eggs with double yolks are perfectly fine for eating but not so good for hatching, although in rare cases, two chicks have been hatched from an egg.

Occasionally, the egg yolk color will look odd. It can be from something a hen ate, but chickens must consume a lot to change the yolk color. Lots of greens and foods with orange or red pigments make the yolk appear deeper yellow to orange.

Gray, mottled, or brownish yolks probably indicate the egg isn't fresh, so discard them. Cooked egg yolks occasionally look green or dark around the outside, but the color is a normal part of the oxidation process and is harmless.

Storing and Handling Eggs

When you've collected all that glorious fresh hen fruit, you need to know how to store and handle it to maintain that farm-fresh flavor. Even a few hens that are good layers can provide you with more eggs than you need at some times, and we discuss ways to deal with excess eggs, too.

How to store eggs

Always store clean, dry eggs in the refrigerator at about 40 degrees. We discuss washing eggs earlier in the chapter. Fresh eggs remain safe to eat for several days at room temperature, but they lose quality quickly.

Modern refrigerators with a frost-free feature quickly draw moisture from eggs. This feature causes a change in texture and makes them lose that fresh taste quickly. Always store eggs in covered containers that retain some moisture. Store the eggs with the large end up. Never freeze eggs in the shell. The shells will crack, and bacteria can contaminate the inside of the egg.

Because an eggshell is porous, eggs can pick up flavors if they're stored with strong-smelling foods like onions, cabbage, and fish. And as anyone who's ever dyed a hard-boiled egg knows, the shell can absorb colors if it's placed on colored paper or other surfaces that bleed color.

The egg carton was designed to keep eggs fresh and safe and home flock owners can use cartons to store eggs too. You can recycle cartons from store-bought eggs or you can purchase cartons from poultry supply catalogs and feed stores. Egg cartons allow you to stack the eggs to save room, and they allow you to easily mark the carton with a date so you know when you collected the eggs. They protect eggs from breakage and loss of moisture. Other containers, such as plastic containers, covered bowls, recycled butter dishes, and plastic ice cream boxes, also work for storing and protecting eggs.

If you're producing farm-fresh eggs, letting them get old in storage doesn't make sense. Mark the date eggs are collected, and rotate your eggs so that you're using the oldest eggs first. Eggs remain edible for a couple months if properly stored.

You may want to use your oldest eggs when you hard-boil eggs for salads or deviled eggs. Fresh eggs are difficult to peel the shell off after cooking. After two weeks or so, your eggs will peel like store-purchased eggs after cooking because they will have lost moisture and the inner membrane will have shrunk away from the shell.

The easiest way for homeowners to store eggs for a long time is to freeze them. Before freezing, crack them into a bowl with a pinch of salt or sugar (discard the shells) and then lightly beat them. You can combine the eggs in twos, threes, or other amounts you find useful and freeze them in baggies or small containers. Frozen in this way, the eggs remain good for about 6 months. You can use them in cooking or making scrambled eggs after you thaw them.

Eggs to discard

Always discard eggs that are cracked, soft, or thin-shelled and eggs that look or smell bad. Eggs that are really dirty have a heavy bacterial load and probably should be discarded. After 3 months in the refrigerator and 6 in the freezer, discard the eggs.

Also discard any eggs that have been in contact with any poisonous substance because eggshells are porous. If you're feeding hens any medications, carefully read and follow the label directions on when you can safely eat eggs from the hens. Don't feed these eggs to pets.

If you have many ill chickens or you see a lot of sudden deaths in your flock, it may be wise to discard eggs until you consult with a veterinarian or other expert. Don't feed these eggs to pets, either.

What to do with excess eggs

A time may come when you have way too many eggs to eat at home and you don't want to part with any hens. Or you may want a small home business and think that selling eggs is a good idea. If the excess eggs are temporary, as an early spring bonanza, consider freezing some, as we mention in "How to store eggs," earlier in the chapter. But if you have many hens (or you want many hens) and they're good layers, you may want to consider selling eggs or pursuing some other options.

Selling eggs for eating

If you have a lot of excess eggs, you may be able to sell some. Don't expect to get rich this way — it takes a lot of hens and a lot of work caring for them to earn a living from eggs. Maybe your excess eggs can support your chicken hobby. But before you get excited about this prospect and order 500 production-type chicks, do your homework.

The USDA allows people to sell eggs without USDA inspection if they have fewer than 3,000 laying hens, sell less than 30 dozen eggs at a time, and sell only eggs they produce on their farm.

However, almost all states add regulations for eggs sold without USDA inspection. Before you sell eggs, check to see what the laws in your state and local area say about home egg sales. Most states require that home producers sell eggs in new cartons, identify where they were produced, and sell only clean, uncracked eggs that were kept at 45 degrees Fahrenheit or less before sale. Some states require that eggs be washed in a certain way or even that they be candled. You can find your state's rules for selling eggs — remember to check county and city/township rules, too: Check `http://nerous.org/state-laws-regulations/egg-laws-by-state/` for info.

Some places may not allow home farm stands or allow you to sell anything from your home. You may not even be able to put up signs advertising eggs in some areas. You may be able to find a farm market in these areas where you can rent a booth to sell eggs. Some people deliver eggs to customers' homes.

Before beginning sales, check around to see who's already selling home-grown eggs and where. You may be able to develop a niche — maybe you'll be the only person selling all blue eggs, for example — but if your area has a lot of competition, expect egg prices to be lower — maybe too low to break even.

Don't expect to sell eggs to retail stores or restaurants. Most retail stores and restaurants have to use USDA-inspected eggs. Getting eggs inspected is a complicated and expensive process, not a project for a home flock owner with a few dozen eggs.

You'll need to purchase new egg cartons to sell eggs in many states. Generally, you'll be required to label those cartons with your name and address. If you're allowed to reuse cartons, you must carefully block out the old seller's information and add your own. Egg cartons are now easy to find at farm stores and from online or mail-order poultry supply places. You can purchase labels or make them from almost any computer.

You may need additional refrigerator space if you're going to sell eggs. Some states require that you keep eggs for sale refrigerated. Eggs will remain fresh for some time at room temperature, but chilling eggs lessens your chances of making someone ill from salmonella and your chances that someone will open a spoiled egg and never buy from you again.

Wash your eggs before you offer them for sale. Many states require that you do so, but if they don't, you still want to wash them. Eggs don't need crusted manure on them to prove they're organic or free range; dirty eggs are a big turnoff to most sensible buyers. You can read about washing eggs earlier in this chapter. You may want to candle eggs for sale, to look for blood spots or foreign bits of things inside them. We discuss candling in Chapter 13. Sort your eggs by size, but don't worry about grading them like eggs in retail stores. Sell only top-quality, fresh eggs if you want return sales.

If you have eggs for sale at a market or farm stand, don't allow customers to handle them. Not only may the eggs be cracked in the process, but you don't know what kind of bacteria and viruses those sellers' hands may leave on the eggs.

Finding other uses for excess eggs

You can sell eggs for hatching if you produce fertilized eggs from purebred chickens. Usually you'll have more luck selling this kind of egg if you own rare breeds or colors of chickens. You may get more for your eggs this way than you would from selling eggs for eating. Advertise on chicken forums online, in poultry magazines, or in newspapers to sell hatching eggs. Schools may also purchase fertile eggs for incubation demonstrations.

Eggs for hatching are handled a bit differently than eggs for eating. You don't wash or refrigerate them. You can store them for only a few days, and you must store and handle them carefully. For more information on storing and handling eggs for hatching, see Chapter 13. You need to decide whether you will sell hatching eggs only locally or whether you will try to sell them by mail.

If you don't want to bother selling eggs, consider donating them to a food kitchen or organization that feeds the hungry. A local animal shelter may also want the eggs to feed to pets that have special nutritional needs. Your neighbor who puts up with the rooster crowing and escaped hens on his porch may really enjoy some eggs. Your own pets and even the hens will also enjoy any excess eggs. Cook eggs before feeding them to pets and chickens.

Dealing with Production Problems and Bad Habits

In this section, we discuss why hens may not produce eggs at all, or at least less often than you like. We also address the problems of hens that lay eggs outside the nest boxes, hens that prefer to sit on eggs than lay them, and hens that consider eggs a tasty treat.

When we talk about bad habits here, we mean the ones in connection with laying eggs, not habits like pulling up all your germinating peas or tasting your strawberries as they ripen. In most small home flocks, few hens display unladylike behavior, but occasionally a rebel arises.

Addressing the failure to lay

Home flock owners who have the space and means to keep a large flock of hens may not really care whether all the birds are producing eggs at an optimum rate. But if you're limited to just a few hens, you're more likely to want their egg production rate to remain reasonably good. Ensuring good egg laying involves two processes: determining why your hens aren't laying and figuring out which hens aren't laying.

Assessing the most common reasons

A hen may not lay for several reasons. Just as in any type of animal, a hen may suffer from a genetic problem or physical injury that prevents laying. But in most cases, lack of laying indicates a management problem. The following are the most common causes of primary laying failures:

- **You've got the wrong sex.** You need hens to get eggs. Don't laugh. We've been consulted many times by people who aren't getting any eggs, only to find that they have all roosters. Chickens can be difficult for new owners to sex.

  If you need help determining which of your chicks are pullets, turn to Chapters 2 and 12 for ways to sex chickens, or ask an experienced chicken owner to help you. Any bird that's crowing definitely isn't going to give you any eggs.

- **Your hen is too young.** Chickens come in many breeds, and they all mature at different rates. Although hens from egg production breeds can begin laying at 18 weeks, many breeds take longer to mature. Be patient with young pullets.

- **Your bird is too old.** Hens don't show obvious signs of aging, so you need an experienced eye to detect them. Many people are sold laying hens that are simply too old to lay. After 3 years of age, many hens stop laying or lay few eggs.
- **Your hens are suffering from poor nutrition.** If a hen has to struggle to meet her basic metabolism requirements, she probably won't lay. If the feed doesn't have the correct balance of vitamins and minerals, her hormones probably aren't functioning correctly and hormones control the laying cycle. Check out Chapter 8 for more information about chicken nutrition.
- **Your hens are plagued by poor health.** Hens that are ill either don't begin laying or stop laying altogether. New chicken owners may have a hard time telling whether their chickens are sick. Birds try to hide the symptoms of illness until they're near death. And some diseases produce few symptoms other than a lack of vigor and a lack of laying. Check out Chapter 10 for tips on how to keep your flock healthy.
- **Your birds are stressed out.** After you eliminate the source of stress or the chickens adjust to the situation, hens generally return to normal laying. See the section "Keeping up a routine and minimizing stress" for details on how to reduce the stress in your chickens' lives.
- **You haven't provided supplemental lighting and daylight is decreasing.**

If you have hens that aren't laying and you're sure it's not because of the aforementioned reasons, you may want to have an experienced chicken owner or a vet who's familiar with poultry look over your flock. Your county Extension office also may have an agriculture agent who can help you.

Figuring out which hen has the problem

If you have a flock of six hens and you never collect more than three eggs a day, one of them probably isn't laying. How do you know which bird that is? And if you're offered laying hens to buy, how do you know whether they're still laying?

If you look carefully at a laying hen next to a nonlaying hen, you can usually tell which one is laying. The comb, ear lobes, and wattle of a laying hen are shiny, plump, and brightly colored. The fleshy parts of a nonlaying hen look small, dry, pale, and shriveled. The *vent* (anal area) looks large and moist in laying hens, whereas it looks dry and small in nonlaying birds. If you carefully feel the pelvic bones, you can see that the bones of the laying hens are spaced farther apart (with about two fingers' width between them) than those of nonlayers.

If a nonlaying hen is healthy, she may look fat and sleek next to her more productive sisters because she's not expending energy on making eggs. If she's ill or old, she may look fluffed up and drowsy and may move more slowly.

You can also watch the behavior of hens in your flock — if you can reliably tell them apart. Hens that aren't laying don't spend much time around the nest boxes — unless they've become broody, in which case they spend all their time in a nest. Hens cackle loudly when they lay, so if you see a hen near a nest box cackling in the morning, she's probably laying.

If you still can't tell which hens aren't laying, you can take turns penning the hens with a nice nest box until about noon. Remember, even a good-laying hen doesn't lay an egg every day, so give the hen several chances before you decide she isn't laying.

If you have a large coop or your hens have free range, you may be missing eggs because the hens aren't laying in the nest box and are hiding their eggs instead. Take a look around the yard for hidden nests if your egg count drops. You may want to confine the hens until midday to see whether egg count improves.

Bringing order to hens that lay all over the place

The most annoying problem that crops up in small flocks is hens laying outside the nest box. In any large group of hens, you're going to get the occasional egg laid in the wrong place. But if all your hens, or many of them, seem to be missing the box, something is wrong.

Some hens never seem to learn where to lay eggs. They lay them off a perch or on the floor. If your nest boxes are comfortable and clean, and you have enough of them, the hens that don't lay in the boxes are probably lacking certain hormones.

But hormones are only part of the problem. Laying outside the nest is also partially due to habit, so you need to work on both parts of the problem. Confining your hens in their shelter until afternoon and making sure plenty of comfortable nest boxes are available usually halts this problem.

Up to an hour before an egg is to be laid, a rise in hormones causes the hen to suddenly have motherly feelings. She wants to seek out a box and make a nest. Soon after the egg is laid, in most cases, the hormones drop, and the hen loses interest in the nest box.

A hen can delay egg-laying for a while if a suitable nest isn't available, either because the nest boxes are all in use or she's waiting for that special box. But after a certain amount of time, the hormones drop anyway and she no longer cares where the egg is laid, as long as she gets rid of it. She may lay it from the roost or as she's walking around in the outdoor run.

If a hen is always laying willy-nilly, she's probably lacking that maternal hormone. This problem occurs more with high-production egg layers than other breeds. After all, they've been selected to have fewer hormonal urges.

Keep your hens locked up until most of the laying is done in the morning (most hens are done laying by 10 a.m.). You can usually allow the hens to use an enclosed outside run, but don't turn them out to free-range until they've laid. If your hens are using the outside run too often for egg-laying, confine them to the shelter or check to see if something is wrong with your nest boxes.

Because habit also plays a part in where hens lay eggs, we recommend that you not give young pullets free range until they've established good egg-laying habits. They may decide that under the porch is the proper place to lay eggs, and you may not even know that they're laying there. Also keep new hens in the shelter until later in the day until they become used to laying in your nests.

The one time habit and good nest boxes may not keep a hen laying in the right place is when she has decided she wants to raise a family. If the hormones don't drop after the egg is laid, the hen may want to stay on the nest, or go broody. Broodiness can occur whether a rooster is present or not, and the eggs she's sitting on don't have to be fertile eggs.

If a hen is allowed to free-range, she'll sneak off and try to find a secret spot to lay in. If there's a rooster in the flock, he may aid and abet her in this decision. When she first lays eggs, she may not sit on them all day and may come back to the shelter at night to roost.

Hens can be very sneaky, making sure no one is watching them coming and going, thus making these nests hard to find. Not only are you unable to collect the eggs, but the hen is also vulnerable to predators when she stays out all night. We discuss natural incubation and what to do to help a hen hatch eggs in Chapter 13.

If hens are confined and can't choose another place, they may just commandeer a nest box and remain on it all day. They may allow other hens to lay in it, too, but they'll likely attack your hand fiercely as you try to collect eggs.

If you want your hens to produce eggs, a broody hen isn't good. We discuss how to discourage a hen from going broody in the next section.

When an egg hunt isn't fun

Rob says: "My brother bought some adult hens at a fair that were supposed to be layers. He brought them home and let them range freely in his yard. After a month without eggs, he was beginning to think he had made some bad choices. Then while cleaning the yard, he stumbled upon a treasure trove — about 20 eggs hidden by some old tires. Pen up those new hens to avoid having egg hunts year-round."

Getting a broody hen to go back to laying

A hen continues to lay eggs until she has about ten eggs; then she stops laying. At this point, she feels she has a large enough *clutch* (group of eggs) and begins incubating the eggs in earnest, leaving the nest only long enough to eat and drink. She doesn't start laying again until the eggs have hatched and the chicks are raised or until the eggs don't hatch in the normal time (about 21 days).

In laying breeds, the urge to brood eggs almost never happens. In dual-purpose breeds, it's somewhat more likely, and in some breeds, it's quite frequent. We recommend choosing breeds that don't go broody if you're keeping chickens for egg-laying.

Many old books recommend several cruel methods of discouraging a hen from sitting on eggs. However, going broody is a hormonal urge, and the hen isn't consciously choosing to sit on a nest. Throwing cold water on her doesn't stop the hormones and may even make her sick. Exposing her to bright lights, putting her under a bushel basket, and scaring her off the nest don't work well, either. If you block her from sitting in one box, she'll often just move to another. The hormones have to come down before she'll stop.

If your hens go broody, just keep collecting the eggs — never leaving the eggs for more than a few hours, if you can. Let the hen continue laying eggs to replace the ones you remove until she stops. If she still remains on the empty nest, you can just let her sit it out — about 3 weeks will pass before she gives up. Make sure you check each day to be sure other hens haven't laid in the box.

When you remove the eggs, the hen keeps laying for a while to try and replace the snatched eggs. Then one of two things happens:

- She abandons the nest, stops feeling broody, and goes back to normal egg production after a short break.
- She continues to sit on the nest, which other hens may deposit eggs in each day, until the time comes when her eggs would've hatched — in about 3 weeks.

If the hen starts another broody cycle right away, you have three choices: Get rid of her and replace her with a laying-breed hen, let her raise a family if you have a rooster, or just humor her until she gives up.

Handling hens that break and eat eggs

Some hens break and eat their own eggs or the eggs of other hens, although this problem pops up less frequently than the other causes of low egg production. After one hen begins eating eggs, however, the behavior quickly spreads through the flock because hens are quick to copy rewarding behavior. Make sure that you know it's the hens that are breaking and eating eggs before you accuse them, though. Many animals love chicken eggs. Dogs, coons, opossums, skunks, and even rats eat eggs. Except for dogs, these animals are generally nocturnal, so you want to pick up your eggs by nightfall. But when these predators are feeding young or food is scarce, some of them will raid the henhouse whenever eggs are available.

Make sure that if you feed eggshells to your hens, they're crushed into small pieces so they don't resemble eggs. Promptly clean up any eggs that get dropped or are smashed in the nest box. Don't feed raw, discarded eggs to the hens; cook them and mash them instead.

Hens that are allowed to roam in a large enclosure or free-range are less likely to hang around in the shelter and eat eggs than hens that are bored and confined to a small area. Feeding a balanced laying ration and providing crushed oyster shells helps hens get the protein, vitamins, and minerals they crave.

Some people have successfully broken the habit of eating eggs by putting plastic or ceramic eggs in the nest. When the hens can't break them, they give up. This plan doesn't work in all cases — hens that eat eggs may need to be eaten by someone else.

Chapter 16

Raising and Butchering Meat Birds

In This Chapter

- Caring for meat birds
- Knowing when to butcher
- Deciding how to butcher your birds
- Killing and cleaning a chicken

Not everyone is interested in raising animals for meat. But if you're interested in producing some of your own meat, raising meat chickens is a good way to start. Chickens are small and easy to handle — the average person who has a little space can successfully raise all the chickens he and his family want to eat in a year. You can be eating fried homegrown chicken 10 weeks after you get the chicks, or even sooner — so, unlike raising a steer or pigs, you can raise your own meat in less than 3 months and see whether you like it. Plus, chickens don't require large, upfront outlays of cash and time.

Meat birds are managed a bit differently from chickens kept for other reasons, though. In this chapter, we give you the information you need to properly raise meat chickens. We also explain the process of butchering the chickens you've raised — whether you hire someone for the job or you decide to do it yourself.

Raising Meat Chickens

If you're considering raising your own meat chickens, don't expect to save a lot of money unless you regularly pay a premium price for organic, free-range chickens at the store. Most homeowners raising chickens for home use wind up paying as much per pound as they would buying chicken on sale at their local chain grocery store.

But saving money isn't why you want to raise chickens. You want to raise your own chickens because you can control what they're fed and how they're treated. You want to take responsibility for the way some of your food is produced, and you want to feel that sense of pride that comes from knowing how to do it yourself.

Raising meat chickens isn't easy, especially at first, but it isn't so hard that you can't master it. For most people, the hardest part of raising meat chickens is the butchering (which we cover later in this chapter), but the good news is that almost every area of the country has people who will do that job for you (for a fee). So if butchering is the only thing stopping you from raising your own meat chickens, don't worry — you can get all the benefits of raising your own meat chickens without having to handle the butchering yourself.

Before you start raising meat chickens, it's a good idea to investigate places that will custom butcher your finished birds to see if this option makes cents, er, sense. See the section "Hiring Out the Butchering," later in this chapter, for details on fees, transportation, and so on.

You can raise chickens that taste just like the chickens you buy in the store, but if you intend to raise free-range or pastured meat chickens, expect to get used to a new flavor. These ways of raising chickens produce a meat that has more muscle or dark meat and a different flavor. For most people, it's a *better* flavor, but it may take some getting used to.

Looking at the three main approaches to raising meat birds

You have three main approaches to consider when raising meat birds:

- **Getting the meat fast and cheap:** This approach means using Cornish cross chicks, confining them inside, and using commercial feed. You push them to gain weight as fast as possible, and your meat tastes like supermarket chicken.
- **Raising chickens humanely but conventionally:** This approach may mean using Cornish cross chicks or other breeds of chicks. You give the chicks commercial feed and confine them, but you also allow them *some* access to the outdoors and more time to grow before you butcher them.

The controversial commercial chicken

Almost all chickens found at your local supermarket are a Cornish Rock hybrid. They were raised in huge buildings packed wing to wing and fed 24 hours a day. They were *not*, however, given growth hormones, as many people think. The fast growth and heavy meat production comes from genetics, not growth hormones. The use of growth hormones in chickens isn't permitted, and it would be too expensive, anyway. The chickens may have been given preventive antibiotics in their feed or water, although there are regulations for discontinuing these before butchering, and poultry is inspected frequently to detect antibiotics.

Many people object to the way commercial meat birds are produced because of the way they're crowded during growth and the assembly-line manner in which they are butchered and cleaned. The conditions in many of these processing plants are horrible, for both the chickens and the humans working in them.

Some people also object to the taste and texture of commercially raised chicken meat, which is softer and fatter than naturally raised chicken and has a bland taste.

- **Raising organic, free-range, or pastured birds:** This approach involves using chicks from breeds that are active foragers and feeding them only organically grown feeds or organic pasture plants. You also give the birds unrestricted access to the outdoors. The meat tastes different from supermarket chicken, but you may like it better.

You can use blends of these three outlooks, too. For example, you can confine birds and feed them an organic feed. You can use Cornish cross chicks on pasture, but they'll take longer to grow, and you may lose more of them than if you use a more active breed.

After raising many batches of meat chickens in various ways, we've concluded that the most important factor when raising your own meat is that you have control over the environment in which the chickens are raised, what they're fed, and how they're harvested. Raise your chickens in a way that feels right to you.

Choosing the right chickens

When you're choosing chickens for meat birds, choosing the correct breed to match the way you want to raise your meat is important. You can, of course, eat any chicken, any breed, any sex, any age. But for quality meat that's economical to raise, certain breeds stand out. Sex can be important in some cases, too. We discuss those issues in this section.

Breed

The Cornish-White Rock hybrid, which is a cross between those two breeds, goes by various names in chick catalogs, such as Vantress, Hubbards, and simply Cornish X, which means it's a hybrid. This breed produces meat the fastest, and it's the closest to supermarket chicken. Some hatcheries advertise a dark-colored version of the Cornish-Rock hybrid, for people who prefer colored birds.

Most of the chicken consumed in the United States comes from Cornish-Rock hybrids. These chickens have been bred to grow quickly, utilize feed efficiently, and produce lots of breast meat. Commercial and home chicken producers both use these breeds. They're not good birds for pasture raising, however, because they don't gain weight well on pasture and aren't active birds.

You can use any of the dual-purpose or heavy breeds that are fairly fast growing for meat: Plymouth Rocks of any color, Cornish of any color, Orpingtons, Faverolles, Wyandottes, Chanteclers, Hollands, and Delawares are good examples. These breeds are better for pasture raising and can also be raised in confinement. Recently, some hatcheries have been offering meat-type birds that are better suited to free-range or pasture conditions, such as Red Rangers, but these breeds aren't easy to find in all locations. The very heavy breeds — like Brahmas and Jersey Giants — take a long time to mature.

Producing your own meat-type crosses isn't too difficult to do, either. For example, you could have a flock of Buff Orpington hens laying eggs for you and keep a Plymouth Rock rooster with them. Of course, you'd need to incubate the eggs the hens produce to get your meat chicks (see Chapter 13 for more on incubation). For best results, cross two breeds that have good meat qualities.

Technically, you can eat any chicken, but you want to avoid certain breeds because they won't give you much of a meal for your time and money. Breeds to avoid are any of the egg production breeds — such as Leghorns, Rhode Island Reds, Isa Brown, and Cherry Eggers — which have a light frame. Also avoid the small ornamental breeds.

Also avoid the hatchery offerings termed "frypan special" or "fryer mix" or any other term for groups of male chicks from lightweight layer or ornamental breeds. You can eat these, and the chicks will be very cheap, but you'll spend a lot of time and money raising these birds, with very little meat as a result.

Sex

If you're going to raise the Cornish-White Rock hybrid chicks, we feel it's fine to order "as hatched" or "straight run" (which means no sexing, just sending chicks as they're grabbed) because both sexes make good meat birds, with little difference in maturation rate. If you're ordering other breeds, you may want to order *cockerels* (males). They grow a little faster and larger than females, and they may be cheaper.

Two specialty types of meat chickens require that the sex be determined:

- **Capons:** *Capons* are castrated male chickens. Castration used to be a popular method of producing tender, fat chickens more quickly. If you want to go this route, you need to order male chicks — probably White Rocks or another heavy breed — and learn how to caponize them. Castrating a chicken requires invasive surgery because a chicken's testicles are inside the body. We don't believe there's any advantage in caponizing cockerels over using a Cornish-Rock hybrid, as far as meat taste and texture go, and the process is a bit complicated and inhumane.
- **Cornish Game hens:** These birds are a gourmet menu offering. Cornish Game hens are young, small Cornish *pullets* (females). You also can use Cornish-Rock hybrid pullets, with no difference in taste.

Choosing the right time of year to raise chickens

You can raise meat birds at any time of the year, and you can usually find the Cornish-Rock hybrid broiler chicks available somewhere year round. Some other breeds, however, may be hard to find in late fall and winter. You can incubate chicks from your own flock any time you have fertile eggs.

Fertility goes down in most breeds in the fall and winter.

Your climate and your housing method are two factors to take into consideration when choosing a time to raise meat chickens. Cornish-Rock hybrids take about 8 to 10 weeks from the day you get the chicks to butchering day. (Other breeds need up to 20 weeks to mature.) They come out of a brooder at 4 to 5 weeks. (See Chapter 14 for information on raising chicks in a brooder.)

Traditionally, meat animals are slaughtered in the late fall when temperatures are low and few flies are present, and fall is a good time to plan your first butchering. Cold weather reduces the smell, fewer bugs are present, and you won't have feathers stuck to your bare, sweaty skin. Plus, if the temperatures are comfortable, you won't be as inclined to hurry through things.

Meat birds, especially the Cornish-Rock broilers, are very susceptible to heat stress. If you live where summers are hot and humid, you may want to raise them early or late in the year, when conditions are cooler.

If you're going to raise free-range or pastured meat birds, it's best to have them out of the brooder and ready for their pens when the weather is normally dry and mild in your area. Your pasture needs to be growing well then for pastured chickens; avoid muddy times of the year.

You can raise meat birds in a dry, enclosed area in the winter as long as ventilation is good. Cold won't hurt them unless it's really severe — near 0 degrees Fahrenheit — for long periods, but they'll have to eat more feed to make each pound of meat than birds raised in milder weather, and your costs for lighting will also increase.

Consider your time also when planning your meat bird project. If you're very busy at certain times of the year, you may want to schedule the project for when you have more time.

Deciding how many chickens to raise

If you're a beginner with raising meat chickens, start with 10 to 25 meat chicks. That number is a good-size batch for the typical family. After you've raised and butchered a batch or two, evaluate how well your family liked the meat, how you liked the raising process, and how the butchering went before purchasing a larger batch (say, 100 meat chicks). It really doesn't make sense to raise only three or four meat chickens — it's just not efficient in terms of time and money.

For that small, trial batch of meat birds, you may want to share a hatchery order with a friend or order from a farm store that combines orders. Generally, a minimum order of 25 chicks is required for shipping. If you also want layers or fancy chickens, you can combine chicks from those breeds with meat chicks to meet the minimum.

How many meat chickens you want to raise in a year depends on how many your family wants to eat. If your family eats two chickens a week, then 100 chickens will be just about right for the year. Although you can raise 100 chickens in one large batch, you're better off dividing them in several batches over the year. Here's why:

- **You're putting all your eggs in one basket (so to speak).** Although large batches provide some economy in time and in larger purchases of feed, bedding, and chicks, if something goes wrong (such as a predator getting into the pen), you lose a year's supply of meat.

- **Meat chickens don't hold well — you need to butcher them before they get too old.** For most meat chickens, that means before 20 weeks if you like tender meat. You don't want to keep a bunch of them in a pen and just go out each Sunday and kill one. People did that before they had freezers, but then, they expected to eat tougher, stronger meat, too.

  If you do want to do this for some reason, don't use the Cornish-Rock cross meat birds. The longer they live, the more likely they are to die suddenly from a heart attack — and it's not from seeing you pick up the ax.

- **For a large batch of meat birds, you need a large housing unit, a large cash outlay, and a large freezer.** The housing unit will stand empty for a long period of time each year. If you send the birds out to be butchered, you'll need a larger supply of cash to get it done than with a small group. Plus, if you butcher all at once, you need a lot of freezer space. And as meat gets older, it loses some quality — even when packaged correctly for the freezer.

  Smaller groups of meat birds need smaller quarters, and they put less of a strain on your freezer.

If you do your meat birds in batches, you can have one bunch going in the brooder as the time to butcher the first batch nears, or you can skip several months between batches, to give yourself a break.

Caring for meat chickens

Caring for meat birds is a bit different than caring for other types of chickens. Their nutritional requirements are different, so you want to separate your meat birds from the rest — at least after they leave the brooder. Plus, because the meat birds won't be around forever, it doesn't make sense to get them situated in a pecking order only to have that all change when they leave.

The big, slow Cornish-Rock broiler-type chicks often get picked on by more active chicks in a mixed situation, and that stress may even be enough to kill them. Even if you're raising a traditional breed with the intent of eating the males and keeping the females for eggs, you want to separate them when you can tell the sex.

Males without hens fight less and spend more time eating. You can also feed them like other meat birds and grow them a little faster without growing the hens too fast.

Housing

Meat birds need about 2 square feet of floor space per bird. More space is nice for active breeds, but the Cornish-Rock broilers aren't very active. They don't need roosts or nest boxes. Because meat birds are here on a temporary basis, the housing can be temporary, too, but it must protect them from predators and weather. See Part II for housing ideas.

Wire floors are sometimes used for meat birds for cleanliness, but they can cause breast sores in the broiler-type birds. Use a deep litter to cushion these birds, and keep it clean and dry. Because meat birds are prone to leg problems anyway, don't use slippery flooring like tile, paper, and metal.

Meat birds, particularly the Cornish-Rock broilers, benefit from having their housing lighted from 18 to 22 hours a day. The large birds need to eat frequently to keep their metabolism going, and chickens don't eat in the dark. The most recent research, however, shows that they need a short dark period, from 6 to 2 hours, in which they won't eat. This schedule gives the bird's metabolism a break and helps prevent sudden death syndrome. In cold weather, birds require more food, so keep the dark period shorter. Using a timer on your lighting is the best way to achieve this.

Birds on pasture are subjected to natural daylight and darkness, which is one reason they grow a little slower. (Remember, chickens don't eat in the dark.) When days are long, this fact may not make a big difference, but when you're raising meat chickens on pasture in early spring and late fall, it can make a big difference in the time needed to get the size birds you want.

If you're putting meat birds in movable pens on pasture, you must take great care not to crush or run over birds when moving the pens. The Cornish-Rock crosses often move very slowly.

Nutrition

From the time they start eating, meat chicks need a high-protein feed. The protein percentage should be 22 to 24 percent for the first 6 weeks and a minimum of 20 percent after that. Broiler feed is also high in energy (fat), and using it reduces the pounds of feed required per pound of gain, compared to other types of poultry feed. To feed your meat birds, you can purchase commercial meat bird, game bird, or broiler feed. (It goes by different names in different places.) *Starter ration* or *broiler starter* refers to the feed with the higher protein content, and *grower* or *finisher ration* refers to the feed with the slightly lower protein content.

If broiler-type birds don't get enough protein, their legs and wings may become deformed, and they may become unable to walk. They will need to be disposed of then because they won't eat well.

If your brooder houses meat birds as well as other types of birds, feed the higher-protein meat bird feed to all the chicks rather than use a lower-protein feed.

Some meat bird starter feed contains antibiotics. One bag of medicated feed is okay to start chicks, especially if they seem weak from shipping. After that time, you can change to unmedicated feed if the medication worries you. No residue will remain in the meat when the chickens are butchered. Some people prefer to keep their chicks on medicated feed throughout the growth period, and some new medicated broiler feeds allow this feed to be fed right up to butchering. But check the label to see if the medication used requires that you stop feeding it a few days before butchering; if it does, switch the meat birds to an unmedicated feed the last few days.

In some areas, vegetarian and organic feeds are available for purchase. If they're not available where you live, you may be able to have feed specially made for you, although certified organic feed has to be ground under special conditions and from organic grain. Generally, to have your own feed blend mixed for you, you must order a fairly large amount, usually 1,000 pounds. This amount can be hard to transport, store, and use before it goes stale and loses nutrition. For more about chicken feed, read Chapter 8.

We prefer to feed our meat birds pellets, crumbles, or mash, because it blends all the ingredients and prevents birds from picking and choosing what they like and wasting the rest. Feeding whole grains and scraps isn't a good idea for meat birds.

Pastured and free-range poultry

If you have a well-managed pasture (not just a grassy spot in the yard), the pasture can furnish a large part of your meat birds' diet. The birds will probably need some high-protein feed on the side, but the amount of feed will be greatly reduced.

For help in determining what type of pasture grasses to grow in your area, consult with your local county Extension agent. You'll need the proper machinery to plant this pasture and maintain it. In principle, you're producing a crop on the pasture and your animals are harvesting it.

It may take a little longer to get pasture-raised birds to a good eating size. Also, because pasture-fed birds move around a lot more than confined ones, their carcasses have more dark muscle meat and less fatty breast meat. The skin of the birds may be a bit more yellow due to pigments in the grass.

Free-range meat chickens that are just given the run of your land with no particular care taken regarding the type of vegetation they eat will vary tremendously in how fast they grow and how tender they are when butchered. The more exercise the bird gets, the tougher the meat will be.

If you want a good rate of growth in free-range situations, you need to provide some broiler feed. We don't really recommend free-range conditions for meat birds. Managed pasture is much better if you want birds raised on grass.

Chickens must be frequently moved to clean pastures. How often depends on the weather, the rate of vegetation growth, and the number of birds. Move them before they eat all the grass to the roots or the pen gets too dirty.

When moving pastured chickens, you must take care not to harm them or stress them too much. Keep water available at all times, and make sure the birds have a shady place to go when the sun is too hot.

Stress management

Stress affects how fast your meat birds grow and may even kill them. Stress can come from temperature, crowding, disease, predators, or too much noise and confusion in the immediate area. All birds eat less if stressed and, as a consequence, grow more slowly. Even the stress of butchering should be kept to a minimum: The meat tastes better if the animal isn't subjected to a lot of stress just before death. The Cornish-Rock hybrid broiler birds may keel over dead if their stress levels get too high.

Broiler-type birds are subject to *sudden death syndrome.* The chickens are usually found on their backs, feet in the air. You may also notice a convulsion just before death. This condition results from a combination of a genetic weakness, electrolyte imbalances, and heart arrhythmia, and it happens more likely when birds are stressed by heat or other factors. It is also more frequent in males.

Don't eat any meat birds that you find dead. There's always the chance that the bird died of disease, and a carcass that sits around — even for a short time — may be contaminated by flies and other insects.

Home chicken raisers should avoid using medication to treat the stress-related problems of meat birds. Instead, they should improve management techniques to minimize stress. Don't use antibiotics as a preventive for diseases — use them only to treat any diseases that come up. One of the reasons for raising your own meat is to avoid meat contaminated with antibiotics.

If you need to medicate your meat birds for any reason — whether by injection, with medication in the water or feed, or any other method — make sure you read and follow the exact label directions on the product regarding how long you must wait before butchering the birds.

You don't have to rely on medication to keep your meat birds free of stress. Here are some tips for raising stress-free meat birds naturally:

- **Place your meat-bird pens in a place that's out of the way of noisy, busy conditions.** Chickens may seem to adjust to busy, noisy locations, but their bodies still feel stress, and that translates to less weight gain and more problems with illness and sudden death.
- **Provide your meat birds with a clean space, and keep the litter dry.** Large, heavy meat birds like the Cornish-Rock cross are prone to developing sores or blisters on the breast from the amount of time they spend sitting. Even birds in clean conditions develop them, but wet litter makes them more common.

  Usually the blisters are found after butchering; most of the time, you can cut out this area after butchering, and the bird will be safe to eat. But if breast blisters appear infected or raw areas get bigger than a quarter on live birds, you may want to destroy them; we don't recommend eating these birds.
- **Make sure your birds' environment has a comfortable temperature.** Heat is a big killer of heavy meat birds, so make sure your meat-bird area is well ventilated. You may need to run fans in really hot weather.
- **Develop a feeding routine, and stick to it.** Routines reduce stress because the birds know what to expect.
- **Don't let pets and kids scare the birds.** They may literally scare them to death.
- **Keep predators out of the pens.** Predators, obviously, are another source of instant death.

Planning for D-Day

Butchering time is often the hardest part of raising any type of meat. If you go into the project with the firm resolve that you're raising chickens for meat, it's easier. The first time is the hardest; after that, it does get easier.

These days, people who eat meat rarely participate in the killing of the animal that produced it. Don't feel bad if you're conflicted about killing a living creature that has been in your care to provide food for yourself and your family. It's an issue that we all must deal with in our own way.

You can make the butchering process a bit easier for everyone involved by taking a few simple steps:

- **Raise chickens bred specifically for meat.** All the birds are destined to die from the beginning — you don't have to pick and choose who lives and who dies. Your purpose for raising the animals is for meat, and you're committed to this goal from the time you buy the chicks.
- **Keep in mind that broiler-type chickens are an end product — they aren't meant to grow to adulthood.** If you did spare their lives, they probably wouldn't live very long anyway.
- **Approach the day of the killing in a calm, orderly manner, with the proper planning and equipment.** It's a solemn, purposeful occasion, not a fun-filled day.
- **Keep very young children or kids who've gotten emotionally attached to the chickens away on killing day, especially the first few times you butcher.** After you've gotten some experience, you're sure of what you're doing, and you can butcher calmly and neatly, it's okay to introduce older, interested children to the process so they, too, have a sense of what happens when meat is produced.
- **If you have children or other sensitive people in the household, we don't recommend teasing them about the fate of the birds at butchering time or when they're about to consume the harvest.** Make it clear from the beginning what the chickens' purpose is. But never, ever make someone watch the butchering or cleaning of chickens if that person doesn't want to.

Some people can never kill anything unless it's a life-or-death situation. If you fall into this category, send your birds out to be butchered. Be understanding if anyone, including yourself, doesn't want to eat the chickens the day of butchering. After a short freezer stay, the eating is easier.

Families with children or sensitive people may find it easier to hire out the butchering and have the chickens returned home in a plastic bag ready to freeze. It may be easier for you, too — and that's okay.

Knowing when your birds are ready

A chicken can be eaten at any stage of its life, but most of us want at least a little meat on our chickens. If you're raising broiler-type birds on a good commercial feed, you can expect at least some of them to be ready to butcher at 8 weeks, and others soon afterward. Pastured, free-range chickens and breeds other than the Cornish-Rock hybrid take longer to reach a decent size.

How large you like your chickens to be for eating is a personal choice. A meat-type bird usually dresses out about 2 pounds lighter than when it was alive, so a 6-pound live bird yields a 4-pound carcass. Other breeds usually dress out at somewhat less.

If you like tender, small fryers, butcher birds at about 4 pounds live weight. If you like larger birds for broiling and roasting, butcher around that 6-pound mark. In any group of meat birds, some will be bigger than others. Males usually grow a bit faster and larger.

You can butcher the biggest birds first, especially if you're planning on butchering in batches anyway, and let the small ones grow, or you can butcher all at the same time and use them according to size.

Don't wait too long to butcher. Broiler hybrids need to be butchered by 14 weeks, or their huge size will start causing health problems. Other types of chickens just get tougher as they get older. When roosters start crowing or hens begin laying, they've finished growing, for the most part and won't get any bigger — just tougher.

Deciding whether to hire a butcher or do it yourself

When your birds are ready for butchering, the biggest decision is whether to do the butchering yourself. You're not alone if you dread this part of the chicken-raising process. And don't feel like you've failed in your job if you decide to hire someone to handle it for you —nothing is wrong with hiring a butcher. That's the approach we take ourselves.

In this section, we explain some of the reasons you may want to hire a butcher. Then we walk you through what you need if you decide to do it yourself.

Don't try to do the butchering yourself if you don't have everything you need to do it right.

Why you may want to hire the job out

You may have many good reasons for hiring someone to butcher your meat birds. One of the main reasons is time: A skilled person can do the complete job — killing, plucking, and cleaning — in a very short time, whereas it may take you hours to do your first few meat birds.

A bird can be plucked by hand, but a machine does it faster and cleaner, and most professional butchers have plucking machines. You can buy or make them too, but if you're doing only a few birds a year, it's a big expense and one you're not likely to want to incur.

As you may expect, butchering is messy and smelly. It creates a lot of disgusting waste that has to be disposed of — something some people have a hard time doing on a small piece of property. And if you live close to your neighbors, they may object to butchering.

We find it worth the small amount of money it costs per bird to have a professional kill and clean our meat birds for us. No mess, no fuss — and we still have good, wholesome meat that we raised ourselves.

What's involved in doing it yourself

Many people want to experience the whole process of producing meat at least one or two times, from the chick to the chicken fryer. If you believe you can do it, you probably can, even though you may be a little clumsy and slow at first. If you've butchered game or other livestock, you probably have a good idea of what you're getting into; if not, read the rest of this chapter.

In a few urban locations, butchering may be prohibited. If you're an urban chicken-keeper who has managed to produce some meat birds, you may want to check your municipality's laws.

To do your own butchering, you need the following:

- **The proper location:** To butcher your chickens at home, you need to have the right location — and that isn't your kitchen! You may finish the cutting and packing inside, but you need to kill and clean your birds outside your home. (For more on choosing the proper location, see "Choosing the location," later in this chapter.)
- **Plenty of water:** Having abundant, clean water available at your butchering site is important. You use the water for filling the tubs to scald birds, to wash birds as you clean them, to fill containers to chill birds, and to clean up after the butchering is done. If water is available at the site, you'll be less tempted to reuse water when you shouldn't and will clean the birds more thoroughly.
- **Someplace to dispose of the waste:** Wastewater has blood, manure, and bits of feathers and other things in it. If it's legal to run the discharge to a sewer, you could do that; some people route it into the septic tank, and if you're butchering ten or fewer chickens, that would probably work. (Taking a lot of butchering waste puts a big strain on home septic systems.)

  If you can't route the waste into your septic tank, the wastewater needs to flow into the ground somewhere. You may want to have a screen at the end to catch larger pieces for burial.

Don't direct the wastewater onto vegetable crops or into ponds, lakes, streams, or old wells. Doing so is generally illegal.

You'll also have to dispose of solid waste, feathers, heads and feet, organs, and so on from cleaning the birds. Put a big, lined garbage can near your cleaning station, with a lid to discourage flies and wasps. If you have room on your property, bury this solid waste 3 feet deep as soon as possible. Remove it from plastic bags first. If you can't bury it, you need to wrap it in several plastic bags for disposal at a landfill. Make sure it's legal to do so first.

Don't try to compost feathers and guts. It can be done, but it's very smelly and attracts pest animals. Bury it in the garden, but don't spread it on the soil surface. And don't try to put it into the garbage disposer. You'll only clog it — imagine explaining that to a repairman. Plus, this waste shouldn't be anywhere near your kitchen sink, for health reasons.

- **Time:** Butchering your first chickens will take you quite a bit of time, but when you get the hang of it, it should take about 15 to 20 minutes to get each chicken freezer ready. Plan for a lot more time than you think you'll need so you don't feel rushed. Your first few chickens may take an hour each, so plan accordingly and start small.
- **Labor:** Standing, bending, lifting (maybe heavy lifting to empty chilling or washing containers), and repetitive cutting are involved in chicken butchering, but the average person should be able to handle the labor. Because the average home butchering involves small numbers of chickens, you probably won't have to worry about carpal tunnel from handling the knife, but people with limited hand function from arthritis or injury may need some help.

  If you have difficulty standing for long periods of time, you may be able to adapt your cleaning station to a seated position.

Hiring Out the Butchering

If you're unsure of how to butcher a chicken or you just don't want to, you can have your meat and eat it, too. In most areas, someone has made a job out of butchering poultry for other people. Don't be embarrassed or ashamed if you don't want to do the butchering yourself.

You may want to watch someone do it just the first time. Some people may even let you help them and teach you how to do it right. Even if you never intend to do it yourself, you never know when you may need the skill someday.

Finding a butcher

To find a butcher, you can ask at feed stores or ask neighbors with chickens who they use. A 4-H poultry leader or your county Extension office may provide a name. In more urban locations, you may have to drive a bit farther to get the job done.

In some areas, you can hire someone to come to your home to butcher, as long as you have the right facilities. Amish, Mennonite, and some ethnic communities often have people willing to do this.

In some areas, chicken-keepers band together and fund mobile processing units, usually in a converted trailer. You make an appointment to rent the trailer for half a day or so. It's furnished with hookups for electricity and water and has all the proper equipment needed for butchering and cleaning poultry. This option can be inexpensive or expensive, depending on how the unit is funded and who's allowed to use it. Your local county Extension office is the place to ask whether such facilities are available.

Places that butcher large animals rarely handle chickens, but they probably can refer you to someone who does.

Although in some areas you can find processors that offer USDA-inspected butchering, most of these places aren't USDA inspected. Facilities that are inspected rarely accept birds except from contracted growers. USDA inspection really isn't necessary anyway if you're raising chickens for your own meat; it's important only if you want to sell meat. So don't let the lack of inspection stop you from using a certain processor.

Knowing what to expect

If you want to hire out the butchering of your chickens, arrange to take a quick tour of the butchering facilities. They need to be clean, without piles of waste or blood-spattered walls, and you shouldn't see hundreds of flies buzzing around. It shouldn't smell too bad (although even the cleanest butchering facilities have some smell if they've been used recently). Ideally, the facilities will be located in an enclosed building with running water. The best places have these features:

- Cement floors
- Walls that are easily cleaned
- Stainless steel sinks and counters

- Good lighting
- Provisions for heating water to scald birds
- Plucking machines
- Vats to chill the birds in ice water

Talk to the processor and ask the following questions:

- **Do you guarantee that I'll get my own birds back?** If so, what precautions does the facility have in place to ensure that you do get your own birds back?
- **How will the birds be packaged when I come to get them?** Some processors ask you to supply bags to take the chickens home in. Others put several chickens in bags they supply, and still others bag each chicken separately. Most processors return whole birds to you, unless you have arranged for them to be cut up.
- **If I want the feet, necks, or other odd parts, can you accommodate me?** Most processors give you the liver, heart, and gizzards along with the carcass, but if you want these delicacies, ask to make sure.
- **What are your requirements for drop-off and pickup?** In some cases, processors want the birds delivered the evening before butchering. They may want you to pick up the processed birds at a certain time the next day. Others ask that you bring the birds at an appointed time.
- **What are the processing fees?** The fees vary widely across the country and according to what services, such as packaging, the processor provides. If you have several processors in your area, you may want to compare prices against facilities, services, and convenience of scheduling.

If your religion requires the chickens to be butchered in a certain way, hire a processor of your faith, or at least someone who is very familiar with the religious restrictions.

Most poultry processors don't freeze the birds before returning them to you the way most large-animal processors do. They may also require that you furnish your own bags for taking the birds home. Smaller operations may ask that you furnish a cooler or large tub to cool your cleaned birds in.

You'll need secure carriers for transporting your chickens and possibly for them to spend the night in (some processors furnish cages or carriers to hold birds at their place). The birds will be sleeping, so they don't need a lot of room, but in hot weather, don't pack them too tightly in carriers. Don't furnish feed or water; the processor wants the crop to be empty. Your cages may be messy the next day when you pick them up, so plan to protect the car, if needed.

Most processors require appointments, and certain times of the year they may be so busy that they can't handle everyone's requests, so plan ahead. Thanksgiving time is usually busy, as is the time just after your county fair.

Your processor should be able to have your chickens butchered, cleaned, and back to you in 24 hours or less. Usually they're killed in the early morning while they're still drowsy and processed soon afterward. They should be kept chilled in ice water or by some other means and protected from flies and other insects until you pick them up. Don't expect the processor to store the birds too many hours unless you make prior arrangements.

The birds should look clean, with few or no pinfeathers remaining, and you should see no more than an occasional small tear in the skin. On rare occasions, your processor may warn you about something he noticed when cleaning the birds, such as tumors, abscesses, or signs of disease. The carcasses of those birds should be separated from the others, and you may be advised to discard those birds.

Some processors may give you other advice. For example, they may advise you to butcher sooner or later or give you other management tips. Listen to them — they're generally trying to help you.

At home, you may want to rinse the birds well, pat them dry with clean paper towels, and package them for freezing. Most chicken processors don't cut up the birds, although some will do it for an additional fee. You can cut them up or freeze them whole. We discuss packaging and storing meat later in this chapter.

Preparing to Do the Deed Yourself

If you're ready to try the butchering yourself, you need to make sure that you have all the proper equipment and that you're prepared in advance. Doing a little planning makes the butchering much more efficient.

Choosing the location

The ideal butchering spot is a small building with good lighting, electricity and water, and some kind of drainage for wastewater. A nice shed, garage, or barn is ideal for butchering, but you can also set up an outdoor butchering site. Look for a level, well-drained spot with drainage away from the site.

If you decide to do your butchering outdoors, make sure the site is beyond the view of neighbors or passersby, however tolerant they seem. You may want to put up a temporary privacy screen.

The butchering site may remain attractive to insects and stay a bit smelly for at least a short time, so it's best not to use the kids' playhouse, the back porch, or the garage where you plan to host a garage sale next week.

The butchering site can be far from the house, but the cleaning spot should be fairly close to the home so you're close to the tools you'll need for proper cleaning. When people become adept at the killing part, they generally prefer to keep the kill spot close to the cleaning area to save steps. Your method of killing will have some bearing on that decision (see "Looking at your options: Killing methods," later in this chapter).

You need a water supply for the butchering process. This setup can be as simple as a garden hose run to the site with some sort of shut-off valve. If you can arrange some sort of sink structure big enough to hold a chicken, you'll improve your butchering facilities tremendously. This sink will have to drain water somewhere besides on your feet. Having electricity available is a big bonus.

Gathering equipment and supplies

Some people may tell you that all you need to butcher is a knife — and, in an emergency situation, that's probably true. You can do most of the work involving butchering a chicken with your hands and a knife. But just like any other endeavor, the right tools make the job easier, quicker, and safer.

You can purchase many of the tools we mention in farm stores, poultry supply places, or even regular retail stores. Sporting goods stores may also be a place to look for some of the tools. Don't forget to look for used tools at garage sales, in your local paper, or on craigslist or other online sales sites. If you have any older farming relatives, they may have some butchering equipment to pass along.

Killing equipment

You can kill a chicken with your hands (see "Looking at your options: Killing methods," later in this chapter), but most people prefer using a knife or an ax. You also need a way to keep that knife or ax sharp. If you use the ax method, you need a solid surface to lay the chicken's head on that's a comfortable distance off the ground for you. Many people find a large stump or use a sturdy stool or small table. You'll be hitting it with the ax, so it must hold up to that.

You may opt to use killing cones (cone-shape metal or sometimes plastic holders with an open bottom; see Figure 16-1). The chicken is inserted head down either to kill it or after it's killed to drain the blood (or for both reasons). The cone keeps the chicken from running around with its head cut off or splattering blood everywhere. The cones are nailed or hung somewhere off the ground. Most homeowners can make cones from thin sheet metal if they can't find them to buy. It's nice to have a couple sizes, to hold large and small chickens. You may want just a few or many, depending on the number of chickens you usually butcher at a time.

Instead of using cones, some people have a rack or a rafter with nails or hooks in it. The dead chicken's legs are tied together with a piece of rope or wire, and it's hung on a nail to bleed out. The rack or cones need to be at least 3 feet off the ground.

You need some sort of bucket under each chicken to catch the drained blood. Any kind of bucket will do, but blood may stain light-colored plastic.

Finally, you need a garbage can to hold waste.

Figure 16-1: Chicken in a killing cone.

Illustration by Barbara Frake

Plucking equipment

Most home chicken butchers pluck their birds by hand, although plucking machines and plans for making them are available. In some areas, you can also rent plucking machines.

The most important tool a home butcher needs for plucking is scalding hot water and a way to keep it hot and clean. Even if you use a machine plucker, you need to scald the birds first.

You need a large metal container big enough to immerse a whole chicken held by its legs without having too much water overflowing. If you're going to heat the water elsewhere and pour it in, you may want to use a plastic or wood container.

The most labor-intensive and dangerous way to bring out the scalding water is to heat it inside and carry it out to your big pot. Most people prefer some way of heating it on-site. You can use a propane grill, propane camp stove, or electric hot plate to heat water. A new method of heating water for butchering is to use one of the turkey fryers on the market.

Many people prefer to heat water in one container and pour it into the container used to dip the birds into. Others use just one unit, dumping the water in the heating unit when it's dirty and adding new water. If you want continuous production, your heating element needs to be strong enough to heat the water to about 140 degrees Fahrenheit in the time it takes to clean one bird.

You need some sort of thermometer to monitor the heat. You can purchase a candy or deep fry thermometer in most large stores. One with a clip to hold it on the pot is ideal.

If your heating unit is propane, you need propane tanks and propane to run them. If your plucking machine or heating element is electric, you need access to electricity. A lot of water is involved in butchering, and water and electricity don't mix, so make sure any extension cords or plug outlets are located where they won't get wet.

Cooling and cleaning equipment

After you scald the chickens and pluck the feathers, you need a table to work on while cleaning the birds. Make sure the table is at a comfortable position for you to work. You can use a picnic table, a folding table, an old countertop, boards on sawhorses, or any other platform that works for you.

Here's a list of all the equipment you need for cleaning your chickens:

- **A good knife:** You need at least one good, sharp knife for cleaning: either a boning knife or a butcher knife. A boning knife with a 5-inch blade and a butcher knife with an 8-inch blade are common choices. You may want both, and a meat cleaver is also handy at times.

  Sporting goods stores sell good knives for cleaning and butchering. Heft them and see what you like. A knife should fit your hands well. A strong, well-made knife is worth the additional cost.

- **A knife sharpener:** Dull knives cause accidents more often than sharp ones. A butcher's steel is a piece of rounded metal that you hold in one hand and stroke the blade of your knife over to keep it sharp. It doesn't sharpen a dull knife, but it helps keep an edge on a sharp one and should be used after butchering each bird.
- **An outside sink or container with a drain:** After you clean the birds, you need to rinse them. Many people use an old kitchen sink or laundry tub. Some fit a drain in the bottom of a big plastic tub or half-barrel. The drain should connect to a pipe or hose that leads wastewater away from the person cleaning to a suitable discharge site. It should be at a comfortable working position for you.

  Some people just use a garden hose and rinse the birds at their cleaning table. This process gets a bit messy underfoot after a while, unless you're on a well-drained surface and you keep that hosed clean, too. Rubber boots may keep your feet dry and keep you from slipping.
- **A container full of ice water:** After the birds are cleaned and rinsed, they need to be cooled. You'll need large containers of ice water for this. Clean plastic tubs, picnic coolers, or barrels will work. You'll need very cold water — preferably ice water — in the containers. So either buy bags of ice or make ice in your freezer ahead of time and store it.
- **Soap:** Be sure to wash your hands frequently during the butchering process to avoid bacterial cross-contamination of the meat. Wash between butchering stages such as plucking and cleaning. Wash if you use the restroom and, of course, before you eat or smoke. And wash well when you finish the job.
- **Plastic gloves:** If you're squeamish, you may prefer to wear gloves.
- **A spray bottle with a mixture of one part unscented chlorine bleach and two parts water:** Use the bottle to clean the table if it gets contaminated with feces from a punctured intestine or contents of the crop.
- **Paper towels:** Paper towels are handy, especially if you don't like wiping blood on the back of your pants.
- **A waterproof apron:** You can get an apron at janitorial supply places, at some feed and sporting goods stores, or from a gardening supply catalog. Wear old clothes that you don't mind getting stained, even with your apron.

Packaging/freezing supplies

Packaging supplies for chickens are generally plastic freezer bags. Some people are now using vacuum-sealed bags. If you want to, too, you'll need a machine that does the vacuum sealing; they're quite common now in small appliance stores. Size the bags according to how you want to store the chickens. A whole chicken generally needs at least a gallon-size bag, and more likely a 2-gallon bag. If you cut up the chickens before freezing, you can use smaller bags.

Buy a waterproof, permanent marker to write the date on each bag. If you cut up the chickens, you can use *butcher's paper*, heavy brown or white plastic-coated paper, to wrap the pieces. You tie it with string or special tape that's used to seal packages.

Following the Play-by-Play of Butchering Day

D-day has arrived. You have all the necessary equipment, your butchering station is set up, and you're ready to begin. This section explains exactly what to do on butchering day.

Beginning with the kill

Most people are afraid that they'll botch the killing, and the chicken will spring to life as they start to clean it. Trust us: It will never happen. And while we can't say there's no pain, the pain is brief. When you butcher at home, you have the chance to end the bird's life in a calm and dignified way, unlike the chaotic mess of a commercial poultry processing plant.

Choosing the best time of day

We recommend that you kill your birds in the very early morning. Adjust inside lighting the night before so that it doesn't come on until you get inside to gather the birds. You'll be able to pick up sleeping chickens very easily. Gather them and kill them while they're still in a drowsy stupor. Kill all the birds you're going to butcher and have them hanging up to bleed out before daylight, if you can.

If the before-daylight kill isn't practical for you, at least keep the catching simple by going out and catching the birds the night before, either picking them up in the pen after dark or catching them just before nightfall and confining them in carriers or cages. Put them in a comfortable, preferably dark place and don't give them food or water overnight. Kill them as soon as you can in the morning.

Catching the birds the night before lets them and you calm down. If you have to chase birds all over, they'll be very agitated, and you'll be hot, tired, and maybe covered with mud and chicken poop. You'll have lost time that you could have spent cleaning and packaging the birds. You may be very ready to kill the birds at that point, but it's better not to do it in revenge. You'll be calmer and less likely to rush things or make mistakes if the chasing and catching are followed by a good night's sleep before the butchering.

Birds held in a confined, darkened area will be calm and easy to handle. Move slowly and talk calmly while removing each bird from the cage, and disturb the others as little as possible. Birds that are frightened will fight back and frantically try to escape. Research has shown that animals killed when they're in a calm state actually bleed out and clean easier — and taste better, too. Providing a calm and relaxed environment is a humane way to do butchering.

Don't feed your chickens the night before you butcher them. Unless it's very hot, don't give them water, either. Cleaning birds is much easier when their crops and intestines are empty.

Looking at your options: Killing methods

Killing methods are hotly debated among home meat producers. You have several good options for killing chickens, and everyone seems to have a method they prefer. Different cultures have different ways to kill chickens, too. Some religions have very strict rules that must be followed both for the killing and butchering of animals.

If you're Jewish, Moslem, or Hindu, speak to a religious leader to find out how you should kill and butcher your chickens.

For the most part, the common methods of killing described here are equal in terms of the amount of pain or distress they cause to the bird. However, some methods may not appeal to *you*. It's actually not too hard to physically kill a chicken, but it's often very hard mentally for the first-time butcher. Either you become accustomed to it or you don't. If it's too hard for you, remember that you can find people who will do it for you.

Some methods you shouldn't use under any circumstances:

- **Using electrocution or a stun gun** doesn't work well and often doesn't kill the bird.
- **Using gas** of various types contaminates the meat.
- **Shooting** a chicken is a senseless waste and hard to do humanely.
- **Drowning** is inhumane.

- **Using power tools** to cut off the head creates a tremendous mess, and these tools aren't designed to cut through skin cleanly — they rip and shred instead.

Using an ax and a stump

You need a stout, sturdy stump or table. Put two large nails in it just far enough apart to hold a chicken's head. Some people prefer to stretch a wire or heavy cord across the stump and slide the chicken's head under it.

You also need a sharp ax. It doesn't have to be a large, heavy ax, especially if you're a small person. A meat cleaver or machete also works, as long as it's sharp.

Here are the steps to take:

1. **Pick up the chicken by its hind legs and hold it upside down for a minute until it stops struggling.**
2. **Pick up the ax in one hand while still holding the bird in the other (still upside down).**
3. **Place the chicken on the stump quickly, with its head between the nails or under your wire.**
4. **Pull back on the legs slightly with one hand, stretching out the neck.**
5. **With the ax in the other hand, strike the neck quickly and decisively just below the head.**

   You need to cut through the major artery, but you don't need to completely cut off the head for the bird to die. Most people, however, remove the head.

In the past, people released the birds' feet, and the birds jumped up and ran around with their heads cut off — yes, they can do that by nervous "memory" until they bleed out. Rest assured, the birds were dead and didn't feel anything. But this is messy — you need an area where you don't mind blood spattered everywhere, and the sight of a chicken running around this way is traumatic to many people.

Instead of letting the birds run around, we prefer to place them neck down in a killing cone (see "Killing equipment," earlier in this chapter) to bleed out, with something under the bird to catch the blood. Or you can tie the legs together and hang them over a hook by the legs. They may twitch or flop their wings a bit, but they're dead: It's just residual nervous activity. This is much cleaner and less traumatic to the observer.

Using killing cones and letting the birds bleed out

In this method, the bird is picked up by the hind legs and inserted into the cone with its head out the bottom.

Don't leave the bird in the cone more than a few seconds before killing it, or it may escape, unless the cone is so deep it can't get its feet over the edge.

You can either slit the neck with a knife or lop off the head with heavy shears:

- **With a knife:** Stretch the neck out and cut with your knife from the front of the neck toward the back, right below the head. The knife must be sharp, and you must use a firm, steady hand. Don't stab; slice instead. You don't have to cut the neck all the way through.
- **With shears:** Stretch the neck and quickly cut off the head with something like heavy pruning shears. You need tension on the neck to do this well.

If you don't remove the head, the bird may experience some distress until it bleeds out, but it will quickly lose consciousness. You may want to tie the legs together if the bird's head isn't removed, unless the cone is deep enough to prevent the bird from getting its feet on the edge to push itself out. The sides of the cone prevent the bird from flopping around and keep it calm. When blood stops flowing, the bird has bled out. This process usually takes several minutes.

Some people insert something like an ice pick or large screwdriver into the chicken's mouth, plunge it through the roof of the mouth into the brain, and twist it before cutting the bird's throat. The object needs to go toward the back of the head, not straight up. This strategy is supposed to immediately cut off any pain or feeling the bird may have. If you do it right, it probably does, but if you botch it, you're probably causing more pain than if you didn't do it. We don't recommend this practice.

Other methods

You may opt for one of these other killing methods:

- **Killing with a stick:** Put a thick stick like a broomstick on the ground, lift one end, and place the neck of a chicken on the ground under it. Hold on to the chicken's feet. Now bring one foot down on the stick quickly, while pulling upward on the chicken's feet quickly and firmly. This should break the chicken's neck. Put it in the killing cone or hang it up by the feet, slit the throat, and let it bleed out.

- **Wringing the neck:** You can hold the chicken's feet in your left hand, letting the bird dangle upside down. Slide your right hand down the neck to just below the head, and grasp the neck firmly. Now jerk the head down and then sharply back up while twisting it at the same time. This method is generally referred to as "wringing the chicken's neck."

  If you do this too hard, you'll pull off the head, which doesn't mean much but can upset some people.

In all the hand-killing methods, you still need to remove the head or slit the throat to let the bird bleed out.

Removing the feathers

You have to pluck the bird (remove its feathers) before you clean the bird. Do the plucking as soon after you kill the bird as possible. Immerse chickens in heated water to loosen the feathers from the skin before plucking.

For a person who kills only 1 to 25 birds at a time, hand-plucking is the easiest and cheapest method (and it's the method we walk you through here). If you're going to process more than that in one day, you may want to invest in a plucking machine. In most of these, the bird rotates around a series of rubber "fingers" in a drum. You can buy the machines in poultry supply catalogs, or if you're willing to put it together yourself, you can buy parts and instructions for building a plucking machine.

In the following sections, we walk you through the procedure for removing a bird's feathers.

Scalding

In a large container big enough to immerse a chicken, heat water or add heated water. The water should be about 140 degrees Fahrenheit. If the water is too cool, the feathers are hard to remove. If the water is too hot, the skin will start to cook, and it will tear as you pluck.

Holding the chicken by its feet, dip it in the water to cover all feathered areas. Slowly count to ten. Remove the bird and try pulling out the wing feathers: If they come out easily, begin plucking all the other feathers. If they don't slide out easily, dip the bird again.

Change the water when it looks dirty or becomes cool. Scalding is quite smelly. Be careful not to drip the water on yourself, spill it, or get your fingers in the pot while dipping the bird.

Hand-plucking

You may need rubber gloves to handle the hot, wet bird. Grasp feathers and pull them out in the direction they were growing, working as quickly as you can. Try not to grab the skin. Keep plucking until you've removed all the feathers.

Some immature feathers, called *pinfeathers*, may be hard to remove. They look like a thick pin in the bird's skin. Most broiler-type pinfeathers are white, although colored birds may have dark pinfeathers. Grasp them with your fingernails and pull them straight out, or if you don't have fingernails, use tweezers or needle-nose pliers. You can also place a flat blade, like that of a butter knife, on one side of the pinfeather and put a finger on the other to pull it out.

Cleaning and inspecting the bird

Take your scalded, plucked bird to a clean table and lay it on its back. With the butcher knife or cleaver, chop off the bird's feet at the first joint. You don't go through bone; you go through the cartilage of the joint, separating the upper and lower leg bones. Discard the feet unless you like to cook them. If so, place them in a container for later skinning.

Checking out the flesh

How much flesh a chicken has and the size of its breast vary by breed and age. If you're used to store-purchased chicken and you're butchering a breed that's not generally raised for meat, the breast of the bird may look small and skinny to you. This appearance is normal.

The drumsticks of birds raised on pasture may look darker and larger than those of Cornish-Rock broilers, which are what commercial chicken comes from. The legs of other meat breeds and nontraditional meat breeds may look longer also. This difference is also normal.

The age a bird was butchered and its feeding program determine how fleshed out it was. Some breeds of birds never achieve a lot of flesh on their bones. Cornish-Rock meat birds should be quite plump and heavy by 8 to 10 weeks; if not, something is wrong with your feeding and management. Most heavy breeds of traditional chickens finish growing at 20 to 25 weeks.

Examining the bird for signs of disease

When you're butchering your birds, you should be examining the carcass for both signs of disease and signs that your management program is working well. Modern meat birds aren't alive long before they're butchered and thus have less chance to pick up a disease or develop other problems, but it doesn't hurt to at least look over your carcasses and their organs carefully.

A healthy carcass has white or yellow skin. Chicken skin is normally loose and thin. The skin and fat of chickens raised free-range or pastured usually have a deeper yellow pigment. A healthy carcass may have bruises on the skin that can happen during the killing process, but a lot of bruises mean your birds aren't being handled correctly.

Breast blisters on meat birds are fairly common. They look like a blister on the skin with clear fluid. They're superficial blemishes and can be cut out, and the meat is safe to eat.

Silkie chickens aren't often eaten by home chicken owners, but in some cultures, they're a delicacy. Their skin is bluish black, no matter what color their feathers were.

Some conditions render a chicken unsafe to eat, so keep an especially careful eye out for the following:

- Abscesses are lumps filled with pus either outside or deep inside a body. Don't eat birds that have abscesses. Abscesses can happen because of an injury or from disease, but either way, the carcass is unsafe to eat. If many birds have abscesses, something is wrong in your management — check your housing for rough objects and don't crowd your birds.
- Tumors can be hard or soft and can be found anywhere on the body or on an organ. Some may contain fluid. Don't eat birds that have tumors. Don't mistake forming eggs for tumors, though. In older hens, some eggs may be forming in the oviduct, near the backbone of the bird. They can range in size from that of a marble to an almost fully formed egg. If you look inside the lump and you see egg yolk, it's a developing egg. In some cultures, these are considered a delicacy and are saved to be eaten, but most people toss them out.
- The liver of a healthy chicken is reddish brown. If it's very pale or mottled looking or it has white spots, the chicken was diseased. Most other chicken organs don't have much to tell you.
- Open sores or wounds mean the chicken isn't good to eat.
- An outpouring of straw-colored fluid when you open the tiny hole in the back of the bird to remove organs during butchering means that the body cavity filled with fluid because of a disease process. Don't eat the bird.
- A butchered bird that was left for more than 1 hour at temperatures above 40 degrees Fahrenheit should be discarded.
- Any bird that was partially cooked from being scalded too long or at too high a temperature should either be completely cooked immediately or discarded.

Don't eat chickens that died of unknown causes or that were killed by predators. Birds that looked or acted ill before butchering are also unsafe to eat.

Removing the head, neck, and oil gland

If the bird still has a head, you need to remove it. What you do with it depends on whether you eat chicken necks:

- **If you like to eat necks:** Carefully remove the crop and the esophagus and trachea tubes before detaching the neck. Chop off the head just below the beak. Try to slide your knife between the vertebrae instead of cutting through bone.
- **If you throw out necks:** You can just cut through the whole neck close to the body without the hassle of removing the head or fussing with the tubes and crop.

If you're using the neck, slit the skin on the underside of the neck and then cut it off near the body. On the underside of the bared neck, you should see two tubes, like small straws. About halfway down the neck, you'll see a small bulge if the crop still has food in it. Don't squeeze or cut through the crop, if you can help it, or the food inside will contaminate the meat. If the crop isn't full, it will be hard to see.

If crop contents get on the meat, take the carcass to the water source and thoroughly rinse it. Clean up the table with the bleach solution in the spray bottle and rinse before proceeding.

At the end of the neck near where the head once was, find the two small tubes. Cut between them and the rest of the neck at the top, and then pull carefully away from the rest of the neck as you slide your knife or finger down behind the tubes to break the connection all the way down the neck. The trachea is stiffer and has ridges. The esophagus, which the crop is attached to, is thinner and flexible: Make sure you follow down the esophagus to detach the crop from the neck flesh.

When you've detached the tubes and crop from the flesh all the way to the shoulder, cut them off as close to the body as you can and discard them. You can the leave the neck on or remove it and put it in a container to save. Don't worry about the ends of the tubes poking through the neck hole; when you pull out the organs, these should pull out, too.

Now you need to deal with the oil gland. It's a bright yellow spot on the back of the tail and the bottom of the spine. Lay the bird on its breast. Either remove the whole tail at the spine and discard it, or, starting from the spine, use your knife to slice under the oil gland, down and past the tail, and cut off the gland, leaving the tail on.

Removing the organs

Your bird should be minus head, neck, and oil gland now. Place it on its back on the table, butt end facing you. Pull up on the flesh and skin over the top of the *vent*, or butt. With a sharp knife, make a shallow slice across the back of the bird, just above the vent. Don't make it too wide at this point, and don't push too deeply — you want to cut through the skin and flesh to make a small hole to the body cavity. Hold the knife across the bird, not pointing down or poking into it.

When you have a small hole, put your fingers in it on both sides and gently pull apart the skin and flesh. If chicken poop comes out at this point because the bird wasn't empty before you butchered it, or if at any time when you're removing the intestines you break them and poop comes out, you must take the bird to the water source at once and thoroughly rinse it. Try to hold the bird downward while doing this so poop or contaminated water doesn't run into the body cavity. This is why running water is better than a bucket of water to rinse with. Many people use a mild liquid soap to wash the carcass when this happens. You must rinse completely if you use soap.

Place the bird in a clean container while you use your bleach spray to clean off the table before continuing. This kind of mess quickly teaches you to starve your birds at least 12 hours before butchering.

Now that your hole has been widened, slide your whole hand, fingers out straight (yes, your whole hand will fit in all but Cornish Game–size hens), into the bird tight up along the backbone as far as you can, preferably to the neck. Slightly curl your hand and rake — don't squeeze and pull — the organs back toward the rear. Most of the organs will come out in a big clump. They will be attached to the vent by the intestines. Move them to the side and carefully cut around the vent in a circle. Then discard the whole mess.

If you're going to cut up the bird for cooking anyway, you can simplify organ removal by splitting the bird open. You can use a knife or a heavy-duty pair of scissors for this. Turn the bird on its breast and, starting at the rear, cut through the backbone all the way to the neck. You must have a steady hand and cut very shallowly — you don't want to cut into the organ mass. Split the bird open with your hands and then remove the organs.

Some people save the heart, liver, and gizzard for eating. If you like them, separate them into a clean container or plastic bag. The liver has two lobes with a small, green, pointed sac between them. This organ is the gall bladder. If it breaks when you clean the bird, any meat it touches will be stained green and will have a bitter flavor. Before saving the liver to eat, carefully cut it out.

Gizzards you save should be slit open. They may have stones, grain pieces, or foreign objects in them to be removed. They also have a yellow lining that you should remove.

Put your hand back up inside the bird to remove what's left. Usually the heart and lungs are still in there, sometimes with bits of other things. The lungs adhere to the ribcage very closely and can be hard to remove. You need to get a finger between them and the ribcage wall. If you like to eat the heart, save it in your clean container. Some people trim off the wing tips and remove fat under the skin near the large opening at this point.

Now take the carcass to your water source and rinse the inside cavity and the outside carcass thoroughly. Place your cleaned bird in cold water for chilling. The water should totally cover the bird so no parts are exposed to insects. Chicken should be chilled as soon as possible after butchering.

Some people believe that aging the bird a day or two in a moderately cold temperature, such as that of a refrigerator, before eating or freezing makes for more tender and flavorful meat. We don't believe that young birds, especially the Cornish-Rock hybrid broilers, need this.

Packaging Home-Butchered Poultry

Whether you butchered your birds yourself or had someone do it for you, you need to properly package that good meat so that you can store it safely until you're ready to eat it. The most popular way to store home-butchered poultry is by freezing it, with canning a distant second. In this section, packaging refers to birds that are going to be stored by freezing.

After you've cleaned your birds, you need to chill them to bring the body temperature down to as close to 40 degrees Fahrenheit as possible. When the birds are cool, you can begin to package them for storage. Packaging the chickens inside your home is fine, and it's probably the easiest way to go. Some people do cut up the birds where they butcher them, and some even package them there, but most put on the finishing touches in the home.

Rinsing and checking the chicken

Whether you bring the birds in from your home-butchering site or back from a processor, you first need to give them another good rinsing and check them over. Before placing the chickens in your kitchen sink, scrub it out with a solution of one part bleach to three parts water; then rinse with hot, soapy water.

Believe it or not, most kitchen sinks have as much bacteria as your toilet bowl. Also scrub and rinse any counters or tables you will be working at.

Use cool running water to rinse the inside cavity of the chickens as well as the outside. Check for remaining bits of organs and pinfeathers, and remove them. Place the birds on some clean paper towels and pat them dry with other clean towels, inside and out. It's better to use disposable towels when working with meat and discard them after each use because cloth towels can spread bacteria.

If you have organs to store, such as the livers, wash them in cool water and drain. If the gizzards weren't opened during butchering and you want to store them, slice them open, remove any foreign bodies inside, and remove the yellow liner inside the gizzard.

Cutting the chicken in a usable fashion

If you don't eat the necks, backs, tails, wings, or other parts, why store them? If you just use them for soup or broth, store them separately. Cutting up chickens conserves space in the freezer and makes preparation for cooking easier. Of course, if you like the look of a whole roasted chicken, you may want to do minimal trimming, such as removing the neck and tail before storing.

Some people prefer to separate parts of the chicken with breasts in one package, legs in another, and so forth. Others cut up a chicken and store all the parts from one chicken in one package. How you like to cook your chicken dictates how you package the parts. You also need to consider the number of parts per package.

You need sharp butcher knives or boning knives for cutting chicken. Try to make your cuts so they separate parts between joints rather than sawing through bones. You have to cut some bones to separate breasts from the back. Put your parts in clean bowls or pans until you're ready to package them. Most people use plastic freezer bags to package chicken for freezing. (For more tips on packaging home-butchered chicken, check out the extra article at www.dummies.com/extras/raisingchickens.)

Avoiding freezer overload

If you're going to butcher many chickens at a time, you'll probably need more than the freezer space that comes with your refrigerator. Chest and upright freezers work equally well. How many cubic feet you need depends on what you intend to store. Buy the most energy-efficient freezer you can afford. In the long run, it saves you money.

Home freezers aren't meant for freezing huge quantities of meat at a time. Don't start with an empty freezer, stuff it, and expect it to freeze your meat correctly. Chicken should be chilled to 40 degrees Fahrenheit or lower before you attempt to freeze it.

Turn on an empty freezer a day before you intend to use it. Check your owner's guide for the recommended number of pounds to freeze at one time. If you try to freeze more than that, the meat may take a long time to reach the correct temperature. This delay gives bacteria a chance to grow, affects the quality of the meat, and, even worse, may burn out the freezer's motor.

So what do you do if you have 100 pounds of chicken to freeze and the directions state a limit of 30 pounds? Put 30 pounds in the freezer and try to spread it out so most of the package surface is exposed. Keep the rest in your refrigerator or in ice chests with ice. When the first batch is frozen hard, which should be in 24 hours, then condense that and add a second batch. Continue until all the chicken is frozen (note, however, that you need to complete this process within 4 to 5 days).

Part VI
The Part of Tens

Want to increase your knowledge of raising chickens even more? Check out the bonus list of resources at www.dummies.com/extras/raisingchickens.

In this part . . .

- Keep the stress out of chicken raising. We give you more than ten tips on keeping your chickens happy and healthy.
- Does a rooster crow only in the morning? Look into this answer and 10 more misconceptions covering all things chicken.

Chapter 17

More than Ten Tips for Keeping Healthy, Stress-Free Chickens

In This Chapter

- Creating a stress-free environment for your birds
- Fulfilling your chickens' most important needs

In addition to basic needs like food and rest, chickens need as little stress as possible to be able to perform well. You're probably aware of how stress affects people and how it impacts their health. Stress can affect chickens' health, too. Stress in chickens may lead to fighting and injury, improper nutrition, and a lowered immune response to disease. Layers may quit laying, and meat birds may die of sudden heart attacks from stress.

What causes stress to chickens? Crowded conditions, chickens frequently being moved in and out of the flock, poor ventilation, heat, cold, irregular lighting, poor-quality feed, lack of water, disease, parasites, and predators can all take their toll.

In this chapter, we divulge the ten most important ways you can keep your chickens happy and healthy and, if you're raising them for eggs or meat, ensure a plentiful supply.

Choose the Right Breed for Your Needs

Not all breeds of chicken perform equally well in your environment. First, decide what you want chickens for — laying eggs, providing meat, showing, or just enjoying. Then carefully study the breed characteristics and choose a breed that seems to fit your needs. If you do this, you're less likely to give up on raising chickens, and your chickens will be healthier and happier, too.

In Chapter 3, we discuss which breeds are good for laying, for meat, and so forth, and we give you some information on the behavior characteristics of the breed and whether they're better suited to certain climates. If possible, visit a poultry show at your county fair or another location to see the chickens and talk to their owners.

Be aware that people who like a certain breed are a little partial to it, but their experience with the breed may not translate into a good experience for you. Ask questions and visit chicken forums like `www.backyardchickens.com` to get a feel for what breed is right for you.

Set Up Suitable Housing

Having the right housing is not only better for the chickens — it's also better for you, so be sure to plan your chicken housing and get it set up before you buy the birds. Plan the size of the housing, how you'll access it to care for the birds and collect eggs, how it will fit into your yard, and how you'll light it. Make sure the housing accommodates your needs as well as the chickens'. Can you collect eggs and clean the housing easily? Will you have enough room for the number of chickens you want? The housing doesn't have to be elaborate, but it needs to be clean and functional.

Chicken housing needs to protect your birds from both the elements and predators. It should keep them dry and out of drafts. Housing usually consists of indoor and outdoor space. Make sure each large-size chicken has at least 3 square feet of indoor space and 3 to 5 square feet of outdoor space for optimum health. The more space you can provide, the happier your chickens will be. And the more functional the housing is for you, the happier you'll be with chicken keeping.

In Part II, we discuss chicken housing in much detail. Having good furnishings in the housing, such as nest boxes, roosts, and feeding equipment, also keeps your chickens happy and healthy and makes things more convenient for you.

Supplement Lighting When Needed

A chicken's life cycle revolves around the amount of daylight or artificial light it receives. Chickens are prompted to lay eggs and mate when the days are long, and they molt when the days start getting shorter. Molting is the process by which a chicken replaces all its feathers, and it's energy intensive. When chickens molt, they usually stop laying.

Chickens have to molt sometime, but you can manipulate the light your chickens receive to keep them laying or have them molt when it's best for you. Supplementing the light of young pullets helps them grow faster and mature sooner and can get them laying in the fall or winter.

To keep chickens laying and prompt young pullets to start laying, supplement the natural light so that they get 14 to 16 hours of bright light per day. Don't reduce the amount of light hens and pullets are getting unless you want them to stop laying and molt.

You don't have to worry about meat birds molting, because they should be in the freezer long before then. But keeping the lights on for 18 to 20 hours a day lets them eat and drink more and grow faster. You do want to keep your show birds from molting before an important show, to keep them looking their best. Supplementing the light for pet birds isn't important.

Chickens prefer to have 14 to 16 hours of daylight and 10 hours of darkness, or at least dim light. You can supplement natural lighting with artificial light to obtain the right lighting conditions. Leaving a small night light on in chicken housing is beneficial because it allows chickens to defend themselves against some predators and avoids nighttime damage when chickens panic over more benign things that go bump in the night.

Control Pests

Pests are creatures like wild birds, rats, mice, and flies that hang around poultry housing. Not only are they offensive to neighbors (and a huge pain for you), but they can also be dangerous to your chickens. Rats; mice; and wild birds like starlings and sparrows, wild geese, and ducks can carry many diseases to your chickens. They also eat a lot of feed, which can become a huge money drain if you aren't paying attention.

Controlling pests means keeping the coop clean, storing feed so that pests can't access it, putting out poison bait or traps when you notice signs of pests, and maintaining secure housing that limits pests' access. You can find out more about controlling pests in Chapter 9.

Protect Against Predators

Besides pests, predators are a big concern for chicken-keepers — after all, people aren't the only ones who enjoy chicken for dinner. If we want to have healthy chickens, we need to protect them; chickens have few defenses of their own. Predator protection works best if you can anticipate problems and protect the chickens with sturdy pens or restricted areas to roam. (Chickens running free, even in urban areas, are at the mercy of all kinds of animals and other dangers, including cars and mowers. Supervised play is best for chickens.)

- **Use roofs, wire, or netting on outside runs to avoid predation from hawks and owls.**
- **Clear away overgrown areas and brush piles around chicken housing.**
- **Place locks on coop doors if two-legged predators are a problem.**
- **Leave a night light on in the coop at night.** Chickens have a better chance to defend themselves if they can see, and they're less likely to panic.
- **Make sure fencing is strong and sufficiently tall.** Chicken wire isn't very sturdy, and even a medium-size dog can rip it apart. (Dogs are one of the top chicken predators, by the way.) We suggest using heavy welded wire on outside runs. Burying it in the ground a foot or so deep to keep out burrowers is also wise.

For more guidance on protecting your flock against predators and dealing with them if they do manage to get a hold of one of your birds, see Chapter 9.

Control Parasites

Parasites not only make birds uncomfortable, but they also can carry disease and lower a chicken's immune system response to disease. Birds carrying a heavy load of internal or external parasites produce fewer eggs, grow more slowly, and eat more feed. Keeping your birds well fed and free of both stress and disease helps their bodies repel parasites and makes them better able to tolerate any they may still contract.

Some parasites, like worms, are hard to eliminate entirely because eggs persist in the environment. If your birds aren't producing eggs well or they look thin and unhealthy, it may be time to check for worms. Lice live on the birds, but ticks and mites may spend most of their time in some part of the housing; to control them, you need to treat the housing as well as the chickens.

Treating for parasites may mean giving chickens medications or spraying either them or the housing with pesticides. Modern products for eliminating — or at least controlling — parasites are readily available in poultry supply catalogs and from veterinarians.

We discuss disease and parasite control in Chapter 10.

Vaccinate

Preventing problems is always better than trying to fix them. When you purchase baby chicks, you're often offered the opportunity to have them vaccinated for a small additional fee. Saying yes is a wise idea.

Vaccines can be given at various life stages of chickens. Many vaccines suggest an optimum age, but if the chicken doesn't get the vaccine then, it can sometimes be administered later. This situation depends on the disease you're trying to prevent. Vaccines can be given by mouth, in the eyes, in the nose, or by injection, depending on the disease they're meant to prevent. Some vaccines prevent disease in one dose; others require several doses.

Many vaccines exist today to prevent chicken diseases. They are reasonably priced, and most home flock owners can administer them. Ask at your county Extension office or at a local vet's office which chicken vaccinations are recommended for your area and get your chickens vaccinated. If you don't want to do it yourself, have a vet do it or ask an experienced friend to help you. For more details on vaccinating your chicks or chickens, see Chapter 10.

Feed a Well-Balanced Diet

Well-fed chickens lay more eggs, grow faster, produce better meat, and have better immune systems to fight off disease. Chickens are like kids, though — you have to supervise their diets. Even if they have a large area of land to forage on, they need at least part of their diet to come from commercial feed so they get all the nutrients they need. Not only will they eat almost anything, whether it's nutritious or not, but a chunk of land just doesn't provide the nutrition that chickens need. Unlike domestic chickens, wild chickens that get all their nutrition from Mother Nature have plenty of space to roam and hunt to meet their needs.

Today's commercial feeds are well balanced, with the correct ratios of protein, minerals, and so on for the type of bird they're labeled for. They come in pellet, mash, or crumb form so the chickens can't pick out their favorite pieces and avoid the rest. If you want organic commercial feeds, these options are now available in many areas.

In Chapter 8, we discuss chicken feed and what constitutes a healthy diet.

Provide Enough Clean Water

Having clean water available at all times is one of the best ways to keep your chickens healthy and productive. Chickens need water available to lay well, grow quickly, and perform all of life's functions. Making sure water is available, even in winter, is essential to their health. If you have trouble keeping water available for your chickens, you may need to use an automatic waterer and a water heater in winter. If you use an automatic waterer, make sure you check it frequently to verify that it's working properly. It's also important to clean and sanitize waterers on a weekly basis.

Chickens can be a bit fussy about water. They don't like water that's too warm or flavored strongly. If they don't drink freely, they don't eat as much, and that starts affecting their production and health. Make sure chickens always have clean, fresh water.

Beware Disease-Transmitting Dangers

Many chicken diseases are carried on clothing, shoes, and hands. When you visit other people's chickens or go to a show, be sure to change your shoes and clothes and wash your hands before tending to your flock. Also think twice about inviting visitors who have chickens of their own to visit your flock. If you have rare or valuable birds, you may want to limit visits. The more visits, the greater the chance you have of exposing your flock to disease. Also be sure to disinfect all borrowed equipment, such as carriers, before and after use.

Use Quarantines Whenever Necessary

One of the easiest but least practiced strategies a home flock owner has for maintaining healthy chickens is to quarantine all new birds and all chickens that come back home from a show or sale. Keep the returning or arriving chickens well away from the rest of the flock for 2 weeks. If you have sick chickens, move them away from the rest of the flock and quarantine them to try to prevent disease spread. Injured birds also need to be quarantined so the others don't pick on them.

Feed and care for quarantined chickens after you take care of the rest of the flock. If the quarantined birds show any signs of disease, you'll need to destroy or treat them, whichever is the most effective method of preventing the spread of disease. To find out how to quarantine birds appropriately, see Chapter 10.

Chapter 18

More than Ten Misconceptions about Chickens, Eggs, and So On

In This Chapter

- Discerning the truth in what you read and hear about chickens
- Allaying fears about bird flu, growth hormones, and antibiotics
- Dispelling myths about what determines egg taste and quality

Oh, the things that are said about chickens! The very word *chicken* brings up the image of a coward, but chickens aren't really cowards. We talk about falsehoods throughout the book, but here's a compilation of the most common myths and misconceptions about chickens and eggs that you may encounter as a chicken owner — or chicken-keeper wannabe. Maybe some of these bits of misinformation are actually keeping you from getting some chickens of your own, so here we clear them up once and for all.

And when you're out there throwing around chicken references like "dumb cluck" or "hen-pecked," remember that while some are based on fact, most are misconceptions about chicken behavior. Knowing a bit about chickens from reading this book and actually observing chickens will help you become a champion for the cause of chicken-keeping.

Bird Flu Is a Risk to Reckon With

Some people want to keep chickens out of cities and suburban areas or are afraid to own chickens because they fear bird influenza. You're more likely to get human flu or West Nile virus than bird flu, or *avian flu.*

The fact is, bird flu has been among us for a long time. Wild birds carry many strains of bird flu, just as humans carry many strains of human flu. Outbreaks of bird flu have occurred among domestic poultry in the United States, but no incident so far has come from the dreaded H5N1or H7N9 strains, the ones responsible for disease and death in humans in Asia and some other parts of the world. Most strains of bird flu don't infect humans.

Avian flu doesn't pass through the air. The H5N1and H7N9 bird flu viruses usually pass from bird to human and not from human to human, except in a few rare cases. Bird flu can be contracted from handling infected poultry, eating raw eggs or meat, or handling something in the environment contaminated by the virus shed in animal secretions.

The U.S. Department of Agriculture (USDA) is carefully monitoring the health of domestic and wild birds in the United States for bird flu of a deadly strain. Because wild birds are the likely source of this virus, and migratory birds could spread it here, home flock owners are advised to keep wild birds away from their chickens. In the case of an outbreak, the USDA would announce steps for home flock owners to take. Otherwise, common sense and good hygiene can keep chicken owners safe.

Chicken owners need to limit visitors who themselves own poultry from handling their birds or going into chicken quarters. This limitation is tough for many proud chicken owners, but it's a major way to keep disease (including bird flu) from spreading between flocks. The virus can be carried on shoes, clothing, and even car tires. If you exhibit birds or buy new ones, put them through a 2-week quarantine period before you allow them to join the rest of the flock.

If most of the chickens in your flock suddenly die within a short period of time and without many symptoms, contact your local county Extension office, your state health department, or a local USDA office. Those experts will either give you advice or tell you who to contact. Always wear gloves when handling dead or ill chickens, and keep your hands washed!

You Can't Raise Chickens If You Live in the City

Chickens aren't just for country folk anymore. Anyone who has a small yard can find a place for a few chickens, even if you live in a bustling urban neighborhood.

We believe chickens can be raised in cities safely without disturbing the neighbors unduly. Pigeons have been allowed as pets in most cities for a long time, and they require similar care. If your city isn't one of the enlightened cities that actually allows keeping chickens, we encourage you to fight for new regulations to allow it. In Chapter 1, we give you some tips on how to get your city to amend outdated laws if yours is still in the dark ages. Fortunately, an increasing number of cities are legalizing urban chicken-keeping.

If chickens are kept clean, they don't smell any more than the flock of Canada geese in the park or the neighbor's three Great Danes. Hens aren't any noisier than a blaring car stereo or leaf blower. Chickens allow urban dwellers to have some neat pets that make breakfast for them, too. Chickens are easier to care for than dogs, and they're quiet at night, unlike the neighborhood cats.

Roosters Crow Only in the Morning

Roosters do greet the sun exuberantly, but they also crow all day long — and sometimes if they're awakened at night, they crow then, too. Roosters crow like songbirds sing, to mark their territory and make the hens aware of their presence. Healthy roosters crow every chance they get, although crowing frequency and sound vary by individual.

You Need a Rooster to Get Eggs

A hen is born with all the eggs she'll ever have, and nature tricks her into laying them regardless of whether a rooster is around. The eggs are equally tasty, nutritious, and abundant even if a rooster isn't present.

Hens don't seem to miss a rooster as long as they have hen friends to chum around with. Of course, none of their eggs can ever become chicks, but many chicken breeds don't care to be mothers anyway. If you can have roosters, though, it's fun to watch roosters escort and care for their hens.

Keeping Chickens Penned Is Inhumane

Chickens like to be able to roam freely, but it isn't always safe for them to do so, even in the country. Most livestock is kept confined in some way for its own safety, and chickens are no exception. Your kids aren't the only ones who like chicken for dinner.

Chickens can be just as happy in a good-sized pen with nutritious food and a warm, dry place to sleep as your dog is confined to the backyard or your horse is confined to the pasture. They can be allowed supervised roaming from time to time, just like your pets. And confined chickens don't annoy the neighbors or damage the flower beds. Confined chickens pose less of a health risk, too, because they aren't as likely to come in contact with wild birds that carry diseases, such as bird flu.

Chickens Are Vegetarians

Chickens love meat, including fried chicken (believe it or not, this is true). Chickens are designed to eat just about anything, and they really need some of the amino acids they get from consuming animal-based proteins. Makers of commercial poultry feed usually add amino acids that are missing from grain-based diets, or they include safe animal sources of protein.

Homemade diets that are based on only grain may not keep your chickens at optimum health, especially in the winter, when they can't dig some maggots out of the litter or catch moths. And pasture-only diets just aren't a good way to grow chickens. Most pasture-based chicken-raising also involves commercial feed.

Big, Brown, Organic Eggs Are Best in Taste and Quality

If you eat your own eggs or buy them locally, they're generally much fresher than store-bought eggs — and they taste better. Farm-fresh eggs are generally brown because breeds that lay brown eggs are easier for most owners of small flocks to care for. But if all eggs are equally fresh, there isn't usually a difference in taste or nutrition. Green and blue eggs also taste the same as brown or white ones. Likewise, small eggs taste like jumbo eggs. Beyond shell color, chickens that have access to greens or that eat marigold flowers, for example, have eggs with deeper yellow yolks, which appeals to some people.

While some people think organic eggs taste better, it's usually because, once again, they're fresher. In general, however, in a blindfold taste test, organic and non-organic eggs are indistinguishable.

Chicken eggs can taste differently if the hens are fed a lot of certain foods, like flax seed, fish, or onions, or if the eggs aren't stored properly. Eggs can also pick up unusual flavors if they're stored next to foods with strong odors.

Nutritional claims about certain eggs vary widely in credibility. Chickens can be fed in a way that results in eggs with less cholesterol and more amounts of certain nutrients, but this strategy is an exacting science that most small flock owners can't practice. Besides, the nutritional gurus are now telling us that the cholesterol we get from eggs isn't the kind that builds up in our blood anyway.

Fertilized and Unfertilized Eggs Are Easily Distinguishable

Only a trained eye can tell fertilized and unfertilized eggs apart, unless they're stored improperly and an embryo begins growing. And blood spots in an egg don't mean it's fertilized; they're simply the result of a vein rupturing as an egg is released from the ovary.

Store-bought eggs are almost always infertile eggs because commercial breeders don't keep roosters with hens. Only a store selling locally produced eggs from a small flock with a rooster has a chance of getting a fertilized egg in there. But if you keep a rooster with your hens, chances are very good that the eggs you eat are fertilized. If that bothers you, don't keep a rooster with your hens — it's that simple.

Fertilized eggs don't taste any different than unfertilized ones. And that tiny bit of chicken sperm doesn't give the egg any nutritional boost, either.

Egg-Carton Advertising Is the Absolute Truth

When buying eggs, beware: "Cage-free" doesn't mean organically raised, and it doesn't mean the hens roam the farm freely. It usually means the birds were housed in large pens with a little room to move around. Growers refer to this environment as cage-free, but really, it's just a giant cage with a lot of chickens crowded into it. It's slightly better than being crowded into cages so small a chicken can't stand up or flap its wings, the way most commercial layers are housed. The cheaper eggs you buy from big-box stores aren't going to come from hens that roam freely outside, no matter what deceptive words are used on the carton.

"Organic" doesn't mean the hens weren't kept in small cages, either — at least, not yet in the United States. It just refers to the feed they were given, not the conditions they were kept in. In Europe, however, eggs labeled "organic" must come from hens that have access to the outdoors.

Buying your eggs locally from hens kept in small flocks — whether or not they're free-ranging or fed organically — gives you the best-tasting eggs, short of collecting them each morning from your own hens. And it probably means that the hens were kept in more humane conditions than commercial, caged layers.

Chickens Are Good for Your Garden

Many people claim that chickens can till your soil, pull the weeds, eat the bugs, and fertilize the soil, but the truth is that chickens ruin your garden. They till the soil, all right — right after you plant that crop of beans. They eat the weeds — along with all the lettuce. And while they eat the tomato worms, they take a bite out of each tomato.

Chickens belong in the garden only in the fall, just before you clean it all out but after you've harvested all you want to eat. They can harvest leftovers and eat bugs then. If you want them to till the soil, fall is the time, long before you plant again.

Chicken manure is good for the garden only after it has been composted. Fresh chicken manure deposited in the garden burns plants and brings the risk of salmonella and E. coli bacteria contaminating your fresh veggies.

Chickens Are Dumb and Cowardly

Most people who have raised chickens for any length of time strongly defend their chickens' intelligence and can tell you many tales of chicken bravery. And remember that sometimes it's smarter to run from danger than to face it, so don't judge the chicken that retreats from danger.

As birds — or animals, for that matter — go, chickens are pretty intelligent. They can learn to count and they understand the concept of zero. They can be trained to do tricks and to recognize colors. They can figure out how to get out of almost any pen you put them in, sooner or later. Chickens and other birds have been observed planning future actions or anticipating reactions to an action they're going to take. And chickens learn by observing and copying other chickens.

Chickens have a well-organized social system that limits strife among a flock. Anyone who has ever watched a rooster coaxing his hens over to some choice food knows that they communicate among themselves.

Although the word *chicken* has come to mean "cowardly," chickens can be very brave when defending their babies or their flock. Hens sometimes sacrifice themselves for their chicks. Roosters often fight to the death, even though most of us would consider that rather stupid behavior. And roosters can be formidable when protecting their girls — just ask anyone who has been chased by an angry rooster!

Index

• Numerics •

• A •

• C •

• D •

• E •

• I •

• J •

• K •

• L •

• M •

• N •

• O •

• P •

• Q •

• R •

• T •

• U •

• V •

• W •

• Y •

• Z •

About the Authors

Kimberley Willis lives with her husband, Steve, on a small farm in the thumb area of Michigan. She is now retired and enjoying her grandchildren, her chickens, and gardening. Kim worked at the MSU Extension office in Lapeer County Michigan for 18 years as a horticulturist and doubled as the resident chicken expert. She has raised numerous breeds of chickens and other poultry for eggs, meat, and showing for more than 40 years. She is a strong advocate of eating locally and shares her bounty of eggs with family and friends. Kim is a proud member of www.backyardchickens.com and is also a member of the Huron County Michigan Fowl Group on Facebook.

In addition, Kim is a garden and country living writer for Examiner.com and has written more than 600 articles online. You can read some of her articles on a wide variety of country and garden topics at www.examiner.com/gardening-in-detroit/kimberley-willis or at www.examiner.com/country-living-in-detroit/kimberley-willis. You can read her weekly garden blog at www.gardeninggranny.blogspot.com.

Rob Ludlow; his wife, Emily; and their two beautiful daughters, Alana and April are the perfect example of the suburban family with a small flock of backyard chickens. What started out as a fun hobby raising a few egg-laying hens has turned into almost an addiction.

Originally, Rob started posting his experiences with chickens on his hobby web site, www.Nifty-Stuff.com, but when he realized how much his obsession with chickens was growing, he concentrated his efforts on a site completely devoted to the subject. Now Rob owns and manages www.BackYardChickens.com (BYC), the largest and fastest-growing community of chicken enthusiasts in the world.

Rob hopes to work with BYC's tens of thousands of members to promote a change of the old concept "A chicken in every pot" to a new version, the BYC vision: "A chicken in every yard!"

Dedication

Kimberley Willis: I would like to dedicate this book to my husband, Steve, who always lends a hand with housework and cooking while I immerse myself in writing; and to the memory of my grandfather, who gave me my first chickens to raise.

Rob Ludlow: To the three most important girls in my life — Emily, Alana, and April — who not only support, but also contribute to my joy of raising backyard chickens.

Author's Acknowledgments

Kimberley Willis: I would like to acknowledge the efforts of editors Jennifer Connolly and Christy Pingleton, who worked on the first edition of this book, and editors Erin Calligan Mooney and Linda Brandon, who guided me on the second edition. I also want to acknowledge and thank Barb Doyen of Doyen Literary Services, who has worked tirelessly to encourage and support me and who has always been a friend as well as an agent. I'd also like to thank Rob Ludlow and the thousands of folks at `www.BackYardChickens.com` for keeping me current with the concerns of chicken owners and for help in making the first edition of this book such a success.

Rob Ludlow: Thanks to my brother, Michael, for getting me started with chickens, and to Mike Baker and Kristin DeMint for their help with the project. Especially huge thanks to the incredibly smart, patient, and helpful staff at www.backyardchickens.com and the thousands of friendly BYC community members.

Publisher's Acknowledgments

Acquisitions Editor: Erin Calligan Mooney

Project Editor: Linda Brandon

Copy Editor: Krista Hansing

Technical Editor: Paul Wylie

Art Coordinator: Alicia B. South

Project Coordinator: Emily Benford

Project Manager: Jennifer Ehrlich

Illustrator: Barbara Frake

Cover Image: ©iStock.com/aluxum